Advances in Experimental Medicine and Biology

Volume 1453

Advances in Experimental Medicine and Biology provides a platform for scientific contributions in the main disciplines of the biomedicine and the life sciences. This series publishes thematic volumes on contemporary research in the areas of microbiology, immunology, neurosciences, biochemistry, biomedical engineering, genetics, physiology, and cancer research. Covering emerging topics and techniques in basic and clinical science, it brings together clinicians and researchers from various fields.

Advances in Experimental Medicine and Biology has been publishing exceptional works in the field for over 40 years, and is indexed in SCOPUS, Medline (PubMed), EMBASE, BIOSIS, Reaxys, EMBiology, the Chemical Abstracts Service (CAS), and Pathway Studio.

2022 CiteScore: 6.2

M. Cristina Vega
Francisco J. Fernández
Editors

Advanced Technologies for Protein Complex Production and Characterization

Volume II

Springer

Editors
M. Cristina Vega
Centro de Investigaciones
Biológicas Margarita Salas (CIB-CSIC)
Madrid, Spain

Francisco J. Fernández
Abvance Biotech SL
Madrid, Spain

ISSN 0065-2598 ISSN 2214-8019 (electronic)
Advances in Experimental Medicine and Biology
ISBN 978-3-031-52192-8 ISBN 978-3-031-52193-5 (eBook)
https://doi.org/10.1007/978-3-031-52193-5

This Springer imprint is published by the registered company Springer Nature Switzerland AG
The registered company address is: Gewerbestrasse 11, 6330 Cham, Switzerland

Paper in this product is recyclable.

Preface

Macromolecular complexes stand at the forefront of contemporary biological research. They carry out most functions in the cell and outside the cell, as proteins have not evolved to operate in isolation [1]. In fact, it has become increasingly challenging to single out proteins or other macromolecules that can perform significant cellular functions in isolation. While proteins can act as monomers, such as many enzymes, they are invariably integrated within a dynamic network where substrate and product fluxes and allosteric regulators coordinate the dozens of distinct proteins that carry out a complete pathway. Multisubunit complexes maintain and build our genomes; read, write, and modify genetic and epigenetic information; execute essential metabolic functions; discern extracellular cues; organize signal transduction processes; and more. Protein and nucleoprotein complexes also comprise viruses and virulence factors secreted by cellular pathogens. Modern biomedicine leverages protein complexes for numerous therapeutic applications, ranging from the systems that produce them to the therapeutic proteins themselves, such as antibodies. In biotechnology, protein assemblies synthesize high-value chemical products and will one day capture sunlight and CO_2 as part of a cleaner, more sustainable economy. Protein complexes are fundamental to transformative processes and innovations across basic biology, biomedicine, and biotechnology.

The first volume of *Advanced Technologies for Protein Complex Production and Characterization* (ATPCPC) originated from work carried out by the editor and collaborators in the context of the ComplexINC collaborative project (EC FP7 2011–2015). This consortium pooled leading expertise to pioneer novel technologies and production tools for complex protein biologics. That first volume provided a tour of some of the most valuable technologies for producing and purifying protein complexes and the methods routinely used for their biochemical, biophysical, and structural characterization [2]. Having achieved over 118,000 accesses and 249 citations, the volume has successfully reached a broad and diverse audience, ranging from university students to junior and senior scientists across academia and industry, in both fundamental and applied research. This second volume of *Advanced Technologies for Protein Complex Production and Characterization* (ATPCPC2) aspires to become a companion to the first one. In ATPCPC2, our focus has shifted from the expression platforms that produce complex protein biologics to techniques for their identification, exploration, characterization, structural determination, and modeling. ATPCPC2 covers a wide range of biochemical, biophysical, and structural methodologies adopted across laboratories worldwide, facilitating the in-depth study of protein-protein and protein-nucleic acid complexes for various purposes and at different scales.

The driving force behind ATPCPC2 stems from the continuous advance of protein complexes as a central research endeavor in biology, biomedicine, and biotechnology, combined with the profusion of methods and techniques that have been developed (and continue to be developed). Since the publication of ATPCPC, some of these methods have led to dramatic leaps forward in our understanding of the structure and function of macromolecular complexes. A case in point is cryo-electron microscopy's "resolution revolution" [3]. With this new volume, we aim to both complete and expand upon the original volume's scope, with a particular emphasis on the biochemical and structural elucidation of protein complexes.

This volume opens with two chapters dedicated to biochemical methods targeting ribonucleoprotein complexes. The first describes techniques involving immunoprecipitation (Chap. 1), while the second describes methodologies for assembling and purifying RNA-protein complexes, in vitro and in vivo (Chap. 2). The following two chapters explore cutting-edge mass spectrometry approaches to study protein complexes, including protein complex composition, subunit stoichiometry, and protein-protein interactions (Chaps. 3 and 4). Chapter 5 delves into methods to discover and investigate peptide linear motifs, underscoring their pivotal roles in mediating the assembly and regulation of protein complexes. Chapters 6 and 7 focus on biophysical techniques with a strong track record for characterizing macromolecular complexes. Specifically, Chap. 6 introduces biolayer interferometry (BLI) as a label-free technique to measure the kinetic parameters and affinity constants for biological interactions, including those involving proteins, peptides, and nucleic acids. Analytical ultracentrifugation (AUC) is the topic of Chap. 7, a versatile technique with a broad field of applications that can elucidate the size, shape, stoichiometry, and binding affinity of macromolecular complexes in solution. Lastly, Chaps. 8 through 13 provide an overview of contemporary applications of structural biology methodologies, spanning from nuclear magnetic resonance and X-ray crystallography to cutting-edge techniques like X-ray free electron lasers, small-angle X-ray scattering, and high-resolution cryo-electron microscopy.

In the concluding remarks of the introductory volume, we posited that "Proteins, their complexes, and their activities are central to modern biology, biotechnology, and biomedicine, and their heterologous production is often a vital prerequisite for discovering their function in health and disease." Within this volume, we have underscored state-of-the-art methodologies and strategies for characterizing the proteins and protein complexes that researchers produce. We have endeavored to incorporate up-to-date and authoritative contributions yet sufficiently accessible to be helpful to the broadest possible audience: from students to seasoned researchers, spanning academic and industrial environments and bridging fundamental science to translational biomedicine and biotechnology. The advancements over the past decade in techniques for expressing, purifying, and analyzing proteins and their complexes have undeniably catalyzed monumental progress in the life sciences for the benefit of all.

Madrid, Spain M. Cristina Vega

Francisco J. Fernández

References

1. Alberts B (1998) The cell as a collection of protein machines: preparing the next generation of molecular biologists. Cell 92:291–294. https://doi.org/10.1016/S0092-8674(00)80922-8
2. Vega MC (2016) Advanced technologies for protein complex production and characterization. Springer International Publishing, Cham
3. Kühlbrandt W (2014) The resolution revolution. Science 343:1443–1444. https://doi.org/10.1126/science.1251652

The original version of the book has been revised. A correction to this book can be found at https://doi.org/10.1007/978-3-031-52193-5_14

Contents

1 Immunoprecipitation Methods to Isolate Messenger Ribonucleoprotein Complexes (mRNP)

Hassan Hayek, Lauriane Gross, Fatima Alghoul, Franck Martin, Gilbert Eriani, and Christine Allmang

Abstract

Throughout their life cycle, messenger RNAs (mRNAs) associate with proteins to form ribonucleoproteins (mRNPs). Each mRNA is part of multiple successive mRNP complexes that participate in their biogenesis, cellular localization, translation and decay. The dynamic composition of mRNP complexes and their structural remodelling play crucial roles in the control of gene expression. Studying the endogenous composition of different mRNP complexes is a major challenge. In this chapter, we describe the variety of protein-centric immunoprecipitation methods available for the identification of mRNP complexes and the requirements for their experimental settings.

Keywords

Protein-RNA interactions · RNA immunoprecipitation · Crosslinking · Messenger ribonucleoprotein complexes

H. Hayek · L. Gross · F. Alghoul · F. Martin · G. Eriani · C. Allmang (✉)
Architecture et Réactivité de l'ARN, Centre National de la Recherche Scientifique, Institut de Biologie Moléculaire et Cellulaire, Université de Strasbourg, Strasbourg, France
e-mail: c.allmang@unistra.fr

1.1 Introduction

Expression of protein-coding genes in eukaryotes is a complex and coordinated mechanism that involves many steps and relies on multiple factors and several molecular machineries. Cellular RNAs are associated with RNA-binding proteins (RBPs) to form ribonucleoprotein (RNP) complexes of highly dynamic compositions [1, 2]. Throughout their life cycle, mRNAs are part of multiple successive mRNP complexes that play critical roles in their biogenesis, cellular localization, translation and decay [3, 4]. The dynamic composition of mRNP complexes allows both post-transcriptional and specific translational regulation mechanisms to take place [5, 6]. Some trans-acting factors can interact stably with specific mRNPs but structural remodelling of mRNPs also involves transient interactions with protein and RNA chaperone complexes and numerous regulatory factors that are difficult to trap [7–9]. Studying the endogenous composition of different mRNP complexes is therefore a major challenge. The advent of new and high-throughput approaches has enabled the genome-wide determination of mRNP composition using techniques such as affinity purification-mass spectrometry (MS) as well as protein–RNA UV crosslinking approaches combined with RNA deep-sequencing (RIP-seq) [10–12]. Two types of approaches to study RNA–protein complexes have emerged: protein-centric and RNA-centric approaches. In protein-

M. C. Vega, F. J. Fernández (eds.), *Advanced Technologies for Protein Complex Production and Characterization*, Advances in Experimental Medicine and Biology 1453,
https://doi.org/10.1007/978-3-031-52193-5_1

centric approaches, mRNAs associated with RBPs are identified after the immunoprecipitation of the RBPs by microarray analysis or RNA-seq [13–15]. In RNA-centric approaches, RBPs are recovered by RNA pull-down methods and subsequently identified by mass spectrometry (MS). Several independent laboratories have developed breakthrough methods for the mRNA interactome discovery. Based on protein-mRNA interactome capture protocols these studies now allow genome wide cartography of mRNPs in yeast, mammalian, invertebrates or plant cells, under various conditions [16–20]. These unbiased methods have emerged as powerful tools to identify new RBPs; they nevertheless need to be complemented by individual mRNA/protein detection methods in order to map pairwise interactions between the RBPs and their mRNA targets and to identify RBPs with overlapping target specificity. Some mRNA centric methods are limited to the study of polyadenylated mRNAs and thus exclude mRNA categories such as histone mRNAs that lack poly-A tails. Global analysis can also oversee low abundance mRNAs or mRNAs containing non-canonical codons subjected to translation recoding. Targeted mRNA immunoprecipitation experiments are particularly useful in these cases. In this chapter, we will focus on standard protein-centric immunoprecipitation methods that allow the identification of native mRNPs containing specific mRNAs of interest. These methods can be used as initial targeted approaches or to confirm and consolidate high throughput or global analysis results. We have applied the panel of immunoprecipitation methods described herein to analyse the implication of various chaperone complexes and modification enzymes that act transiently in the assembly and biogenesis mechanisms of selenoprotein mRNAs as well as the role of translation initiation complexes in histone H4 mRNA translation.

1.2 Endogenous mRNP Purification Using RNA Immunoprecipitation Methods

mRNA form complexes with partner proteins in ribonucleoprotein particles (mRNP) from the onset of their transcription. Their assembly, transport and subcellular localization and degradation involve interactions with numerous intermediate protein complexes. Protein-centric immunoprecipitation methods allow to unravel the fate of mRNAs and their uptake by different known protein complexes. Various methods have been developed for the recovery of endogenous mRNPs under native physiological conditions or after stabilization of the interactions by chemical or UV cross-linking (Table 1.1).

1.2.1 Native Immunopurification

Native RNA immunoprecipitation (RIP) allows the identification of RNA–protein interactions under physiological conditions by using a protein-specific antibody and the detection of interacting RNAs. mRNA interactants can be identified by targeted-RNA detection methods like quantitative Real Time-PCR (qRT-PCR), northern blot but also by genome-wide methods such as microarray (RIP-chip) or deep-sequencing (RIP-seq). Typical RIP experiments for the study of eukaryotic mRNPs are outlined in Fig. 1.1a, mainly applying to HEK293 or HeLa cells. These methods preserve native complexes present in the cells and most often reflect *in vivo* associations. Nevertheless it was demonstrated that reassociation of RNA-binding proteins can occur after cell lysis [21]. Interactions that were prevented *in vivo* due to differential compartmentalization might occur during the lysis and immunoprecipitation experiment. Several protocols have therefore been optimized to minimize RNP rearrangements during the process of immunoprecipitation [13, 22]. For instance, purification of cytoplasmic mRNPs requires mild lysis conditions in order to leave nuclei intact [23–25]. Specific conditions for nuclear mRNP have also been developed [26, 27]. Another important issue is the specificity of the interactions detected by these methods. Indeed, specific interactions that occur with low abundance mRNAs *in vivo* could be masked by the non-specific interactions of abundant transcripts. Several precautions need to be taken in the experimental procedures to avoid this (see below).

Table 1.1 List of protein-centric methods for the purification of mRNP complexes

	Method of mRNP purification	Applications	Limitations
In vivo RIP	Native RIP	Endogenous RNP complexes purification	Possible RNP rearrangements Possible nonspecific interactions Requires highly specific anti-RBP antibodies
	Epitope tagging and RIP	Confirmation of native RIP Analysis of low abundance target proteins and their target mRNA Efficient RNP affinity purification	Possible impact of the epitope tag on the target protein: misfolding, loss of function, mislocalization Potential disruption of functional protein complexes Nonspecific interactions with the epitope tag
	Formaldehyde cross-linking and RIP	Capture of RNA-protein interactions Stabilization of transitory interactions and indirect RNA-binding factors	False positives due to excess of cross-link False negatives due to low cross-linking efficiency
	UV cross-linking and RIP	Identification of direct mRNA-protein interactions Trapping of functional interactions Genome-wide analysis: PAR-CLIP HITS-CLIP; iCLIP; iCLAP CRAC	False negatives due to low cross-linking efficiency
Cell free mRNP purification	GST pull-down	Reconstitution and screening for mRNA-protein interactions	Possible artefacts due to protein misfolding, lack of post-translational modifications Nonspecific interactions with the GST-tag
	Cross-linking	Identification RBPs using *in vitro* transcribed mRNAs: 4-thioU internally labelled mRNAs 6-thioG capped mRNAs, tracking of cap-binding proteins	Optimization of *in vitro* 4-thioU incorporation and 6-thioG cap labelling
Direct mRNA IP	m^6A Me-RIP m^5C RIP	Global identification of mRNA transcriptional modifications patterns	Specificity of the antibodies
	TMG-IP	Detection of endogenous $m_3{}^{2,2,7}G$-capped mRNAs	Low recovery levels of mRNAs possibly due to their abundance, stability, short-half life

RIP RNA immunoprecipitation, *PAR-CLIP* photoactivatable ribonucleoside-enhanced cross-linking and immunoprecipitation, *HITS-CLIP* high-throughput sequencing CLIP, *iCLIP* individual nucleotide resolution CLIP, *iCLAP* individual-nucleotide-resolution UV-cross-linking and affinity purification, *CRAC UV* cross-linking and analysis of cDNA, *GST* (glutathione-S-transferase) pull-down, *m^6A Me-RIP* anti N^6-methyladenosine RIP, *m^5C RIP* anti 5-methylcytidine RIP, *TMG-IP* anti $m_3{}^{2,2,7}G$ IP

1.2.1.1 Key Requirements for RIP

To prepare high-quality extracts that contain intact pre-mRNPs or mature mRNPs several factors need to be optimized. In particular extracts have to be kept cold and snap-frozen immediately after lysis and centrifugation. To limit degradation, inhibitors of proteases and ribonucleases (RNases) have to be added at all steps. Extracts should be diluted only after thawing and during the immunoprecipitation steps. The total amount of cell extract used in RIP experiments needs to be adapted based on the abundance of the RNA-binding protein or complex targeted by the antibodies as well as the subsequent method of RNA detection. Typically, extracts generated from 3 to 5 × 10^6 mammalian cells will be required when the RNA is detected by qRT-PCR or 5 to 20 × 10^6 when using microarrays.

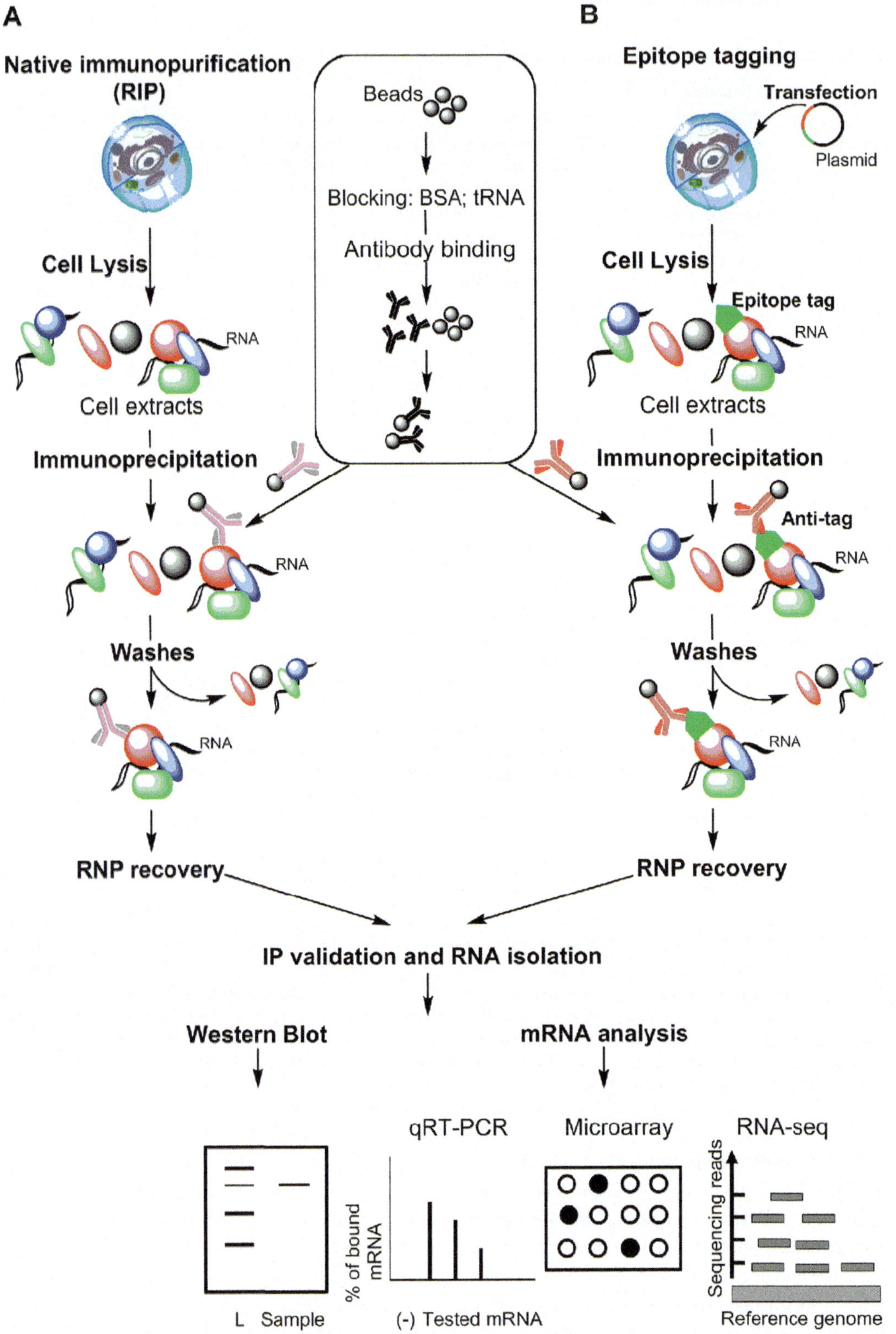

Fig. 1.1 Schematic representation of RNA immunoprecipitation (RIP) methods for the identification of endogenous protein-mRNA interactions. Native mRNPs can be recovered directly from HEK293 or HeLa cell lysates (**a**) or after transfection and expression of the epitope tagged protein (**b**). In both cases RNAs bound to specific proteins

It is essential to limit potential nonspecific interactions of the mRNAs of interest with the immunoprecipitation matrix. Various affinity matrices and supports can be used successfully for the recovery of mRNP complexes after incubation of the extracts with a specific antibody. These include protein A-sepharose, agarose but also wide range of magnetic beads. It is advisable to compare different matrices because the nonspecific binding of a given mRNA can vary depending on the immunoprecipitation support. It is also recommended to pre-clear the extracts in contact with non-coated beads before immunoprecipitation and to dilute the lysate in order to remove nonspecific binders. When the targeted mRNA will subsequently be detected by methods such as qRT-PCR, pre-coating of the beads with purified BSA and yeast total tRNA can considerably reduce nonspecific background (Fig. 1.1). For standard RIP, extensive washing of the immunoprecipitated pellet is required and the stringency of the washing buffers needs to be adapted to the stability of the RNP complex analysed. The mRNP is ultimately released and dissociated into RNA and proteins. This is mostly performed under denaturing conditions. mRNP affinity purification can also be performed using anti-peptide antibodies against the target protein. In this case bound mRNP complexes can be eluted in native conditions by the peptide used to generate the antibodies. This is particularly interesting when studying low abundance mRNPs. This method allowed us to co-immunoprecipitate selenoprotein mRNAs and protein complexes associated to SECIS binding protein 2 (SBP2), a key protein that interacts with all selenoprotein mRNAs and recruits translation and assembly factors to the mRNP [28, 29]. In any case, validation steps of the RIP will include verification of optimal protein pulldown by Western blot and evaluation of the integrity of protein complexes associated to the mRNP by standard mass spectrometry.

1.2.1.2 mRNA Identification and Validation of the RIP Experiment

After release of the RNP components, the ultimate goal is to isolate and characterize the mRNA from the immunoprecipitated pellet (Fig. 1.1). For proper interpretation of the data it is essential to compare the RIP result to a negative control (IP with pre-immune serum or non-RBP protein target). For standard RIP, qRT-PCR remains the method of choice to measure RNA levels as it allows to normalize the immunoprecipitated mRNA to its abundance in total lysate [30]. To determine if the interaction of given RBP can be generalized to an entire family of mRNAs the microarray analysis of the RIP content is a useful strategy. Result differences can be observed between the two detection methods, most likely because qRT-PCR is more sensitive and specific than microarrays for the detection of the endogenous population of mRNAs [31]. Deep-sequencing analysis of RIP experiments allows the genome-wide characterization of RNAs bound to a given RBP and the mapping of protein-RNA regulatory networks using appropriate bioinformatics tools [32, 33]. mRNA interactants identified using native purification methods often require further experimental validation using multiple distinct experimental approaches. A way to confirm the validity of the

Fig. 1.1 (continued) of interest are immunoprecipitated from cell lysates in native conditions by using either a protein-specific antibody (**a**) or anti-tag antibody (**b**). After several washing steps the RNA-protein complexes (RNPs) are recovered by elution in either denaturing or native conditions. The validity of the immunoprecipitation experiment is verified by western blot. The RNA is extracted and interacting mRNAs are identified by targeted RNA detection methods (qRT-PCR, microarray) or global high throughput sequencing analysis (RNA-seq). The insert represents the preparation of the affinity matrix before immunoprecipitation. This includes pre-coating of the beads with purified BSA and yeast total tRNA to reduce nonspecific background before binding the antibodies

mRNA interactant is to compare the RIP results obtained against an endogenous protein under normal and RNAi inhibition of the target protein [34]. Alternative immunoprecipitation approaches based on epitope tagging or cross-linking methods can be used, and they will be discussed below. Functional validation of the mRNA-protein interactions unveiled by the RIP experiment will have to include a wide panel of methodologies such as subcellular localization experiments (FISH) [35], yeast-two hybrid interaction tests (Y2H) [36], electrophoretic mobility shift assays (EMSA) [37, 38], and more.

1.2.2 Epitope Tagging and RIP

The use of an epitope-tagged RBP for mRNP isolation is a useful complementary approach to confirm endogenous RIP data. It can also be used when the endogenous target protein is not detected by standard RIP because of its low cellular abundance or the inaccessibility of the epitope to the immunoprecipitating antibody (Fig. 1.1b). After transfection and expression of the epitope tagged protein in cells, usually HEK293 or HeLa, mRNAs are coprecipitated using antibodies directed against the epitope tag. Some of the most commonly used epitope tags include FLAG, HA, His, Myc, GST, GFP and V5 [39, 40]. Limitations linked to the use of epitope tags are the possibility for the tag to impact the folding of the fusion protein, to disrupt functional protein complexes and to interact non-specifically with cellular proteins. Large globular tags may also affect the function and subcellular localization of the fusion protein. Among various tags the green fluorescent protein (GFP) has proven particularly efficient for affinity purification of RNP complexes and proteomic analysis, despite its size, because it shows minimal nonspecific binding to mammalian cell proteins [41]. We have used epitope tagging successfully to analyse the contribution of different chaperone complexes and modification enzymes in selenoprotein mRNA assembly and translation. This allowed us to demonstrate interactions between selenoprotein mRNAs and components of the HSP90 chaperone complex [28]. Our findings also establish that a FLAG-tagged version of trimethylguanosine synthase 1 (Tgs1) interacts with selenoprotein mRNAs for cap hypermethylation [31]. We also showed that the ubiquitous Survival of MotoNeurons (SMN) chaperone and the methylosome complex devoted to sn- and snoRNP maturation contribute to recruiting Tgs1 to selenoprotein mRNPs [29]. In this case we used GFP-tagging as a method of choice to decipher the role of individual SMN and methylosome components in selenoprotein mRNP assembly and translation.

1.2.3 Formaldehyde Cross-linking and RIP

When studying RNA-protein interactions *in vivo* it is essential to minimize RNP reassortments that can occur during cell disruption. Chemical treatments can be employed for this purpose in order to cross-link proteins with their associated mRNAs prior to the cell lysis and immunoprecipitation steps (Fig. 1.2a). Formaldehyde is a useful reversible crosslinking agent capable of capturing interactions between various macromolecules mainly by formation of covalent bonds between amino groups that are in close proximity (around 2 Å) [42]. It has been used extensively as a fixative to maintain the structural integrity of cells and in chromatin immunoprecipitation (ChIP) experiments to study DNA-protein interactions and cellular networks [43, 44], but also to delineate RNA-protein and protein-protein interactions [45]. Formaldehyde cross-linking is particularly useful to detect transitory interactions and to stabilize indirect RNA-binding factors that are part of higher-order RNPs. It was used successfully in various studies including the identification of RNAs binding to the CTD of RNA polymerase II, the binding of the histone acetyltransferase Elongator to nascent mRNAs [46, 47] as well as the interaction of eukaryotic Translation Initiation Factor 3 (eIF3) with various mRNAs [48]. The success of the RIP assay depends on the degree of cross-linking. The cross-linking procedure therefore needs to be optimized to avoid false

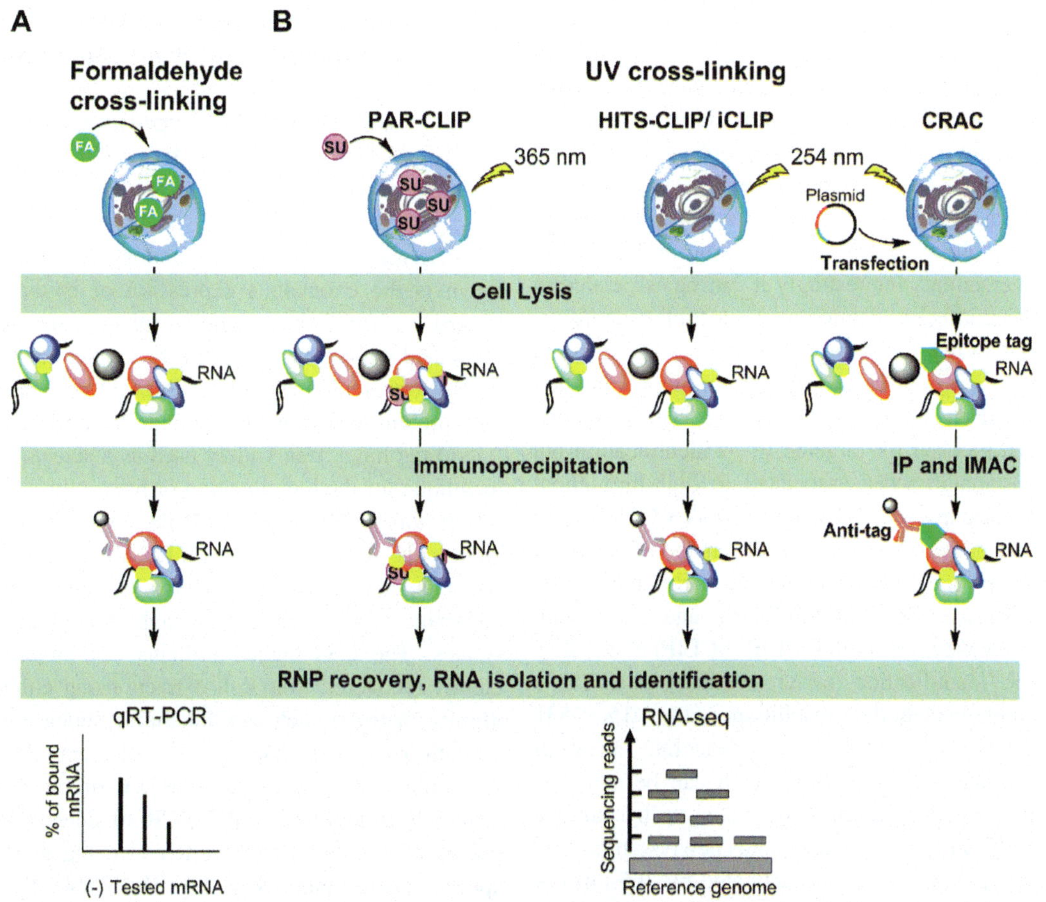

Fig. 1.2 Outline of chemical and UV cross-linking immunoprecipitation methods. (**a**) Formaldehyde (FA) cross-linking. (**b**) UV cross-linking methods coupled to high-throughput analysis are represented from left to right: photoactivatable ribonucleoside-enhanced crosslinking and immunoprecipitation (PAR-CLIP), High-throughput sequencing CLIP (HITS-CLIP) and individual-nucleotide resolution CLIP (iCLIP), individual nucleotide crosslinking and cDNA analysis (CRAC). PAR-CLIP uses 4-thioU ribonucleosides; UV at 365 nm induces zero length cross-links between the photoactivatable 4-thioU moiety of the RNA and RNA-binding proteins (RBPs), while the other three methods utilize UV cross-linking at 254 nm. Isolation of RNP complexes is achieved by immunoprecipitation (IP) or double affinity purification by IP and immobilized metal ion affinity chromatography (IMAC) in the case of CRAC. FA represents formaldehyde and SU and represents 4-thioU nucleotides respectively. Yellow dots symbolize cross-links between RNA and proteins

positive interactions and to achieve optimal RNA-protein complex recovery. The concentration of formaldehyde and the duration of cross-linking need to be adjusted [49]. Optimal formaldehyde concentration typically ranges between 0.1% and 1.0% and the duration of fixation varies between 5 min and 1 h. Cross-linking reactions are quenched by the addition of glycine (pH 7). Interestingly formaldehyde crosslinking can be reversed by incubation of the IP pellets at 70 °C for the characterization of immunoprecipitated components.

1.2.4 UV Cross-linking and RIP

Unlike formaldehyde cross-linking, which is able to capture interactions with protein complexes,

UV cross-linking exclusively identifies direct RNA-protein interactions. UV irradiation at 254 nm generates short-lived radicals that react with nucleic acids and amino acids such as lysine, cysteine, phenylalanine, tryptophan and tyrosine located in close proximity (zero distance) forming covalent bonds [50]. *In vivo* UV cross-linking thus allows trapping functional protein-RNAs interactions. Interestingly it can be associated to denaturing purification methods for the identification of the interacting RNA and has given rise to the cross-linking and immunoprecipitation (CLIP) technique [11, 12, 51, 52]. This technique can be used for targeted RNA identification but has mainly been associated to high-throughput sequencing of cDNA library (HITS-CLIP). Several powerful variants of the methodology exist such as PhotoActivatable Ribonucleoside enhanced CLIP (PAR-CLIP) and individual-nucleotide resolution CLIP (iCLIP) that allow the identification of cross-linking sites at a single-nucleotide resolution [14, 15, 53]. PAR-CLIP employs the nucleotide analogue 4-thiouridine (4-thioU), which can be added to the growth medium and is taken up by cultured cells and incorporated into newly-synthesized RNAs [15]. 4-thioU is activated by 365-nm UV irradiation, instead of UV 254 nm used in CLIP, to generate covalent crosslinks between proteins and RNAs. In both CLIP and PAR-CLIP experiments the RNA is fragmented by RNase treatment and immunoprecipitation allows the recovery of co-precipitated RNA fragments. Other variations make use of double epitope-tagged RBPs in cross-linking experiments followed by purification of the RNPs by tandem affinity purification (iCLAP) or immunoprecipitation coupled to immobilized metal-ion affinity chromatography (IMAC) in the CRAC method [54, 55]. Some *in vivo* methods combine UV cross-linking, GFP-based immunoprecipitation and quantification of co-isolated polyadenylated [poly(A)] RNAs with fluorescent oligo (DT) probes [56]. The main high-throughput methods are summarized in Fig. 1.2b and Table 1.1.

1.3 Cell-Free Extract mRNP Immunoprecipitation Methods

1.3.1 GST Pull-Down Experiments Using Total RNA from Cell Extracts

An alternative approach to endogenous pull-down is the exogenous expression of epitope-tagged recombinant proteins that can be incubated with cell extracts for the reconstitution of RNA-protein interactions. Purified recombinant proteins carrying a glutathione-S-transferase (GST) tag have been widely used as a screening technique for the identification of protein-protein interactions. GST-tagged proteins are also efficient tools to uncover protein-mRNA interactions when incubated with total RNAs isolated from cell extracts, native extracts or rabbit reticulocyte lysates. The GST-tagged proteins can be efficiently recovered from cell extracts using either affinity matrices such as Glutathione sepharose, Glutathione magnetic beads or anti-GST antibodies. The co-precipitated RNA can be subsequently analyzed by qRT-PCR to detect the presence of specific target mRNAs using gene-specific primer pairs (Fig. 1.3a and Table 1.1). Glutathione-S-transferase (GST) pull-down procedures have been developed for the purification of eukaryotic mRNAs using a mutant version of the mRNA 5′ cap-binding protein (eIF4E) with increased affinity for the m^7GTP moiety of the cap [57]. Using this method we could demonstrate that several selenoprotein mRNAs are not recognized efficiently by translation initiation factor eIF4E and have an alternative cap structure [31].

1.3.2 Crosslinking of Total Cell Extracts with *In Vitro* Transcribed 4-thioU or 6-thioG Labelled mRNAs

The 4-thiouridine PAR-CLIP approach can be adapted to study RNA-protein interactions using

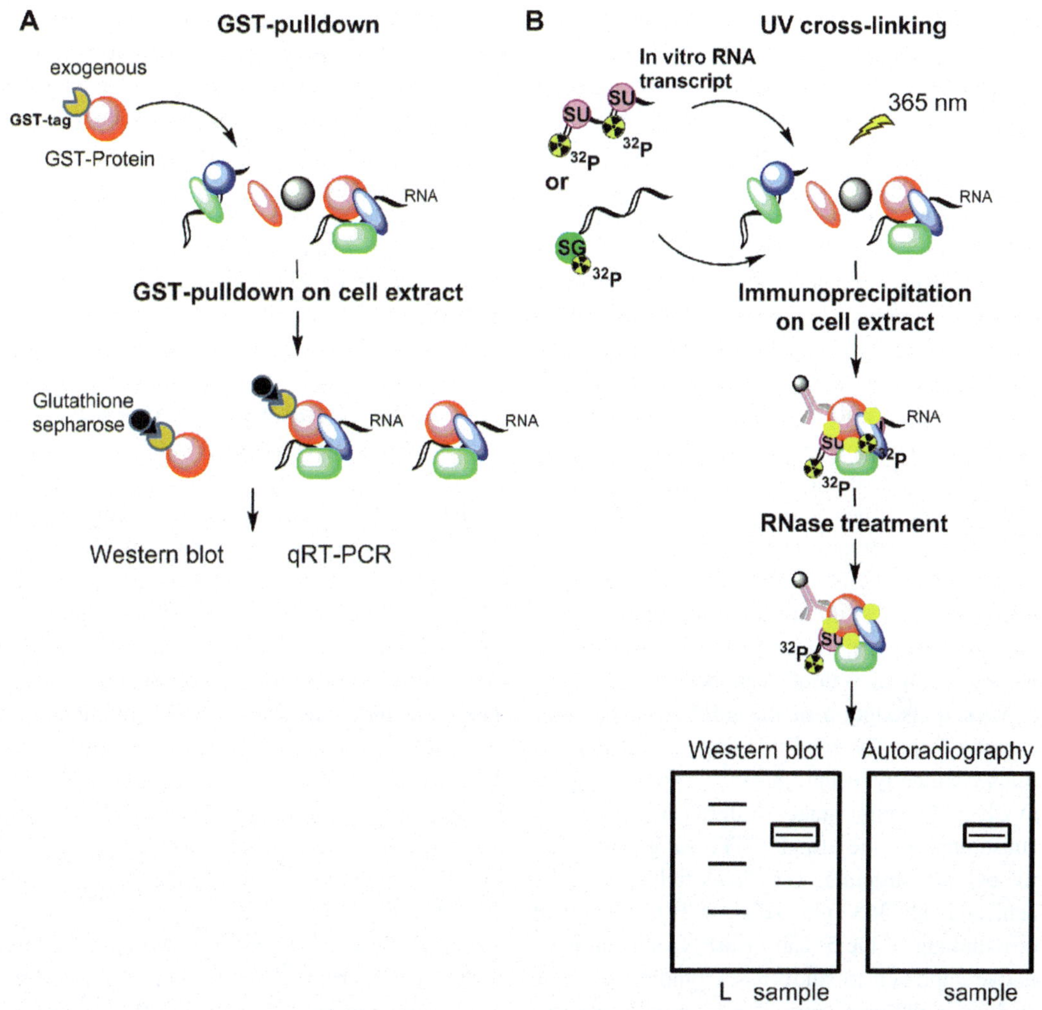

Fig. 1.3 Cell-free extract mRNP immunoprecipitation methods. (**a**) GST pull-down. An exogenous GST-tagged recombinant protein produced in *E. coli* is incubated with cell extracts to reconstitute RNA-protein interactions. After affinity purification using Glutathione sepharose beads, the co-precipitated mRNA can be identified by target mRNA identification methods such as qRT-PCR. (**b**) Illustration of cross-linking methods using exogenous 4-thioU or 6-thioG labelled mRNAs. 4-thioU and ^{32}P labelled transcripts or 6-thioG/^{32}P cap modified mRNA transcripts are incubated with cell free extracts and cross-linked to RBPs at 365 nm. Partial RNase digestion allows transferring the fragments of radiolabelled mRNA to the interacting protein. After immunoprecipitation comparison of Western blot and Phosphorimager analysis of an SDS-gels allows to identify the 5′-^{32}P-labeled protein. SU and SG represent 4-thioU and 6-thioG nucleotides respectively. Cross-links are illustrated as in Fig. 1.2

an exogenous *in vitro* transcribed ^{32}P-labelled mRNA containing 4-thioU nucleotides that is incubated in the presence of total cell extracts (Fig. 1.3b). After UV-cross linking at 365 nm, the crosslinked RNAs is digested by RNase T1 or A [58]. Alternatively *in vitro* transcribed ^{32}P-labelled mRNA devoid of 4-thioU can also be UV cross-linked directly to proteins from cell extracts at 254 nm. Subsequent immunoprecipitation and separation by denaturing gel electrophoresis allows resolving RNase-protected ^{32}P-labelled mRNA fragments cross-linked to specific RNA binding proteins. Overlay of Western-blot and autoradiography analysis of the SDS–PAGE gel can allow the detection of pairwise protein-RNA interactions. Critical to the success of these experiments is the preparation of the *in vitro* 4-thioU labelled RNA, in particular the optimization of the 4-thioUTP: UTP ratio during transcription [59, 60]. UV crosslinking using mRNAs containing 4-thiouridine have allowed the characterization RNA–protein interactions between various mRNAs and ribosomal proteins as well as numerous translation initiation factors including eIF3, eIF4GII [61–63]. These experiments have also revealed the mRNA binding function of Gemin5, a component of the SMN chaperone complex, and its role in translation [64]. We have recently developed UV and chemical crosslinking methods specifically dedicated to track proteins bound to the 5′ cap structure of an mRNA of interest during the translation process in cell-free extracts [65]. Site-specific incorporation of 6-thioG can be achieved after *in vitro* transcription of the mRNA at the m^7G capping step. The modified cap is added post-transcriptionally by the Vaccine Capping Enzyme (VCE). At this stage 6-thioGTP (s6G) is substituted to the usual GTP, to generate an m^7s6G-cap structure. UV cross-linking can be achieved at 365 nm (Fig. 1.3b). Chemical crosslinking of the 5′ cap of mRNAs is an alternative approach to study RNA binding proteins such as initiation factors, it was initially used to identify the cap-binding protein eIF4E [66]. The target mRNA transcripts containing a radioactively labelled $m^7[^{32}P]G(5')pppG$ cap, is oxidized with sodium periodate and incubated in cell-free extracts leading to the formation of Schiff bases between the 5′ cap and amino groups of interacting proteins [65]. Subsequent RNase A digestion allows to leave the radiolabelled cap covalently linked to its protein partners that can be identified directly by Western blot or after immunoprecipitation. These methods allowed us to follow the histone H4 mRNA cap binding proteins during H4 mRNA translation with the use of specific translation inhibitors for different steps [65, 67].

1.4 Direct mRNA IP: Anti-m^6A, m^5C and Anti-TMG IP

While mRNP assembly and remodelling is essential to gene expression control, mRNA post-transcriptional modifications have emerged as an additional layer of gene expression regulation [68]. Different types of nucleotide modification have been documented in eukaryotic mRNAs. These include internal mRNA modifications such as 5-methylcytosine (m^5C) [69], and N^6-methyladenosine (m^6A) [70], that play regulatory roles, but also m^7G and $m_3{}^{2,2,7}G$ 5′ cap modifications. Mammalian mRNAs, synthesized by RNA polymerase II (polII), are generally characterized by the presence of a 7-methylguanosine (m^7G) cap structure at their 5′ end [71]. The m^7G cap required for mRNA processing, translation initiation, mRNA transport, splicing and degradation [72–77]. We showed that several selenoprotein mRNAs bear a trimethylated $m_3{}^{2,2,7}G$ cap (Fig. 1.4a) and are thus subjected to a non-conventional translation initiation mechanism [31]. Only a limited number of mRNA

Fig. 1.4 (continued) polysome fractions. Cytoplasmic extracts are fractionated onto 7–47% (w/v) linear sucrose gradient after cycloheximide treatment in order to block translation elongation. A typical absorbance profile is represented in the top panel and the positions of the polysomes, 80S ribosome as well as free RNAs are indicated. The middle panel represents relative mRNA abundance in each fraction measured by qRT-PCR in the case of the GPx1 selenoprotein mRNA and a control mRNA. Vertical bars mark the position of the polysome as well as RNP and free fractions that are pooled and can be analysed by TMG-IP. The amount of RNA immunoprecipitated from the polysome and RNP pools are measured separately by qRT-PCR and normalized to 100%. Error bars represent standard deviations of an average of two independent experiments

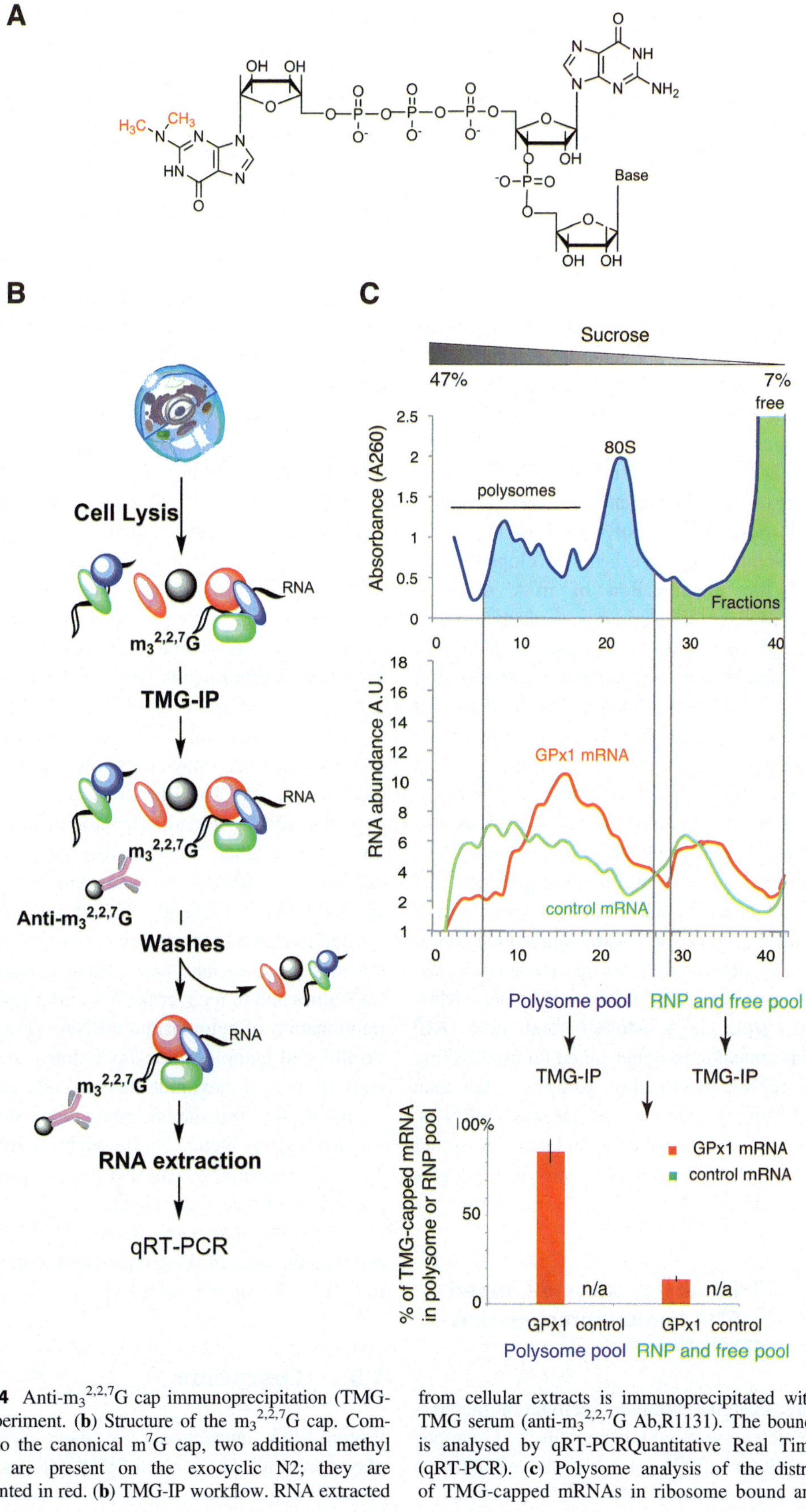

Fig. 1.4 Anti-$m_3^{2,2,7}$G cap immunoprecipitation (TMG-IP) experiment. (**b**) Structure of the $m_3^{2,2,7}$G cap. Compared to the canonical m^7G cap, two additional methyl groups are present on the exocyclic N2; they are represented in red. (**b**) TMG-IP workflow. RNA extracted from cellular extracts is immunoprecipitated with anti-TMG serum (anti-$m_3^{2,2,7}$G Ab,R1131). The bound RNA is analysed by qRT-PCRQuantitative Real Time-PCR (qRT-PCR). (**c**) Polysome analysis of the distribution of TMG-capped mRNAs in ribosome bound and free

immunoprecipitation methods allow to directly recover endogenous mRNAs using antibodies directed against specific nucleotide modification; these include detection of $m_3^{2,2,7}$G-cap, m^6A and m^5C (Table 1.1).

1.4.1 m^6A Me and m^5C RIP

N^6-methyladenosine (m^6A) is the most prevalent internal (non-cap) modification present in the messenger RNA of higher eukaryotes [78–80]. This RNA methylation is reversible and may dynamically control mRNA metabolism. RIP methods called anti-m^6A Me-RIP [81] or m^6A-seq [82], based on antibody-mediated capture followed by either qRT-PCR or massively parallel sequencing have been developed. They allowed the identification of m^6A modified mRNAs [83] and transcriptomes (methylome) in human cells and mouse tissues and revealed m^6A enrichments within long exons and around stop codons [84, 85], further suggesting fundamental regulatory roles of m^6A. 5-methylcytidine (m^5C) is another mRNA modification that plays a role in gene regulation [86, 87]. Anti-m^5C RIP is based on the use of monoclonal antibodies that specifically bind 5-methylcytosine [88] and that have been broadly used in pull-down experiments of modified DNA molecules. These antibodies were raised against 5-methylcytosine nucleotide conjugated to ovalbumin without the ribose or deoxyribose sugar and are blind to the DNA/RNA context. Both m^6A Me-RIP and m^5C RIP methods appear to be better suited for establishing global mRNA modification patterns rather than for the specific recovery of targeted mRNAs, and they are usually associated to other epitranscriptome characterization technologies (for a review see [89]).

1.4.2 Trimethylguanosine-Capped mRNA Immunoprecipitation (TMG-IP)

We have designed immunoprecipitation methods for the detection of endogenous $m_3^{2,2,7}$G-capped selenoprotein mRNAs from total cell extracts or after polysome fractionation of cytoplasmic extracts [90]. The detection of the $m_3^{2,2,7}$G-capped RNAs can be performed by immunoprecipitation experiments using a highly specific anti-$m_3^{2,2,7}$G cap R1131 serum (or anti-TMG serum) that was demonstrated not to recognize monomethylated caps [91, 92]. On the contrary, antibodies developed against m^7G-cap recognize both m^7G and $m_3^{2,2,7}$G modifications. Experiments based on the use of anti-TMG serum are called TMG-IP and have been used successfully to characterize the $m_3^{2,2,7}$G-capped small non-coding RNAs such as snRNAs and snoRNAs. We have adapted this assay coupled to real time quantitative PCR for the detection of endogenous $m_3^{2,2,7}$G-capped selenoprotein mRNAs *in vivo* from cultured cells (Fig. 1.4b) [31, 90]. The fraction of the $m_3^{2,2,7}$G-capped selenoprotein mRNAs recovered by TMG-IP does not exceed 5–15%, whereas this level can reach close to 100% for $m_3^{2,2,7}$G-capped snoRNAs. Determinants such as the larger size and lower abundance as well as stability and shorter half-life of mRNAs compared to non-coding RNAs may contribute to their lower recovery in the TMG-IP. The contribution of potential mRNA folded structures in the 5′UTR must also be taken into consideration in the recognition processes of individual mRNAs by the antibody. The TMG-IP method can also be applied to evaluate the ability of $m_3^{2,2,7}$G-capped mRNAs to associate with actively translating ribosomes and to recover mRNAs after polysome fractionation of cytoplasmic mRNAs (Fig. 1.4c). To this end cytoplasmic extracts must be recovered from cycloheximide-treated cells in order to block the translation elongation step and fractionated on linear 7–47% sucrose gradients. TMG-IP experiments can then be performed on pooled fractions that contain either free or non-polysome associated RNAs. This allows to estimate the ratio of ribosome bound compared to free $m_3^{2,2,7}$G-capped mRNAs [31].

1.5 Conclusion

Protein-RNA interactions and their modularity play a central role in the mRNA fate including their assembly, modification, transport, translation and degradation. While many methods have been developed to examine mRNP complexes, there are still significant challenges that need to be addressed. The precise nature of the protein complexes that interact with most RNAs in the cell is still poorly understood and exploring low abundance transcripts remains a difficult task. The advent of high-throughput methods as well as new computational approaches to address protein–protein and protein-RNA networks is transforming our understanding of mRNP architecture as well as biology. In this chapter we have presented protein-centric techniques designed to reveal and validate targeted mRNA–protein interactions.

Acknowledgments The authors are supported by institutional funds from the French Centre National de la Recherche Scientifique (CNRS), the French Agence Nationale de la Recherche [grant ANR H4translation to GE, FM and CA], Doctoral grants of the University of Strasbourg (to LG and FA), and a Lebanese Doctoral grant to HH.

References

1. Singh G, Pratt G, Yeo GW et al (2015) The clothes make the mRNA: past and present trends in mRNP fashion. Annu Rev Biochem 84:325–354
2. Moore MJ (2005) From birth to death: the complex lives of eukaryotic mRNAs. Science 309:1514–1518
3. Mitchell SF, Parker R (2014) Principles and properties of eukaryotic mRNPs. Mol Cell 54:547–558
4. Björk P, Wieslander L (2017) Integration of mRNP formation and export. Cell Mol Life Sci 74:2875–2897
5. Müller-McNicoll M, Neugebauer KM (2013) How cells get the message: dynamic assembly and function of mRNA–protein complexes. Nat Rev Genet 14:275
6. Keene JD, Tenenbaum SA (2002) Eukaryotic mRNPs may represent posttranscriptional operons. Mol Cell 9: 1161–1167
7. Singh RN, Howell MD, Ottesen EW et al (2017) Diverse role of survival motor neuron protein. Biochim Biophys Acta – Gene Regulat Mech 1860: 299–315
8. Massenet S, Bertrand E, Verheggen C (2017) Assembly and trafficking of box C/D and H/ACA snoRNPs. RNA Biol 14:680–692
9. Chen C-YA, Shyu A-B (2014) Emerging mechanisms of mRNP remodeling regulation. Wiley Interdiscip Rev RNA 5:713–722
10. McHugh CA, Russell P, Guttman M (2014) Methods for comprehensive experimental identification of RNA-protein interactions. Genome Biol 15:203
11. Marchese D, de Groot NS, Lorenzo Gotor N et al (2016) Advances in the characterization of RNA-binding proteins. Wiley Interdiscip Rev RNA 7:793–810
12. Li X, Song J, Yi C (2014) Genome-wide mapping of cellular protein-RNA interactions enabled by chemical crosslinking. Genomics Proteomics Bioinformatics 12: 72–78
13. Tenenbaum SA, Carson CC, Lager PJ et al (2000) Identifying mRNA subsets in messenger ribonucleoprotein complexes by using cDNA arrays. Proc Natl Acad Sci U S A 97:14085–14090
14. König J, Zarnack K, Rot G et al (2010) iCLIP reveals the function of hnRNP particles in splicing at individual nucleotide resolution. Nat Struct Mol Biol 17:909–915
15. Hafner M, Landthaler M, Burger L et al (2010) Transcriptome-wide identification of RNA-binding protein and microRNA target sites by PAR-CLIP. Cell 141:129–141
16. Kastelic N, Landthaler M (2017) mRNA interactome capture in mammalian cells. Methods 126:38–43
17. Castello A, Horos R, Strein C et al (2013) System-wide identification of RNA-binding proteins by interactome capture. Nat Protoc 8:491
18. Castello A, Frese CK, Fischer B et al (2017) Identification of RNA-binding domains of RNA-binding proteins in cultured cells on a system-wide scale with RBDmap. Nat Protoc 12:2447
19. Ryder SP (2016) Protein-mRNA interactome capture: cartography of the mRNP landscape. F1000Research 5:2627
20. Köster T, Marondedze C, Meyer K et al (2017) RNA-binding proteins revisited – the emerging arabidopsis mRNA interactome. Trends Plant Sci 22: 512–526
21. Mili S, Steitz J (2004) Evidence for reassociation of RNA-binding proteins after cell lysis: implications for the interpretation of immunoprecipitation analyses. RNA 10:1692–1694
22. Keene JD, Komisarow JM, Friedersdorf MB (2006) RIP-Chip: the isolation and identification of mRNAs, microRNAs and protein components of ribonucleoprotein complexes from cell extracts. Nat Protoc 1:302–307
23. Mayeda A, Krainer AR (1999) Preparation of HeLa cell nuclear and cytosolic S100 extracts for in vitro splicing. Methods Mol Biol 118:309–314
24. Kataoka N, Dreyfuss G (2008) Preparation of efficient splicing extracts from whole cells, nuclei, and cytoplasmic fractions. Methods Mol Biol 488:357–365
25. Abmayr SM, Yao T, Parmely T et al (2006) Preparation of nuclear and cytoplasmic extracts from mammalian cells. Curr Protoc Pharmacol 12:3
26. Dignam JD, Lebovitz RM, Roeder RG (1983) Accurate transcription initiation by RNA polymerase II in a soluble extract from isolated mammalian nuclei. Nucleic Acids Res 11:1475–1489
27. Webb C-HT, Hertel KJ (2014) Preparation of splicing competent nuclear extracts. Methods Mole Biol (Clifton, NJ) 1126:117–121

28. Boulon S, Marmier-Gourrier N, Pradet-Balade B et al (2008) The Hsp90 chaperone controls the biogenesis of L7Ae RNPs through conserved machinery. J Cell Biol 180:579–595
29. Gribling-Burrer A-S, Leichter M, Wurth L et al (2017) SECIS-binding protein 2 interacts with the SMN complex and the methylosome for selenoprotein mRNP assembly and translation. Nucleic Acids Res 45: 5399–5413
30. Bustin SA, Benes V, Garson JA et al (2009) The MIQE guidelines: minimum information for publication of quantitative real-time PCR experiments. Clin Chem 55:611–622
31. Wurth L, Gribling-Burrer A-S, Verheggen C et al (2014) Hypermethylated-capped selenoprotein mRNAs in mammals. Nucleic Acids Res 42:8663–8677
32. Wheeler EC, Van Nostrand EL, Yeo GW (2018) Advances and challenges in the detection of transcriptome-wide protein-RNA interactions. Wiley Interdiscip Rev RNA 9:e1436
33. Zambelli F, Pavesi G (2015) RIP-seq data analysis to determine RNA–protein associations. Methods Mol Biol 1269:293–303
34. Selbach M, Mann M (2006) Protein interaction screening by quantitative immunoprecipitation combined with knockdown (QUICK). Nat Methods 3:981
35. Meyer C, Garzia A, Tuschl T (2017) Simultaneous detection of the subcellular localization of RNAs and proteins in cultured cells by combined multicolor RNA-FISH and IF. Methods 118–119:101–110
36. Van Criekinge W, Beyaert R (1999) Yeast two-hybrid: state of the art. Biol Proced Online 2:1–38
37. Setzer DR (1999) Measuring equilibrium and kinetic constants using gel retardation assays. Methods Mol Biol 118:115–128
38. Hellman LM, Fried MG (2007) Electrophoretic mobility shift assay (EMSA) for detecting protein-nucleic acid interactions. Nat Protoc 2:1849–1861
39. Trinkle-Mulcahy L (2012) Resolving protein interactions and complexes by affinity purification followed by label-based quantitative mass spectrometry. Proteomics 10:1623–1638
40. Gerace E, Moazed D (2015) Affinity pull-down of proteins using anti-FLAG M2 agarose beads. Methods Enzymol 559:99–110
41. Trinkle-Mulcahy L, Boulon S, Lam YW et al (2008) Identifying specific protein interaction partners using quantitative mass spectrometry and bead proteomes. J Cell Biol 183:223–239
42. Hoffman EA, Frey BL, Smith LM et al (2015) Formaldehyde crosslinking: a tool for the study of chromatin complexes. J Biol Chem 290:26404–26411
43. Orlando V (2000) Mapping chromosomal proteins in vivo by formaldehyde-crosslinked-chromatin immunoprecipitation. Trends Biochem Sci 25:99–104
44. Sutherland BW, Toews J, Kast J (2008) Utility of formaldehyde cross-linking and mass spectrometry in the study of protein-protein interactions. J Mass Spectrom 43:699–715
45. Au PCK, Helliwell C, Wang M-B (2014) Characterizing RNA–protein interaction using cross-linking and metabolite supplemented nuclear RNA-immunoprecipitation. Mol Biol Rep 41:2971–2977
46. Kaneko S, Manley JL (2005) The mammalian RNA polymerase II C-terminal domain interacts with RNA to suppress transcription-coupled 3′ end formation. Mol Cell 20:91–103
47. Gilbert C, Kristjuhan A, Winkler GS et al (2004) Elongator interactions with nascent mRNA revealed by RNA immunoprecipitation. Mol Cell 14:457–464
48. Jivotovskaya AV, Valášek L, Hinnebusch AG et al (2006) Eukaryotic translation initiation factor 3 (eIF3) and eIF2 can promote mRNA binding to 40S subunits independently of eIF4G in yeast. Mol Cell Biol 26:1355–1372
49. Niranjanakumari S, Lasda E, Brazas R et al (2002) Reversible cross-linking combined with immunoprecipitation to study RNA–protein interactions in vivo. Methods 26:182–190
50. Pashev IG, Dimitrov SI, Angelov D (1991) Crosslinking proteins to nucleic acids by ultraviolet laser irradiation. Trends Biochem Sci 16:323–326
51. Jensen KB, Darnell RB (2008) CLIP: crosslinking and immunoprecipitation of in vivo RNA targets of RNA-binding proteins. Methods Mol Biol 488:85–98
52. Kishore S, Jaskiewicz L, Burger L et al (2011) A quantitative analysis of CLIP methods for identifying binding sites of RNA-binding proteins. Nat Methods 8:559–567
53. Hussain S, Sajini AA, Blanco S et al (2013) NSun2-mediated cytosine-5 methylation of vault noncoding RNA determines its processing into regulatory small RNAs. Cell Rep 4:255–261
54. Bohnsack MT, Martin R, Granneman S et al (2009) Prp43 bound at different sites on the pre-rRNA performs distinct functions in ribosome synthesis. Mol Cell 36:583–592
55. Granneman S, Kudla G, Petfalski E et al (2009) Identification of protein binding sites on U3 snoRNA and pre-rRNA by UV cross-linking and high-throughput analysis of cDNAs. Proc Natl Acad Sci 106:9613–9618
56. Strein C, Alleaume AM, Rothbauer U et al (2014) A versatile assay for RNA-binding proteins in living cells. RNA 20:721–731
57. Choi YH, Hagedorn CH (2003) Purifying mRNAs with a high-affinity eIF4E mutant identifies the short 3′ poly(A) end phenotype. Proc Natl Acad Sci U S A 100:7033–7038
58. Lee ASY, Kranzusch PJ, Cate JHD (2015) eIF3 targets cell proliferation mRNAs for translational activation or repression. Nature 522:111–114
59. Stade K, Rinke-Appel J, Brimacombe R (1989) Site-directed cross-linking of mRNA analogues to the Escherichia coli ribosome; identification of 30S ribosomal components that can be cross-linked to the mRNA at various points 5′ with respect to the decoding site. Nucleic Acids Res 17:9889–9908

60. Bélanger F, Baigude H, Rana TM (2009) U30 of 7SK RNA forms a specific photo-crosslink with Hexim1 in the context of both a minimal RNA-binding site and a full reconstituted 7SK/Hexim1/P-TEFb ribonucleoprotein complex. J Mol Biol 386:1094–1107
61. López de Quinto S, Lafuente E, Martínez-Salas E (2001) IRES interaction with translation initiation factors: functional characterization of novel RNA contacts with eIF3, eIF4B, and eIF4GII. RNA 7:1213–1226
62. Pisarev AV, Kolupaeva VG, Yusupov MM et al (2008) Ribosomal position and contacts of mRNA in eukaryotic translation initiation complexes. EMBO J 27:1609–1621
63. Lee ASY, Kranzusch PJ, Doudna JA et al (2016) eIF3d is an mRNA cap-binding protein that is required for specialized translation initiation. Nature 536:96
64. Pacheco A, de Quinto SL, Ramajo J et al (2009) A novel role for Gemin5 in mRNA translation. Nucleic Acids Res 37:582–590
65. Gross L, Schaeffer L, Alghoul F et al (2018) Tracking the m^7G-cap during translation initiation by crosslinking methods. Methods 137:3–10
66. Sonenberg N, Shatkin AJ (1977) Reovirus mRNA can be covalently crosslinked via the 5′ cap to proteins in initiation complexes. Proc Natl Acad Sci U S A 74: 4288–4292
67. Martin F, Barends S, Jaeger S et al (2011) Cap-assisted internal initiation of translation of histone H4. Mol Cell 41:197–209
68. Batista PJ (2017) The RNA modification N6-methyladenosine and its implications in human disease. Genomics Proteomics Bioinformatics 15:154–163
69. Squires JE, Patel HR, Nousch M et al (2012) Widespread occurrence of 5-methylcytosine in human coding and non-coding RNA. Nucleic Acids Res 40:5023–5033
70. Desrosiers R, Friderici K, Rottman F (1974) Identification of methylated nucleosides in messenger RNA from Novikoff hepatoma cells. Proc Natl Acad Sci 71: 3971–3975
71. Shatkin AJ (1976) Capping of eucaryotic mRNAs. Cell 9:645–653
72. Topisirovic I, Svitkin YV, Sonenberg N et al (2011) Cap and cap-binding proteins in the control of gene expression, Wiley interdisciplinary reviews. RNA 2:277–298
73. Houseley J, Tollervey D (2009) The many pathways of RNA degradation. Cell 136:763–776
74. Kohler A, Hurt E (2007) Exporting RNA from the nucleus to the cytoplasm. Nat Rev Mol Cell Biol 8: 761–773
75. Fortes P, Inada T, Preiss T et al (2000) The yeast nuclear cap binding complex can interact with translation factor eIF4G and mediate translation initiation. Mol Cell 6:191–196
76. Wilusz CJ, Wormington M, Peltz SW (2001) The cap-to-tail guide to mRNA turnover. Nat Rev Mol Cell Biol 2:237–246
77. Izaurralde E, Lewis J, Gamberi C et al (1995) A cap-binding protein complex mediating U snRNA export. Nature 376:709–712
78. Tuck MT (1992) The formation of internal 6-methyladenine residues in eucaryotic messenger rna. Int J Biochem 24:379–386
79. Wang X, Lu Z, Gomez A et al (2013) N6-methyladenosine-dependent regulation of messenger RNA stability. Nature 505:117–120
80. Jia G, Fu Y, He C (2013) Reversible RNA adenosine methylation in biological regulation. Trends Genet 29: 108–115
81. Warda AS, Kretschmer J, Hackert P et al (2017) Human METTL16 is a N^6-methyladenosine (m^6A) methyltransferase that targets pre-mRNAs and various non-coding RNAs. EMBO Rep 18:2004–2014
82. Dominissini D, Moshitch-Moshkovitz S, Schwartz S et al (2012) Topology of the human and mouse m6A RNA methylomes revealed by m6A-seq. Nature 485:201
83. Liu N, Zhou KI, Parisien M et al (2017) N6-methyladenosine alters RNA structure to regulate binding of a low-complexity protein. Nucleic Acids Res 45:6051–6063
84. Dominissini D, Moshitch-Moshkovitz S, Salmon-Divon M et al (2013) Transcriptome-wide mapping of N6-methyladenosine by m6A-seq based on immunocapturing and massively parallel sequencing. Nat Protoc 8:176
85. Meyer KD, Saletore Y, Zumbo P et al (2012) Comprehensive analysis of mRNA methylation reveals enrichment in 3′ UTRs and near stop codons. Cell 149: 1635–1646
86. Delatte B, Wang F, Ngoc LV et al (2016) Transcriptome-wide distribution and function of RNA hydroxymethylcytosine. Science 351:282–285
87. Yang X, Yang Y, Sun B-F et al (2017) 5-methylcytosine promotes mRNA export—NSUN2 as the methyltransferase and ALYREF as an m5C reader. Cell Res 27:606
88. Edelheit S, Schwartz S, Mumbach MR et al (2013) Transcriptome-wide mapping of 5-methylcytidine RNA modifications in bacteria, archaea, and yeast reveals m5C within archaeal mRNAs. PLoS Genet 9: e1003602
89. Li X, Xiong X, Yi C (2016) Epitranscriptome sequencing technologies: decoding RNA modifications. Nat Methods 14:23
90. Gribling-Burrer A-S, Eriani G, Allmang C (2018) Modification of selenoprotein mRNAs by cap tri-methylation. Methods Mol Biol 1661:125–141
91. Luhrmann R, Appel B, Bringmann P et al (1982) Isolation and characterization of rabbit anti-m3 2,2,7G antibodies. Nucleic Acids Res 10:7103–7113
92. Tycowski KT, Aab A, Steitz JA (2004) Guide RNAs with 5′ caps and novel box C/D snoRNA-like domains for modification of snRNAs in metazoa. Curr Biol 14: 1985–1995

Purification of *In Vivo* or *In Vitro*-Assembled RNA-Protein Complexes by RNA Centric Methods

2

Aurélie Janvier, Hassan Hayek, Fatima Alghoul, Lauriane Gross, Christine Allmang, Franck Martin, and Gilbert Eriani

Abstract

Throughout their entire life cycle, RNAs are associated with RNA-binding proteins (RBPs), forming ribonucleoprotein (RNP) complexes with highly dynamic compositions and very diverse functions in RNA metabolism, including splicing, translational regulation, ribosome assembly. Many RNPs remain poorly characterized due to the challenges inherent in their purification and subsequent biochemical characterization. Therefore, developing methods to isolate specific RNA-protein complexes is an important initial step toward understanding their function. Many elegant methodologies have been developed to isolate RNPs. This chapter describes different approaches and methods devised for RNA-specific purification of a target RNP. We focused on general methods for selecting RNPs that target a given RNA under conditions favourable for the copurification of associated factors including RNAs and protein components of the RNP.

Keywords

RNA protein complexes · RNP ribonucleoprotein · Ribosomes · Affinity purification · Mammalian cell-free extracts · Complex assembly · RNA aptamer · RNA tag

2.1 Introduction: Issues and Challenges of RNP Purifications

The use of small, genetically introduced affinity tags for the production and purification of recombinant proteins and their complexes with other proteins has greatly improved our knowledge on their function in a variety of research fields. The usual protein affinity tags include polyhistidine (poly-His), the hemagglutinin epitope (HA-tag), myc epitope, glutathione S-transferase (GST-tag), Maltose-Binding Protein (MBP-tag), Strep-tag, FLAG epitope (Flag-tag), protein A, etc. These affinity tags all bind with high affinity to a ligand that can be immobilized on a chromatography resin for further purification. In most cases, the bound complexes can either be released from the resin by competitive elution or cleaved off by a protease with a recognition site incorporated in the fusion protein. The widespread use and success of protein affinity tags has led to the development of comparable tags for nucleic acids leading to new RNA-centric methods for affinity-purification of

A. Janvier · H. Hayek · F. Alghoul · L. Gross · C. Allmang · F. Martin · G. Eriani (✉)
Architecture et Réactivité de l'ARN, Centre National de la Recherche Scientifique, Institut de Biologie Moléculaire et Cellulaire, Université de Strasbourg, Strasbourg, France
e-mail: g.eriani@unistra.fr

M. C. Vega, F. J. Fernández (eds.), *Advanced Technologies for Protein Complex Production and Characterization*, Advances in Experimental Medicine and Biology 1453,
https://doi.org/10.1007/978-3-031-52193-5_2

RNA-binding proteins (RBDs) or ribonucleoparticles (RNPs). RNA-centric methods target RBPs or RNPs binding to a single RNA of interest. The majority of the existing methods use tagged RNAs as baits to capture and study complexes bound to it. During these last years, particular emphasis has been placed on developing new affinity tags for RNA molecules. Between the early studies in the 90 s and today's methods, a large variety of diverse RNA tags have been described with various applications including functional analyses, in addition to mass-spectrometry, cell imaging, or three-dimensional structure determination.

Different approaches have been used to tag RNA molecules. (i) RNAs can be chemically tagged during *in vitro* transcription through the incorporation of modified ribonucleotides that contain biotin, fluorescent dyes, or other compounds. (ii) Well-characterized protein-binding RNA sequences can be incorporated during *in vitro* or *in vivo* transcription. Such natural sequences might be further optimized in size and affinity. (iii) Similarly, artificially selected RNA aptamers can be incorporated during transcription. Because of their small sizes, such aptamer tags are easy to insert into the RNA of interest. The tagged RNAs and assembled RNPs are subsequently purified by affinity-purification on an adapted resin. The limiting step often consists in the elution of the assembled complexes in native conditions since the binding affinities are typically high. (iv) Hybridization of biotinylated oligonucleotides that are complementary to the RNA target is an efficient purification strategy. It requires accessible single-stranded regions and elution can be achieved under denaturing conditions, or displacement by a competitor oligonucleotide, or by targeted RNase degradation. (v) Alternatively, the RNA sequence of interest may be ligated to a biotinylated-DNA oligonucleotide in order to yield chimeric RNA-DNA molecules that can efficiently bind streptavidin beads. Benefits of this procedure include the absence of constraints resulting from the use of a foreign structured tag. In addition, the elution is performed in native conditions by DNase or targeted RNase degradation that releases the RNA-protein complexes without any extra sequence. Each of these methods has particular advantages and disadvantages for the tagging or affinity isolation or elution of RNPs. They will be discussed hereafter.

2.2 First Affinity Purification Methods

Two early studies performed in the 90 s established the foundations of affinity purification based on natural RNA regulatory sequence as bait [1, 2]. The studies focused on the Iron Responsive Element (IRE) that binds the Iron Responsive Element Binding-Protein (IRE-BP) involved in iron homeostasis. In the first study, the RNA transcript containing the regulatory sequences upstream of a biotinylated unstructured tail was added in solution to unfractionated lysates containing the protein target. Then, the high affinity interaction between biotin and avidin was used to attach the IRE/IRE-BP to a solid matrix. For that, the assembled mRNA-protein complexes were bound to biotin-agarose beads through a succinylated-avidin intermediate (Fig. 2.1a). IRE-BP was subsequently eluted from the RNA with 2 M KCl [1]. In the second study, *in vitro* transcribed polyadenylated RNA was hybridized to a poly(U)-Sepharose resin and incubated with pre-purified cellular extracts. The RNA-protein complexes were assembled, and the adsorbed proteins were recovered by elution with 1 M KCl [2] (Fig. 2.1b). Although these two seminal studies reached the mRNP assembly on a specific RNA target, the complex was dissociated during the high-salt elution. Nevertheless, they successfully tested two RNA-tagging procedures. The first was using biotinylated nucleotides inserted in the RNA for binding to the beads, and the second procedure took advantage of the RNA-RNA hybridization to immobilize RNA-protein complexes on a solid matrix.

Over the same period, a more general affinity purification method that allowed the isolation of specific RNAs and RNA-protein complexes was described [3]. The approach, illustrated in Fig. 2.1c, hinges on the specific interaction between the bacteriophage R17 coat protein and a short hairpin found in its genomic RNA. In this

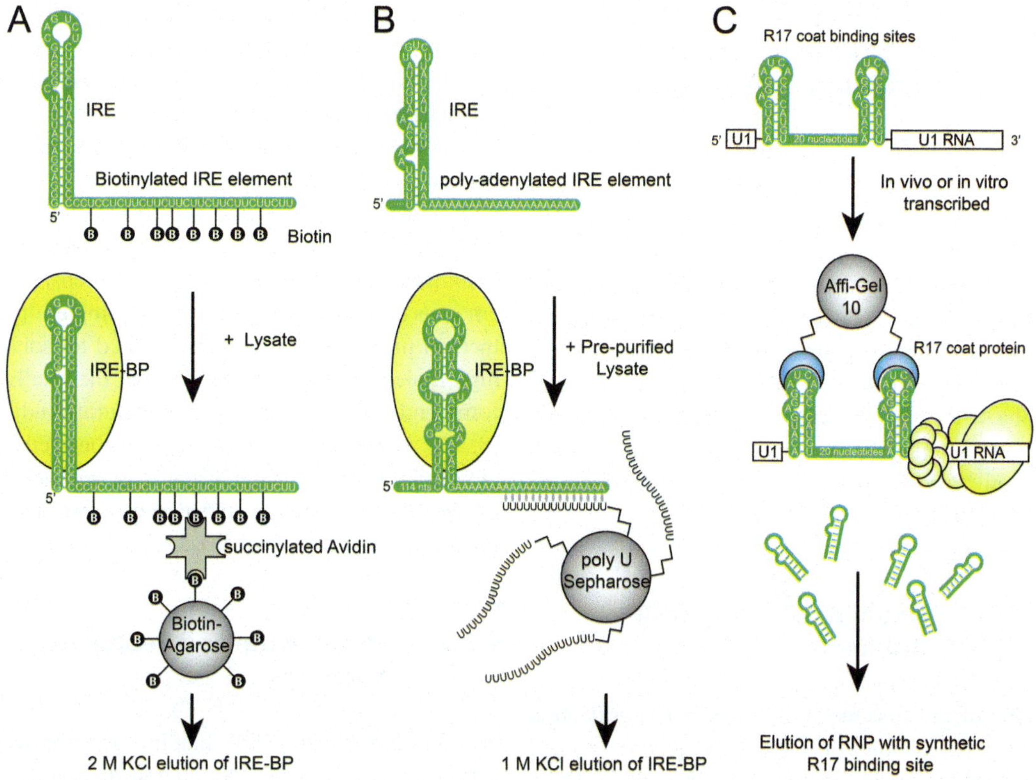

Fig. 2.1 Outline of the three methods of purification of a specific RNA protein complex from a mixture of proteins. (**a**) Schematic representation of the affinity purification of IRE-BP based on Biotin-Avidin affinity. The allylamine uridine triphosphate was randomly incorporated throughout the entire transcript before coupling with succinimidyl-biotin. Bound IRE-BP was subsequently eluted with 2 M KCl. IRE, Iron Responsive Element; IRE-BP, IRE-Binding protein. (**b**) Representation of the affinity purification of IRE-BP based on polyU resin. In vitro transcribed polyadenylated RNA was bound to poly (U)-Sepharose. Prepurified IRE-BP was specifically adsorbed to the IRE and subsequently eluted with 1 M KCl. The IRE here shown is one of the transferrin IRE elements. (**c**) Affinity purification of snRNP U1 based on the affinity of R17 coat protein for its RNA-binding element. A chimeric RNA containing two R17 recognition sites inserted in U1 RNA was prepared either *in vitro*, by transcription with a phage polymerase, or *in vivo*, by cellular transcription of a transfected or injected DNA template. The chimeric RNA binds to appropriate factors in the cell extract. The resulting complexes are selectively retained on a support to which R17 coat protein has been covalently coupled. Specific RNA molecules and any associated factors are eluted with an excess of R17 recognition elements

strategy, the R17 coat protein (CP) was covalently attached to beads to retain RNA sequences containing the short hairpin sequence. To improve the interaction between the CP and RNA hairpin, a high affinity variant of the RNA hairpin was used ($K_a = 3.5 \times 10^{10}$ M [4]). A hybrid RNA containing two R17 recognition sites fused to the RNA sequence of interest was prepared either *in vitro*, by transcription with a phage polymerase, or *in vivo*, by cellular transcription of a transfected or injected DNA template. After incubating the chimeric RNA with appropriate factors in the cell or lysate, the resulting RNA-factor complexes were retained on the resin coupled to the R17 coat protein. The binding of RNAs to the beads turned out to be rapid, efficient, and highly selective. After eluting the column with an excess of R17 recognition sites, the authors could obtain biologically active factors and complexes of interest [4].

2.3 Affinity Purification Based on Artificially Selected or Natural RNA Motifs

Over the years, a large variety of aptamer-based affinity purification approaches have been developed for the isolation of *in vitro*-assembled RNA–protein complexes. These include the use of StreptoTag aptamers [5], streptavidin aptamer [6], Sephadex aptamer [7], tobramycin aptamers [8], and Mango aptamer [9]. Other natural RNA elements have been used such as the bacteriophage R17 MS2 sequence (binding site of the coat protein CP), or the equivalent PP7 RNA from *Pseudomonas aeruginosa* bacteriophage PP7 [10–13].

2.3.1 Antibiotic-Binding RNA Aptamers

One of the first methods for affinity-purification was based on the RNA sequence named StreptoTag aptamer [5]. This 48-nt long RNA element binds with high binding specificity to the antibiotic streptomycin. It was isolated via SELEX and binds to the antibiotic with a dissociation constant (Kd) of around 1 μM (Fig. 2.2a). The StreptoTag was inserted by *in vitro* transcription in a hybrid RNA and was incubated with crude extract. The assembly reaction was then applied to a dihydrostreptomycin-Sepharose column. The resulting bound complexes were washed and specifically eluted upon addition of free streptomycin. Using this purification scheme, properly assembled spliceosomal U1A protein and the bacteriophage MS2 coat protein could be isolated via their appropriate RNA motif [5] as well as 48S ribosomal complexes assembled on an IRES element in milligram quantities [14]. The method turned out to be more specific than the procedures that use poly-U Sepharose [2] or biotin/streptavidin beads [1] that often copurify unspecific proteins [5]. One of the drawbacks of the method consists in the preparation of the dihydrostreptomycin-Sepharose column. Although the protocol for coupling streptomycin to Sepharose 6B is well established, it is a time-consuming process, and the coupled matrix can only be stored for a maximum of 4 weeks. In addition, the quality of the matrix might be subjected to batch-to-batch variation.

A similar affinity-purification procedure was developed using an RNA aptamer selected to bind with 5 nM affinity the aminoglycoside antibiotic tobramycin [15, 16] (Fig. 2.2a). The tobramycin-binding aptamer was used to purify preparative amounts of human pre-spliceosome complex. The method turned out to be ideally suited to isolate high molecular weight complexes in their native form and for functional and structural studies [8, 16]. However, the drawbacks of the method are similar to those of the StreptoTag and mainly concern the preparation and preservation of the tobramycin-coupled Sepharose support.

2.3.2 Streptavidin and Sephadex Aptamers

The S1 Streptavidin-RNA binding aptamer was selected based on a tight binding to a commonly available target molecule in such a way that the ligand–RNP complex can be selectively and gently dissociated afterwards. The S1 Streptavidin tag binds specifically to streptavidin and can be eluted by competition with biotin. Streptavidin is a protein produced by *Streptomyces avidinii*, which binds to biotin with extraordinary high affinity. The K_d for the biotin interaction is one of the strongest non-covalent biological interactions found in nature (10^{-14} M). Streptavidin is commercially available in many forms, either as a purified protein or conjugated with enzymes, dyes, or many supporting matrices. The Streptavidin aptamer, as an affinity tag (Fig. 2.2a), inserted into the RNA component of the large RNA subunit of RNase P, was successfully used to purify the active form of the ribonucleoprotein RNase P [6]. The S1 aptamer was further used for the isolation of an mRNP involved in translation activation [17] and purification of mRNA-interacting proteins from human cells and Drosophila embryo extracts using *in vitro* transcribed RNAs attached to streptavidin via the S1 tag [18, 19]. It was also shown that the

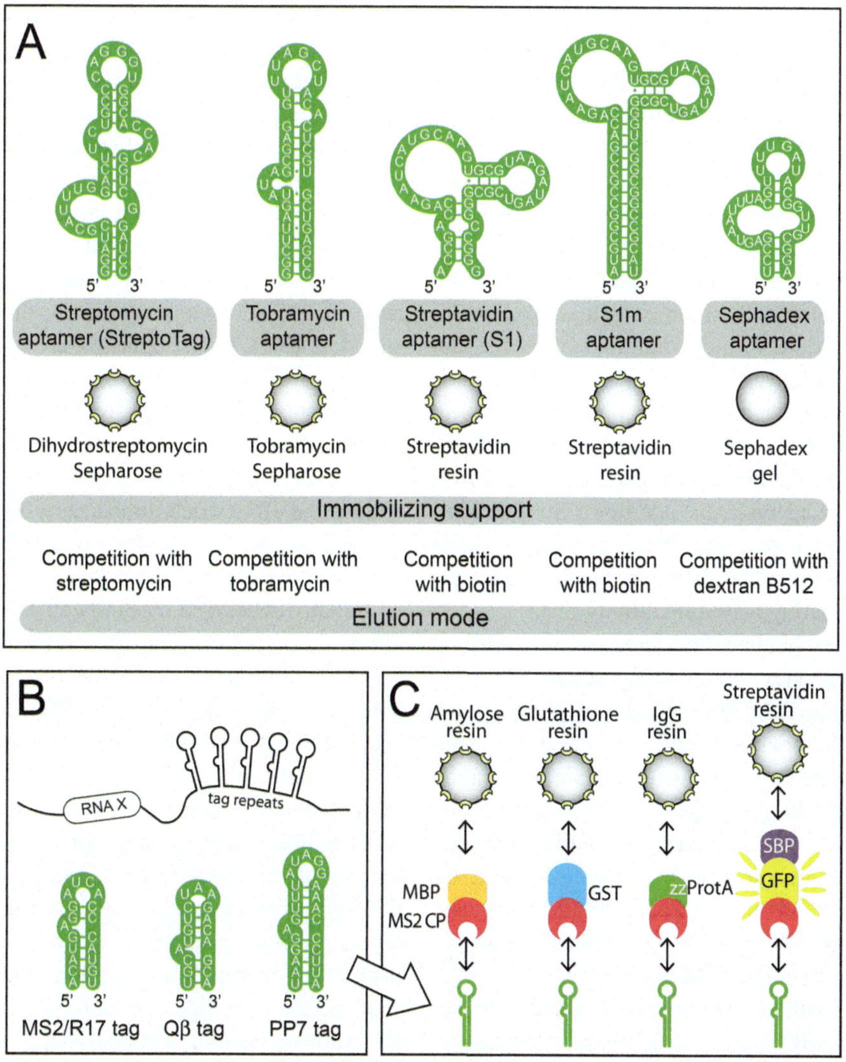

Fig. 2.2 Strategies of purification of RNPs based on artificial RNA aptamers or natural tags. (**a**) Sequences and predicted secondary structures of different RNA aptamers that have been used to develop purification strategies. The streptomycin RNA aptamer sequence (StreptoTag) binds with high affinity to dihydrostreptomycin coupled to a sepharose column matrix. The tobramycin aptamer binds efficiently to tobramycin-derivatized sepharose beads. The streptavidin aptamer (S1) specifically binds to streptavidin conjugated to many commercial supports. S1 was optimized in S1m that can be repeated (4×S1m). Elution is obtained under mild conditions by adding biotin. The Sephadex aptamer is specific to the Sephadex matrix. Elution is performed by dextran B512, which is the base material of Sephadex. (**b**) Several natural sequences from bacteriophages have been used as tags to affinity-purify RNPs. The MS2 hairpin is the most frequently used, with a variable number of copies in the RNA target. (**c**) Since the affinity of the repeated MS2 sequence for the coat protein (CP) is too high for an efficient elution, several fusion proteins have been engineered in order to affinity-bind the MS2 tag on one side (CP) and additional peptides with different binding properties. The Maltose-Binding Peptide (MBP) binds to amylose columns and release of the complexes is obtained by adding maltose or by protease cleavage at specific sites inserted between the fused domains. Other domains have been fused with the CP domain. The GST domain binds to glutathione resin, the zz domain of protein A (zzProtA) binds to IgG coated beads and the fusion with the Streptavidin-Binding Peptide (SBP) allows binding to streptavidin coated resins whereas the GFP domain adds fluorescent properties. In all that cases, co-expression or co-purification of the fusion proteins is needed

addition of a tRNA scaffold to the Streptavidin aptamer increased binding efficiency of about tenfold. Similarly, by optimizing the RNA aptamer S1 in structure and repeat conformation, the affinity for Streptavidin was increased and found to be optimal with a fourfold repeat of S1m (4 × S1m) and be more efficient than the established MS2 and PP7 systems from bacteriophages [20].

The advantages of the Streptavidin tag include an elution step with biotin, which is done under mild conditions. Mild conditions preserve the particle integrity and increase purity since non-specific contaminants that bind to the column support would not elute (on the contrary to stringent elution with high salt, pH changes, or with denaturing agents). In addition, Streptavidin-matrices are widely distributed as agarose beads or magnetic particles. This eliminates the necessity of preparing a specific affinity resin.

Sephadex-binding RNA aptamers were selected against Sephadex, a commonly used matrix in gel filtration (Fig. 2.2a) [7]. Sephadex is widely used in many laboratories and readily available. It is relatively inexpensive, making large-scale purification more affordable. Sephadex is stable and can be regenerated several times. Its basic structure, formed of glucose repeats connected via α-1,6 glucosidic bonds, provides a high binding capacity. Using Sephadex G-100, the Sephadex aptamer could be purified from a complex mixture of cellular RNA, with an enrichment of at least 60,000-fold. Similarly, yeast nuclear RNase P containing one RNA molecule tightly associated with 9 protein subunits could be purified and eluted by competition with soluble dextran B512 [21].

2.3.3 Affinity Purification with RNA Tags Derived from Natural Sequences

A variety of studies have focused on natural RNA tags that are small and form structured RNA motifs that can fold independently of the tagged RNA (Fig. 2.2b). They bind specific ligands with high affinity. They include the MS2 RNA hairpin [10–13, 22, 23], the *Pseudomonas* phage 7 (PP7) RNA hairpin for coat protein [24–28], the 19-nts of the B-box of lambda bacteriophage that bind the antiterminator protein N (22 amino acid RNA-binding domain) [29–31], or the U1 snRNA hairpin II that binds U1 A protein [32].

The most widely used affinity purification with RNA tag is that of the MS2 coat protein (CP) and its cognate RNA. MS2 is a member of a family of closely related enterobacterial viruses that includes bacteriophage f2, bacteriophage Qβ, and R17. The coat proteins of these single-stranded RNA bacteriophages are translational repressors of their viral replicase. They achieve this function by specifically binding an RNA hairpin that encompasses the replicase start codon. This interaction is generally conserved among the RNA bacteriophages with variations in the sequences of the coat proteins and RNA hairpins during evolution.

Initially developed by Bardwell & Wickens in 1990 (see above) with the R17 recognition sequence originating from the phage R17, the MS2 tag of bacteriophage MS2 has similar sequence and binds the coat protein (CP) with similar high affinity. Three copies of the MS2 tag are usually placed in the RNA of interest in order to improve its binding efficiency. Since the affinity of the repeated MS2 sequence for MS2 CP is too high to consider an efficient elution by direct affinity purification, a hybrid protein is used as an intermediate to affinity-bind the MS2 sequences on one side (CP moiety) in fusion with the Maltose-Binding Peptide (MBP moiety) in order to interact with the resin on the other side. Therefore, there is no requirement for the MS2 CP to be immobilized, and the isolation can be performed using the MBP and not the MS2 coat protein. The most used fusion protein involves the N-terminal domain of the MBP combined with the MS2 CP part with a double mutation (V75Q and A81G) that prevents the oligomerization [33]. The MS2 CP-MBP fusion protein is expressed in *Escherichia coli* and after affinity purification the complex can be released from the affinity resin by elution with maltose (Fig. 2.2b, c). Alternatively, the elution can be performed using a protease cleavage site located between the MS2 coat protein site and the protein that binds directly to the affinity resin. In this case, purified complexes are released from the resin on by cleavage with a specific protease. Regardless of whether the elution proceeds via the maltose or by protease cleavage, the MS2 CP-MBP or MS2 CP remain bound to the mRNA-protein complex, which may represent a disadvantage for structural or functional studies. This approach has been successfully used to purify many large molecular weight complexes such as U2 snRNP with the ATP-independent spliceosomal complex [10], ribosomes on IRES elements [12] or native spliceosomes [11, 13, 22, 34].

Variations of the fusion MS2 CP-MBP include the MS2 CP-GST which binds to Glutathione-beads with an elution step by glutathione [35]. An immunoglobulin variant was proposed with the MS2 CP domain fused to the immunoglobulin-binding domain of the *Staphylococcus aureus* protein A (ZZ domain) (Fig. 2.2c). The complex of interest could be purified from a cell lysate using immunoglobulin-conjugated beads [36]. The MS2-BioTRAP method is another RNA tagging system designed for purification of *in vivo* assembled RNPs. In this method, the RNA of interest harbouring multiple MS2 RNA elements is co-expressed with a MS2 CP fused to the HisBio tag (HB tag). The HB tag adds to the MS2 CP two hexa-His tags, a TEV cleavage site, and a signal sequence for *in vivo* biotinylation. Here, the high affinity of biotin for streptavidin is used to isolate the endogenously biotinylated HB-tagged protein associated with the MS2-tagged RNA within the RNPs [37]. In addition, the MS2 aptamer inspired numerous derived methods for targeting RNPs or miRNAs such as RaPID [38], MS2-TRAP [35], and Ribotrap [36]. The MS2-CP was also fused with Green Fluorescent Protein (GFP) to visualize mRNAs bearing the MS2 aptamer *in vivo* [39–41]. MS2-CP-GFP could also be fused with the Streptavidin-Binding Protein (SBP) in order to give a MS2-CP-GFP-SBP, which could be visualized by fluorescence and purified using streptavidin-conjugated beads [38], however these methods go far beyond the scope of this article.

Generally, all these methods are relatively flexible and widely applicable, and in some cases, their *in vivo* applicability allows the study of protein–RNA interactions in physiological conditions. Tagged RNAs can be stably expressed in appropriate mammalian cell lines and RNPs assembled on the tagged RNAs may be purified from extracts using the appropriate affinity purification method. The primary use of such techniques has involved identification of mRNP components by mass spectrometry, but purified RNPs may also be used for downstream functional assays. However, in some cases, the incorporation of a tag into the RNA bait may alter its secondary structures and possibly the formation of ribonucleoprotein complexes. On the other hand, the presence of the exogenous tag may also induce binding of additional nonspecific RNA binding proteins.

2.4 RNP Capture Based on Antisense Biotinylated Oligonucleotides

These methods are using antisense oligoribonucleotides that are complementary to the RNA of interest (mRNA, or RNP). These complexes are subsequently enriched and purified from cell extracts by hybridization. Usually, one or several biotin residues are included at the 5′ end or internally of the oligonucleotide to immobilize it to a Streptavidin column. Following hybridization, binding and extensive washing steps, the RNA of interest is eluted by a displacement oligonucleotide, or by targeted RNase degradation (Fig. 2.3a). In these methods, the RNA of interest does not need a special preparation; the RNA oligonucleotide is chosen to be specific of the RNA sequence. Usually, 2′-O-alkyl (methyl or allyl) oligoribonucleotides are preferred since they are resistant to nuclease degradation in crude extracts containing high levels of endogenous RNase or DNase activity [42]. Elution of bound RNP from the antisense oligonucleotide can be achieved under non-denaturing conditions, using a displacement oligonucleotide, which can form a thermodynamically more stable duplex with the antisense oligonucleotide [43–45]. Alternatively, a DNA oligonucleotide complementary to

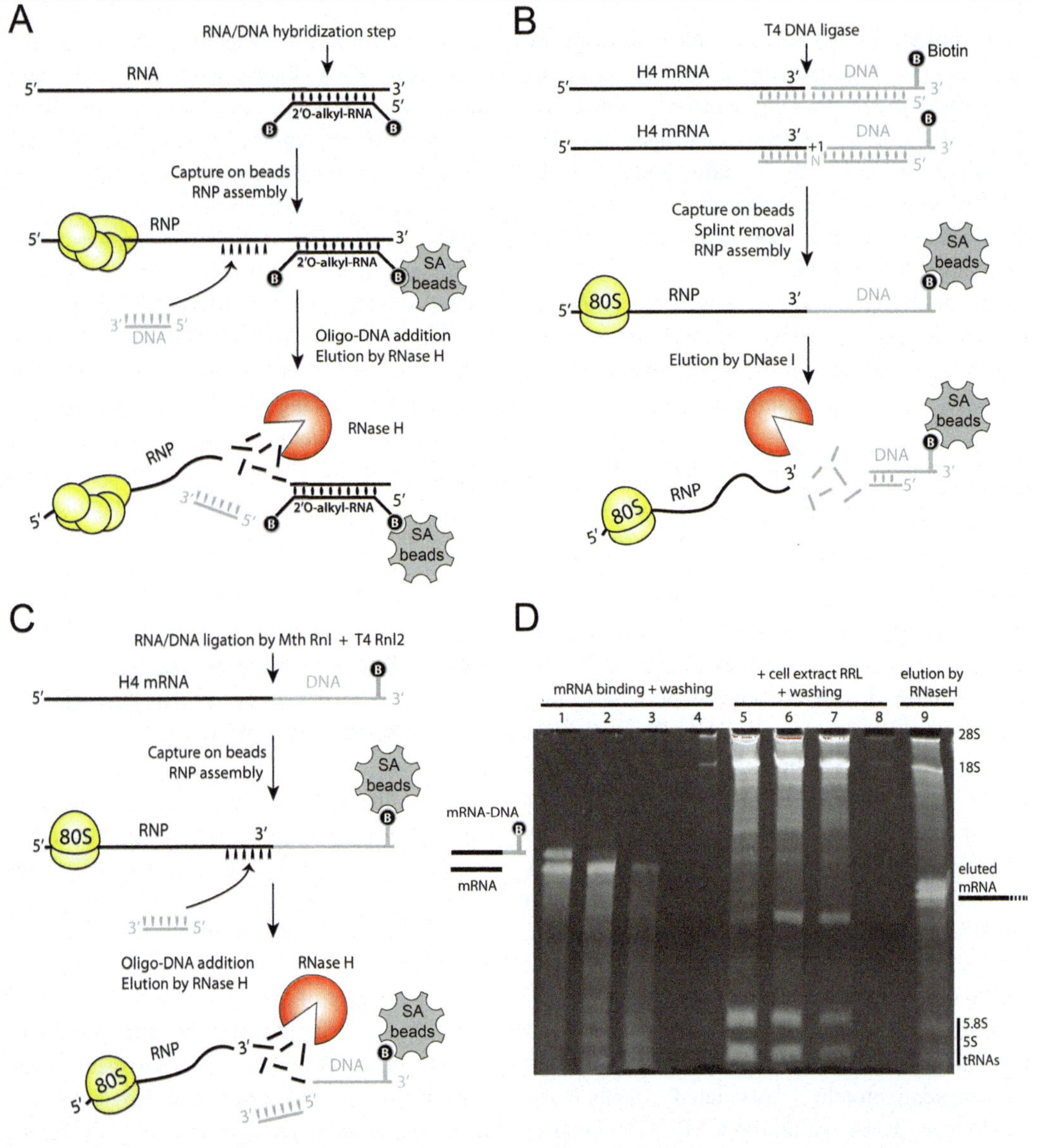

Fig. 2.3 Strategies of purification of RNPs based on hybridization and ligation of biotinylated oligonucleotides. (**a**) Affinity purification by oligonucleotide antisense and elution with RNase H (Lower). The RNP of interest is assembled to the RNA immobilized via an antisense biotinylated RNA oligonucleotide (grey color) to streptavidin beads. The antisense oligonucleotide contains 2′-0-alkyl sugars in order to resist to degradation in cell extracts. After washing, the elution of the RNP is obtained by adding of a DNA oligonucleotide, which forms the cleavage site for RNase H. (**b**) This method is based on the formation of RNA–DNA chimeric molecules that resist to the RNase H activity present in cell extracts. T4 DNA ligase catalyses the ligation of RNA-DNA hybrids if these are joined by base pairing to a "splint oligonucleotide". To circumvent the problem of imperfect pairing due to heterogeneous 3′ ends due to the tendency of RNA polymerases to incorporate *in vitro* non-encoded nucleotides, a degenerated splint carrying a mixture of all four nts is used. After removal of the splint oligo-mixture, RNPs are assembled in rabbit reticulocytes lysate (RRL). Release of the complexes is performed with DNase I or RQ1 DNase, which specifically cleave the DNA part of the chimera. (**c**) In this approach, histone H4 mRNA is ligated to a biotinylated DNA oligonucleotide (in grey) in two successive steps using *M. thermoautotrophicum* RNA ligase (MthRnl) and truncated T4 RNA ligase, forming a chimeric RNA/DNA molecule. Assembly of translation initiation complexes on the mRNA is performed in RRL. Elution is performed with RNase H as in panel A. (**d**) Representative denaturing urea PAGE gel showing the purification of 80S particles assembled on H4 mRNA by the method described in panel C. The chimeric mRNA–biotinylated DNA (lane 1) was loaded on the beads

another part of the RNA of interest is added together with RNase H which will specifically cleave the DNA/RNA duplex and induce the elution of the RNP in native conditions [46, 47].

To successfully apply this technique, an unstructured and accessible region of RNA sequence must be available to hybridize the antisense oligoribonucleotide. Therefore, this method can be difficult for highly structured RNAs and RNPs. The need for duplex formation of a previously single-stranded region of the RNA may tend to destabilize the RNP architecture. On the other hand, compared to methods that incorporate a structured RNA tag, this method avoids folding problems of the RNA due to the foreign sequence. In addition, several different hybridization sites in the RNA target can be quickly tested for accessibility to synthetic oligonucleotide probes.

2.5 RNP Purification Based on Ligation of a DNA-Oligonucleotide Tag

This method brings together several advantages from all the other methods. The RNA sequence of interest is ligated to an unstructured DNA oligonucleotide also called adapter, forming a chimeric RNA-DNA molecule. In this way, standard chemically synthesized DNA oligonucleotides carrying a variety of modifications, including biotin, can be ligated to RNA. The resulting chimera can efficiently bind streptavidin resins or beads where the RNP complexes are subsequently assembled starting from crude cell preparations. Here the absence of a foreign structured tag reduces constraints on RNP formation and interaction with non-specific molecules.

2.5.1 Tagging RNA by Splint-Dependent Ligation with T4 DNA Ligase

In this version of the method, a standard biotinylated-DNA adapter is ligated by T4 DNA ligase, which is able to ligate hybrids of ribo- and deoxyribonucleotide homo-polymers in double stranded regions [48–51]. Therefore, a splint DNA oligonucleotide hybridizing both mRNA and biotinylated DNA oligonucleotide is added to form the double-strand and allow ligation (Fig. 2.3b). Efficient ligation of RNA/DNA duplexes requires stoichiometric or greater concentrations of T4 DNA ligase because this enzyme cannot turnover effectively on RNA-containing duplexes [50]. In addition, since T7 RNA polymerase produces significant amount of N + 1 RNA products by run-off transcription, a second splint oligonucleotide is added with an extra nucleotide (mix of four nucleotides at this position) at the junction site (Fig. 2.3c). Prior to incubation with crude cell extracts, the splint oligonucleotide is removed from the duplex to prevent cleavage by endogenous RNase H (which cleaves RNA/DNA duplexes). To do so, the immobilized chimeric mRNAs is incubated 2 min at 95 °C to unwind mRNA-splint duplexes and then 10 min at room temperature in the presence of a tenfold excess of an antisense splint DNA oligonucleotide containing the complementary sequence to trap the splint and avoid re-annealing on the mRNA. After these washing steps, the immobilized mRNA–DNA chimera is ready for RNP complex capture. It is possible to accommodate 3′-sequence variety of the RNAs of interest by using a combination of splints, whereas the biotinylated oligonucleotide remains constant.

Fig. 2.3 (continued) (7 pmoles are shown, 50% of the mRNA molecules were ligated to the biotinylated DNA oligonucleotide). Unbound mRNAs are shown in the flow-through (lane 2) and washing fractions (lanes 3–4). Translation-initiation complexes were assembled in RRL. Unbound rRNAs and tRNAs are shown in the flow-through (lane 5) and washing fractions (lanes 6–8). The bound 80S particles were eluted with RNase H (lane 9) after adding the DNA oligonucleotide forming the RNase H cleavage site. The positions of the ribosomal RNAs, tRNAs and histone H4 mRNA are indicated

The elution of RNP is performed by DNase I, which degrades the DNA moiety of the chimera (the adapter), releasing in native conditions the RNP assembled on an intact RNA without extra tag.

It is also possible to extend the 3′-end of the RNAs by a poly $(CAA)_n$ unstructured extension that favors hybridizing the splint. For the assembly of translation initiation complexes, the chimeric molecule is obtained by ligating the DNA oligonucleotide to the 3′ end of the mRNA to enable ribosome binding to occur freely on the 5′ end. However, the method can be adapted to isolate RNP complexes assembled on the 3′ end of the mRNA, since nothing prevents ligation of the biotinylated DNA oligonucleotide to the 5′ end of the mRNA.

The method has been used to assemble translation initiation complexes on histone H4 mRNA, hepatitis C virus (HCV) internal ribosome entry site (IRES), β-globin mRNA, and cricket paralysis virus IRES, and these complexes were suitable for cryo-EM and mass spectrometry studies [52–55].

2.5.2 Direct Ligation of 5′-Adenylated Deoxyoligonucleotides Tag by T4 RNA Ligase 2

The advent of deep sequencing has led to the development of new methods to generate very large libraries of DNA molecules. To improve ligation efficiency and limit concatemer formation, the ligation reaction is split in two consecutive steps. ATP is first used to form a 5′-adenylated DNA oligonucleotide (5′--AppDNA) that is the biochemical intermediate of enzymes as DNA- or RNA-ligase. This activated intermediate is a good substrate for various enzymatic reactions, forming a building block to assemble nucleic acids under specific conditions. Then, the 5′-AppDNA molecule acts as substrate during the second step catalyzed by robust T4 ligases that work in the absence of ATP and do not concatenate products or form circles.

The first step of 5′-adenylation of the oligonucleotide is catalysed by a thermostable RNA ligase of *Methanobacterium thermoautotrophicum* (MthRnl) [56, 57]. The enzyme usually catalyzes the intramolecular ligation of single-strand RNA through ligase-adenylate and 5′-AppRNA intermediates. When using 3′-terminal blocked DNA oligonucleotides, and suitable ATP concentration, MthRnl accumulates 5′-AppDNA that can serve for efficient ligation of RNA or DNA molecules.

The second enzyme that ligates the RNA of interest to the activated 5′-AppDNA adapter is a mutated form of truncated T4 RNA ligase 2 (T4 Rnl2tr). T4 Rnl2tr is defective in self-adenylation and readily accepts pre-adenylated substrates for ligation [58]. Adding the R55K and K227Q mutations to T4 Rnl2tr leads to very low background in concatenation and circularization of the RNA molecules [58]. Because the optimized enzyme does not use ATP for ligation but pre-adenylated adapters, up to 75% of efficiency is reached during ligation [58]. As stated above, this procedure is commonly used to optimize adapter ligation for high-throughput sequencing technology. Here, we recently adapted the protocol for the formation of the mRNA-DNA chimera for RNP purification, and more precisely for purification of translation initiation complexes assembled on mRNA. With this method, we are routinely reaching 50% efficiency when ligating a biotinylated DNA adapter to the 3′ OH end of the mRNA (Fig. 2.3c), which is much higher than the result obtained by the splint-dependent ligation with T4 DNA ligase (about 10%). The following steps of RNP assembly and washes basically follow the conventional protocols. At the end, elution is performed with RNase H after adding a complementary DNA oligonucleotide that forms the RNase H cleavage site in the 3′ part of the mRNA. We found that RNase H was less sensitive than DNase I to steric clashes with the matrix beads and larger amounts of RNP are recovered when compared with the DNase elution.

2.6 Outlook

In the past few years, many approaches targeting the assembly of RNPs from cell lysates have been described. The capture of RNPs assembled *in vitro* remains valuable for the structural or biochemical analysis of a large variety of complexes and systems, including the spliceosome, ribosome, and other smaller RNPs such as the telomerase and RNase P. A critical step of these purification processes remains the capture of the RNA sequence of interest to a solid support that will allow separation of the total cell extract. This step relies on the presence, in the RNA target, of an external element with affinity for the solid matrix. Various types or elements have been selected and tested. Synthetic RNA aptamers have been artificially selected to target a specific molecule. Once bound to a solid support this molecule can be used for affinity purification of the RNP. Other elements are natural RNA sequences exhibiting specific binding sites for well-characterized RNA-binding proteins (phage elements). These methods are attractive because RNA aptamers are generally small, directly bind bead matrixes, and are eluted with small molecules. However, to successfully tag a given RNA, the aptamer should be placed in a solvent-accessible region. It is important to consider the steric effect of using such aptamers and the potential interference with the folding, structure, assembly and normal function or interactions of the RNP under study. Conversely, the RNP assembly may disrupt aptamer folding and therefore its ability to bind the solid support. Therefore after designing the hybrid RNA, it may be useful to predict the folding of the sequence. In case attempts to purify RNP complexes all fail, it might be necessary to reiterate the design of the construct by trying different tags, changing their location in the RNA, or adding flexible spacers that would prevent steric hindrance.

The use of biotinylated oligonucleotides, which are complementary to accessible single-stranded regions or ligated to the RNA of interest, is an efficient alternative approach when the RNA structure is too constrained and incompatible with the aptamer approach. Compared to aptamers, biotin is a discrete molecule that only slightly modifies the RNP complex. The biotinylated RNA, in the form of a DNA-RNA hybrid or ligated chimera, exhibits high affinity to streptavidin affinity resins. Efficient assembly and stringent washing steps allow the decrease of the background of contaminating proteins.

In addition, elution of the RNP complexes can be achieved under native conditions using competing oligonucleotides or DNase/RNase which releases the RNP without extra RNA sequence.

In conclusion, we have presented here several general strategies for selecting RNPs *in vitro* assembled in crude cell extracts and exhibiting *in vivo* potential application. Although these methods have been developed for the preparation of specific complexes, most of them have been successfully applied to the preparation of RNP involved in fundamental cellular processes.

Acknowledgments The authors are supported by institutional funds from the Centre National de la Recherche Scientifique (CNRS), the Université de Strasbourg, and grant ANR-17-CE11-0024-01 and ANR-17-CE12-0025-01.

References

1. Rouault TA, Hentze MW, Haile DJ, Harford JB, Klausner RD (1989) The iron-responsive element binding protein: a method for the affinity purification of a regulatory RNA-binding protein. Proc Natl Acad Sci U S A 86:5768–5772
2. Neupert B, Thompson NA, Meyer C, Kuhn LC (1990) A high yield affinity purification method for specific RNA-binding proteins: isolation of the iron regulatory factor from human placenta. Nucleic Acids Res 18: 51–55
3. Bardwell VJ, Wickens M (1990) Purification of RNA and RNA-protein complexes by an R17 coat protein affinity method. Nucleic Acids Res 18:6587–6594
4. Lowary PT, Uhlenbeck OC (1987) An RNA mutation that increases the affinity of an RNA–protein interaction. Nucleic Acids Res 15:10483–10493
5. Bachler M, Schroeder R, von Ahsen U (1999) StreptoTag: a novel method for the isolation of RNA-binding proteins. RNA 5:1509–1516
6. Srisawat C, Engelke DR (2001) Streptavidin aptamers: affinity tags for the study of RNAs and ribonucleoproteins. RNA 7:632–641

7. Srisawat C, Goldstein IJ, Engelke DR (2001) Sephadex-binding RNA ligands: rapid affinity purification of RNA from complex RNA mixtures. Nucleic Acids Res 29:E4
8. Hartmuth K et al (2002) Protein composition of human prespliceosomes isolated by a tobramycin affinity-selection method. Proc Natl Acad Sci U S A 99: 16719–16724
9. Panchapakesan SSS et al (2017) Ribonucleoprotein purification and characterization using RNA mango. RNA 23:1592–1599
10. Das R, Zhou Z, Reed R (2000) Functional association of U2 snRNP with the ATP-independent spliceosomal complex E. Mol Cell 5:779–787
11. Jurica MS, Licklider LJ, Gygi SR, Grigorieff N, Moore MJ (2002) Purification and characterization of native spliceosomes suitable for three-dimensional structural analysis. RNA 8:426–439
12. Ji H, Fraser CS, Yu Y, Leary J, Doudna JA (2004) Coordinated assembly of human translation initiation complexes by the hepatitis C virus internal ribosome entry site RNA. Proc Natl Acad Sci U S A 101: 16990–16995
13. Zhou Z, Sim J, Griffith J, Reed R (2002) Purification and electron microscopic visualization of functional human spliceosomes. Proc Natl Acad Sci U S A 99: 12203–12207
14. Locker N, Easton LE, Lukavsky PJ (2006) Affinity purification of eukaryotic 48S initiation complexes. RNA 12:683–690
15. Wang Y, Rando RR (1995) Specific binding of aminoglycoside antibiotics to RNA. Chem Biol 2: 281–290
16. Hartmuth K, Vornlocher H-P, Lührmann R (2004) Tobramycin affinity tag purification of spliceosomes. Methods Mol Biol 257:47–64
17. Vasudevan S, Steitz JA (2007) AU-rich-element-mediated upregulation of translation by FXR1 and Argonaute 2. Cell 128:1105–1118
18. Iioka H, Loiselle D, Haystead TA, Macara IG (2011) Efficient detection of RNA-protein interactions using tethered RNAs. Nucleic Acids Res 39:e53
19. Dix CI et al (2013) Lissencephaly-1 promotes the recruitment of dynein and dynactin to transported mRNAs. J Cell Biol 202:479–494
20. Leppek K, Stoecklin G (2014) An optimized streptavidin-binding RNA aptamer for purification of ribonucleoprotein complexes identifies novel ARE-binding proteins. Nucleic Acids Res 42:e13
21. Srisawat C, Engelke DR (2002) RNA affinity tags for purification of RNAs and ribonucleoprotein complexes. Methods 26:156–161
22. Deckert J et al (2006) Protein composition and electron microscopy structure of affinity-purified human Spliceosomal B complexes isolated under physiological conditions. Mol Cell Biol 26(14):5528–5543
23. Kastner B et al (2008) GraFix: sample preparation for single-particle electron cryomicroscopy. Nat Methods 5:53–55
24. Lim F, Downey TP, Peabody DS (2001) Translational repression and specific RNA binding by the coat protein of the pseudomonas phage PP7. J Biol Chem 276: 22507–22513
25. Lim F, Peabody DS (2002) RNA recognition site of PP7 coat protein. Nucleic Acids Res 30:4138–4144
26. Hogg JR, Collins K (2007) RNA-based affinity purification reveals 7SK RNPs with distinct composition and regulation. RNA 13:868–880
27. Hogg JR, Goff SP (2010) Upf1 senses 3′UTR length to potentiate mRNA decay. Cell 143:379–389
28. Fritz SE, Haque N, Hogg JR (2018) Highly efficient in vitro translation of authentic affinity-purified messenger ribonucleoprotein complexes. RNA 24: 982–989
29. Baron-Benhamou J, Gehring NH, Kulozik AE, Hentze MW (2004) Using the λN peptide to tether proteins to RNAs. In: MRNA Processing and Metabolism. Humana Press, Totowa, pp 135–154
30. Gehring NH, Hentze MW, Kulozik AE (2008) Chapter 23 tethering assays to investigate nonsense-mediated mRNA decay activating proteins. In: Methods in enzymology. Elsevire, London, pp 467–482
31. Ramanathan M et al (2018) RNA–protein interaction detection in living cells. Nat Methods 15:207–212
32. Piekna-Przybylska D, Liu B, Fournier MJ (2007) The U1 snRNA hairpin II as a RNA affinity tag for selecting snoRNP complexes. Methods Enzymol 425: 317–353
33. LeCuyer KA, Behlen LS, Uhlenbeck OC (1995) Mutants of the bacteriophage MS2 coat protein that alter its cooperative binding to RNA. Biochemistry 34: 10600–10606
34. Bertram K et al (2017) Cryo-EM structure of a pre-catalytic human spliceosome primed for activation. Cell 170:701–713.e11
35. Yoon J-H, Srikantan S, Gorospe M (2012) MS2-TRAP (MS2-tagged RNA affinity purification): tagging RNA to identify associated miRNAs. Methods 58:81–87
36. Beach DL, Keene JD (2008) Ribotrap: targeted purification of RNA-specific RNPs from cell lysates through immunoaffinity precipitation to identify regulatory proteins and RNAs. Methods Mol Biol 419:69–91
37. Tsai BP, Wang X, Huang L, Waterman ML (2011) Quantitative profiling of in vivo-assembled RNA-protein complexes using a novel integrated proteomic approach. Mol Cell Proteomics 10(4): M110.007385
38. Slobodin B, Gerst JE (2010) A novel mRNA affinity purification technique for the identification of interacting proteins and transcripts in ribonucleoprotein complexes. RNA 16:2277–2290
39. Bertrand E et al (1998) Localization of ASH1 mRNA particles in living yeast. Mol Cell 2:437–445
40. Beach DL, Salmon ED, Bloom K (1999) Localization and anchoring of mRNA in budding yeast. Curr Biol 9: 569–578

41. Haim L, Zipor G, Aronov S, Gerst JE (2007) A genomic integration method to visualize localization of endogenous mRNAs in living yeast. Nat Methods 4: 409–412
42. Iribarren AM et al (1990) 2′-O-alkyl oligoribonucleotides as antisense probes. Proc Natl Acad Sci U S A 87: 7747–7751
43. Lingner J, Cech TR (1996) Purification of telomerase from *Euplotes aediculatus*: requirement of a primer 3′ overhang. Proc Natl Acad Sci U S A 93:10712–10717
44. Schnapp G, Rodi HP, Rettig WJ, Schnapp A, Damm K (1998) One-step affinity purification protocol for human telomerase. Nucleic Acids Res 26:3311–3313
45. Kurth I, Cristofari G, Lingner J (2008) An affinity oligonucleotide displacement strategy to purify ribonucleoprotein complexes applied to human telomerase. Methods Mol Biol 488:9–22
46. Namy O, Moran SJ, Stuart DI, Gilbert RJ, Brierley I (2006) A mechanical explanation of RNA pseudoknot function in programmed ribosomal frameshifting. Nature 441:244–247
47. Eliseev B et al (2018) Structure of a human cap-dependent 48S translation pre-initiation complex. Nucleic Acids Res 46:2678–2689
48. Kleppe K, Van de Sande JH, Khorana HG (1970) Polynucleotide ligase-catalyzed joining of deoxyribo-oligonucleotides on ribopolynucleotide templates and of ribo-oligonucleotides on deoxyribopolynucleotide templates. Proc Natl Acad Sci U S A 67:68–73
49. Moore MJ, Sharp PA (1992) Site-specific modification of pre-mRNA: the 2′-hydroxyl groups at the splice sites. Science 256:992–997
50. Moore MJ, Query CC (2000) Joining of RNAs by splinted ligation. Methods Enzym 317:109–123
51. Fareed GC, Wilt EM, Richardson CC (1971) Enzymatic breakage and joining of deoxyribonucleic acid. 8. Hybrids of ribo- and deoxyribonucleotide homopolymers as substrates for polynucleotide ligase of bacteriophage T4. J Biol Chem 246:925–932
52. Prongidi-Fix L et al (2013) Rapid purification of ribosomal particles assembled on histone H4 mRNA: a new method based on mRNA-DNA chimaeras. Biochem J 449:719–728
53. Chicher J et al (2015) Purification of mRNA-programmed translation initiation complexes suitable for mass spectrometry analysis. Proteomics 15: 2417–2425
54. Martin F et al (2016) Ribosomal 18S rRNA base pairs with mRNA during eukaryotic translation initiation. Nat Commun 7:12622
55. Gross L et al (2017) The IRES 5′UTR of the dicistrovirus cricket paralysis virus is a type III IRES containing an essential pseudoknot structure. Nucleic Acids Res 45:8993–9004
56. Torchia C, Takagi Y, Ho CK (2008) Archaeal RNA ligase is a homodimeric protein that catalyzes intramolecular ligation of single-stranded RNA and DNA. Nucleic Acids Res 36:6218–6227
57. Zhelkovsky AM, McReynolds LA (2012) Structure-function analysis of *Methanobacterium thermoautotrophicum* RNA ligase – engineering a thermostable ATP independent enzyme. BMC Mol Biol 13:24
58. Viollet S, Fuchs RT, Munafo DB, Zhuang F, Robb GB (2011) T4 RNA ligase 2 truncated active site mutants: improved tools for RNA analysis. BMC Biotechnol 11:72

3 Peptide-Based Mass Spectrometry for the Investigation of Protein Complexes

Gianluca Degliesposti

Abstract

In the last two decades, biological mass spectrometry has become the gold standard for the identification of proteins in biological samples. The technological advancement of mass spectrometers and the development of methods for ionization, gas phase transfer, peptide fragmentation as well as for acquisition of high-resolution mass spectrometric data marked the success of the technique. This chapter introduces peptide-based mass spectrometry as a tool for the investigation of protein complexes. It provides an overview of the main steps for sample preparation starting from protein fractionation, reduction, alkylation and focus on the final step of protein digestion. The basic concepts of biological mass spectrometry as well as details about instrumental analysis and data acquisition are described. Finally, the most common methods for data analysis and sequence determination are summarized with an emphasis on its application to protein-protein complexes.

Keywords

Protein alkylation · Asp-N · Bottom-up · CID fragmentation · Data dependent acquisition · Protein digestion · ECD fragmentation · ESI · ETD fragmentation · Glu-C · HCD fragmentation · LC-MS/MS · MALDI-TOF/MS · Mascot · MS · Orbitrap · Pepsin · Proteomics · Reduction · Sequest · Trypsin

G. Degliesposti (✉)
MRC Laboratory of Molecular Biology, Cambridge, UK
e-mail: gdeglies@mrc-lmb.cam.ac.uk

3.1 Introduction

Protein identification by mass spectrometry (MS) relies on the determination of its amino acid sequence. Two different approaches, namely top-down and bottom-up, are adopted for this purpose. In a top-down method the identification is achieved by MS fragmentation of the whole protein. On the other hand, the bottom-up (or peptide-based) analyses start with the protein digestion and peptide sequencing followed by protein identification by searching and matching the peptide into databases of protein sequences [1, 2].

The overall process of peptide-based MS can be subdivided into three main steps: sample preparation (Sect. 3.2), data acquisition (Sect. 3.3) and data analysis (Sect. 3.4).

3.2 Sample Preparation

The sample preparation step consists of the transformation of proteins into a mixture of peptides whose mass to charge ratio (*m*/*z*) and

M. C. Vega, F. J. Fernández (eds.), *Advanced Technologies for Protein Complex Production and Characterization*, Advances in Experimental Medicine and Biology 1453,
https://doi.org/10.1007/978-3-031-52193-5_3

fragmentation pattern can be determined by mass spectrometry. A general workflow proceeds through sample denaturation, fractionation, reduction, alkylation and terminates with proteolytic digestion [3].

3.2.1 Sample Fractionation

The digestion of protein complexes composed of many subunits generates significant numbers of different peptides. Sample fractionation aims to improve the MS identification of all the proteins in a complex by reducing sample complexity.

Two commonly used approaches to protein fractionation are gel electrophoresis and size exclusion chromatography. Mono-dimensional and bi-dimensional gel electrophoresis separate proteins in gel bands and spots that after excision are further processed *in-gel*. Conversely, protein separation by size exclusion chromatography (SEC) allows the fractionation and collection of proteins *in-solution* with simultaneous buffer exchange in case optimized buffers for activity and selectivity of proteolytic enzymes are required.

Despite the advantages of reducing the sample complexity and facilitation of buffer exchange, protein fractionation implies sample losses that can lead to biased results in cases of low protein abundance and where quantitative or semiquantitative analysis is required.

3.2.2 Reduction and Alkylation of Cysteine

Cysteine-containing proteins often display disulfide bonds in their tertiary and quaternary structures. Digestion of proteins containing disulfide bonds generates non-linear cross-linked peptides which are more difficult to identify by mass spectrometry. By introducing a preliminary reduction of cysteines, free linear peptides are generated, and their identification simplified. The reduction reaction is usually based on DTT (dithiothreitol), TCEP (tris-(2-carboxyethyl)-phosphine) or 2-ME (2-mercaptoethanol) with comparable performance. Reformation of disulfide bonds is prevented by blocking the sulfhydryl groups with iodoacetamide to generate carbamidomethyl derivatives. Minor side reactions of iodoacetamide are observed to different extents in the following order of frequency: alkylation of N-terminus > glutamic acid > the C-terminus and lysine > aspartic acid > tyrosine > histidine [4].

3.2.3 Protein Digestion

After reduction and alkylation, proteins are usually digested using a proteolytic enzyme. The choice of protease is extremely important to maximize the number of informative and MS detectable peptides. In certain conditions the combination of more than one proteolytic enzyme can increase sequence coverage and identify a higher number of proteins. A quick review of the most commonly used proteases is found below.

Trypsin Trypsin is a serine endoproteinase that cleaves peptide bonds of lysine and arginine residues at the C-terminal side except for the imido bonds formed with proline. The distribution and abundance of lysine and arginine in proteins enables the generation of peptides with a suitable size for MS identification providing good sequence coverage. This feature makes trypsin the most frequently used enzyme in proteomics. Hydrolysis at the C terminus of positively charged residues produces peptides with a neutral balance of charge. This improves their ionization in ESI (electrospray ionization) sources and leads to a better fragmentation in tandem mass spectrometry leading to more accurate peptide identification.

Trypsin activity is optimal in the pH range of 7–9. Mild denaturing conditions such as 1 M urea, 10% acetonitrile and 0.1% SDS are tolerated [5–7].

Lys-C Lys-C is a serine endoproteinase that selectively hydrolyzes the C-terminal amide and imido bond of lysine residues. Lys-C is stable in strong denaturing conditions (*e.g.*, 8 M Urea, 1% SDS and 40% acetonitrile) improving the hydrolysis of structured regions not accessible under non-denaturing conditions. Lys-C specificity for a single amino acid reduces the number of cleavage sites and generates longer peptides than trypsin. By tandem digestion with Lys-C in strong denaturing conditions followed by trypsin digestion in milder conditions, the number of long peptides is reduced, and their MS identification increased [5, 6, 8].

Glu-C Glu-C is a serine endoproteinase that selectively cleaves the C-terminal of glutamic acid residues. The specificity is maximized in ammonium bicarbonate and ammonium acetate buffers. In phosphate buffer selectivity is reduced and hydrolysis occurs at both glutamic and aspartic acid residues. The proline imido-bond generates a missed cleavage site. The catalytic activity of Glu-C is optimal at pH 8 and it is retained in mild denaturing conditions (*e.g.*, 2 M urea, 1 M guanidine chloride, 0.1% SDS and 20% acetonitrile) [9–12].

Due to the high selectivity and the missed cleavages induced by proline at the C-terminal side, Glu-C digestion potentially generates long peptides with weak ionizability. However, if combined with Lys-C and trypsin, Glu-C allows the generation of shorter peptides with improved ionization and fragmentation properties.

Asp-N Asp-N is a zinc metalloproteinase that primarily hydrolyzes peptide bonds at the N-terminal side of aspartic acid and cysteine residues if oxidized to cysteic acid. A secondary cleavage activity towards glutamic acid residues is known but its hydrolysis rate is lower than that at aspartic acid residues. Asp-N activity is maximized at 37 °C in the pH range of 6–8. Mild denaturing conditions such as 1 M urea, 2 M guanidine hydrochloride, 0.1% SDS, 2% CHAPS and 10% acetonitrile are tolerated [13–15].

Pepsin Pepsin is a relatively nonspecific endopeptidase which cleaves predominantly at the C terminus of aromatic residues such as phenylalanine, tyrosine, and tryptophan. Activity at other residues is known but with variable efficiency. The optimal pH is between 1 and 3 and it is permanently inactivated above pH 6. Due to the low selectivity of pepsin the number of short peptides generated is high and for this reason it is used for specific purposes (*e.g.*, Hydrogen Deuterium Exchange experiments). Pepsin can be used free in solution or immobilized on resin supports. The immobilization increases the stability of the enzyme to autolysis preventing the contamination of the sample with pepsin peptides [16–19].

3.2.4 Multiple Enzymatic Digestion

Digestion with more than one enzyme allows the generation of a larger number of shorter peptides. This potentially increases the coverage and the number of the identified proteins. Examples of combined digestion with Lys-C in strong denaturing conditions followed by mild trypsin digestion have been reported as effective methods to improve protein identification in proteomics experiments [20–22].

3.2.5 Chemical Hydrolysis

In certain experimental conditions chemical hydrolysis can be a valid alternative or a complementary method to enzymatic digestion. Examples of chemical and targeted residues are the cyanogen bromide (CNBr) that cleaves at non-oxidized methionine (Met) residues, formic acid selective for aspartic acid and proline dipeptides (Asp-Pro) and hydroxylamine cleaving asparagine and glycine bonds (Asn-Gly) [23].

The combination of enzymatic and chemical hydrolysis can increase the number of detectable peptides. Several protocols based on enzymatic (trypsin) and chemical digestion (CNBr) display an increased sequence coverage and number of identified membrane proteins. This is attributable to a regular and convenient distribution of methionine residues in transmembrane domains [24].

The hydrolytic activity of CNBr requires concentrated acid (*e.g.*, acetic or formic acid) and high temperature of reaction. These conditions promote extensive acetylation or formylation of the sample and should be considered for downstream analysis.

3.3 Sample Analysis

After sample preparation, peptide mixtures are ready for identification. Nowadays, the most commonly employed methods for protein and peptide analysis are based on liquid chromatography and tandem mass spectrometry (LC-MS/MS) and on the indirect analysis of the whole or pre-fractionated samples in solid phase by matrix assisted laser desorption ionization-time of flight mass spectrometry (MALDI-TOF-MS).

3.3.1 LC-MS/MS

The LC-MS/MS methods are commonly based on nano-LC systems coupled to mass spectrometers. These instruments allow flow rates of hundreds of nanoliters per minute and are directly interfaced with the mass spectrometer. The chromatographic separation is usually performed on reverse phase columns with small internal diameters (*e.g.*, 75 μm) which improves the chromatographic resolution enhancing the sensitivity. Elution is carried out using increasing gradients of acetonitrile in water at constant concentration of formic acid. Formic acid protonates the basic groups of amino acid side chains and N-terminal amine groups thus neutralizing negative charge of the carboxylic groups. The elution period is proportional to both sample complexity and mass spectrometry acquisition speed. The amount of sample required for identification depends on the complexity of the sample, the instrumental set-up and the experimental needs [25].

3.3.2 MALDI-TOF-MS

MALDI-TOF-MS is an off-line method based on the co-crystallization of analytes with a light-absorbing matrix (*e.g.*, sinapinic acid and α-cyano-4-hydroxycinnamic acid) which, through laser excitation, induce the analyte ionization. The MALDI ion-source is usually coupled to time-of-flight (TOF) analyzers for the determination of intact mass of proteins and peptides. The simple and rapid operativity of MALDI-TOF-MS found a wide distribution in proteomics laboratories for the analysis of both proteins and peptide. However, the lack of a liquid chromatographic separation step requires the prefractionation of samples before the deposition on the MALDI target [26, 27].

3.3.3 Mass Spectrometers and Data Acquisition

For proteomic investigations aimed at the identification of unknown proteins, the most commonly used mass spectrometers are orbitrap hybrid instruments such as LTQ-Orbitraps or Quadrupole-Orbitraps. Targeted analyses on these

Fig. 3.1 (continued) of the peptide ions is acquired in the Orbitrap analyzer and peaks ranked by intensity. The nth most intense peaks (*e.g.*, the first three in the picture) are further analyzed in the following steps. In sequential events each ion is selected in the quadrupole (MS1) and fragmented in the collision cell. A full scan of the generated fragments is acquired in the Orbitrap (MS2), and MS/MS spectra collected. (**b**) Selected reaction monitor (SRM) on a triple-quadrupole instrument. After electrospray ionization, a list of parent ions is cyclically scanned. Each peptide ion is filtered by the first quadrupole (MS1) and fragmented. Specific fragments of the parent ions are then sequentially monitored by the third quadrupole (MS2) and their intensities recorded (MS/MS spectra). The extract ion chromatogram of each transition is used for quantitative purposes. (**c**) Summary of the theoretical positive ions generated by fragmentation of a positively charged tetrapeptide and their commonly accepted nomenclature [42, 43]

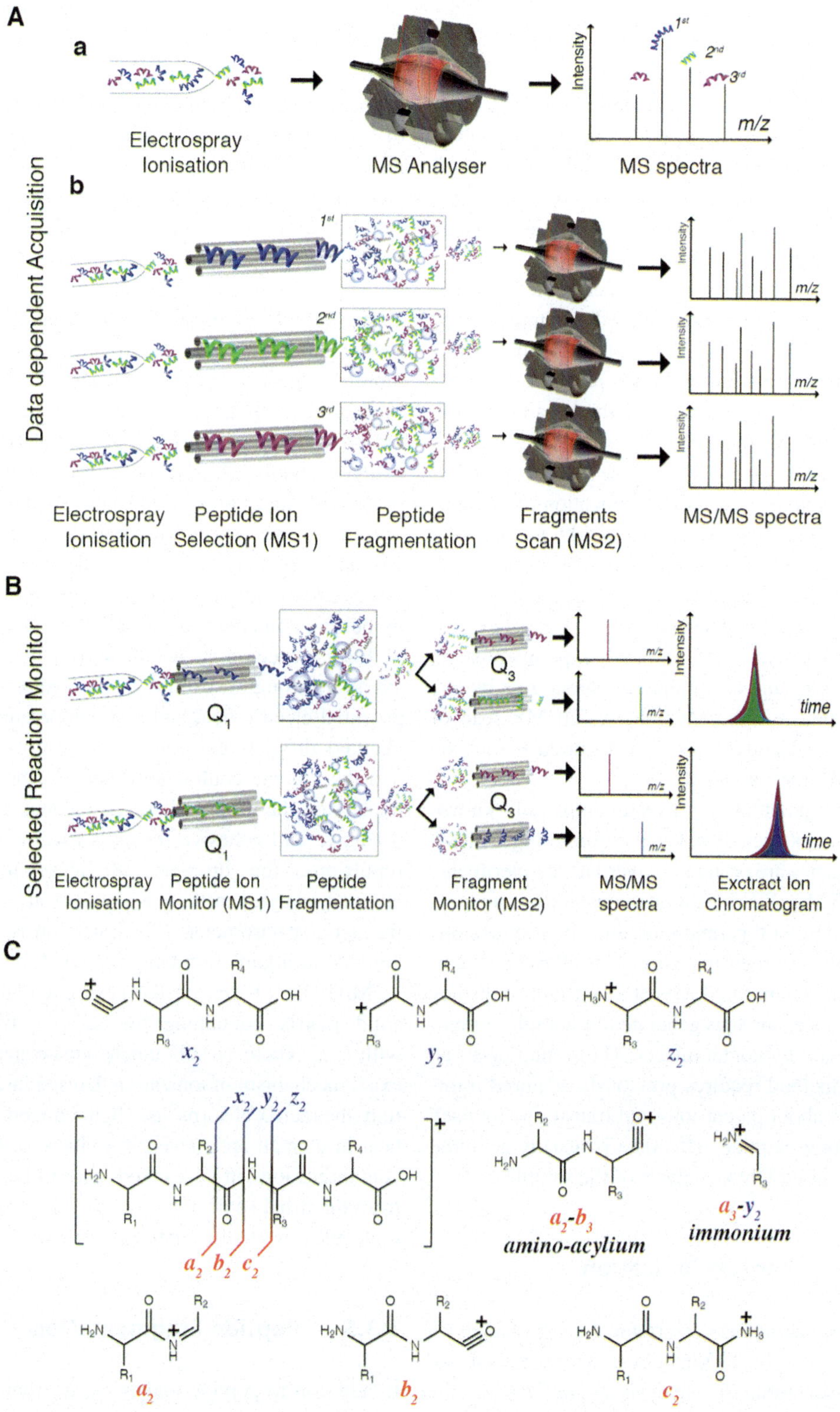

Fig. 3.1 Different methods for data acquisition are available using modern mass-spectrometers. (**a**) Summary of data dependent acquisition (DDA) experiment on a hybrid Quadrupole-Orbitrap instrument. Peptides eluting from liquid chromatography are ionized by electrospray and enter the mass-spectrometer. A high-resolution full scan

instruments enables the acquisition of high resolution and accurate mass data over a wide mass range of unknown peptides and fragmentation products. Triple-Quadrupoles (QqQ) and hybrid Quadrupole-Time-Of-Flight (qTOF) analyzers are commonly used for identification and quantitation of known proteins.

Methods for mass spectrometric data acquisition are designed and tailored to improve the identification of unknown proteins or the quantitative determination of known proteins.

Protein identification methods are usually based on data dependent acquisition (DDA) experiments (Fig. 3.1a). The eluted peptides reach the ion source where ionization and evaporation occur. Ions are accelerated by an electrical field and collected into the mass spectrometer. A full MS scan is acquired, and the first *n* most intense peaks detected are shortlisted for MS/MS analysis. Each MS/MS experiment starts by filtering and accumulating parent ions by *m/z* ratios before fragmentation. A full MS scan of fragmented peptides is then acquired before the next MS/MS step starts.

For quantitative investigations of known proteins, selected reaction monitoring (SRM) approaches are preferable over DDA experiments (Fig. 3.1b). SRM methods constantly monitor a defined set of parent ions and their transitions regardless of their intensity. This allows data collection of monitored ions at a constant frequency and reduces the bias generated by sample composition and by contaminants. The setting up of an SRM method requires previously acquired information about parent ions and transitions for each monitored peptide. This data is usually acquired by an initial DDA analysis of the sample.

3.3.4 Peptide Ionization

Mass spectrometers determine the *m/z* of analytes in the gas phase. Vacuum is a prerequisite for mass spectrometric analysis as well as for the ionization of the analytes.

The MS analysis of peptides in solution or co-crystallized with MALDI matrix requires their transfer to the gas phase from liquid and solid phases, respectively. Macromolecules like proteins and peptides are not volatile. Their evaporation and ionization require a good balance of energy to prevent degradation and loss of intrinsic information. The lack of methods and technologies to achieve this task has for many years been a bottleneck for the MS analysis of proteins and peptides.

After a long development time, in the late 1980s, John B. Fenn and Koichi Tanaka reported two methods for mass spectrometric analyses of biological macromolecules, namely Electrospray Ionization (ESI) and the Matrix Assisted Laser Desorption Ionization (MALDI). These two ionization methods allowed the effective ionization of polypeptides preserving their sequence and introduced mass spectrometry to the field of protein and peptide research. For their revolutionary discoveries Fenn and Tanaka were awarded the Nobel Prize in chemistry in 2002 [28–30].

ESI is based on the nebulization of the sample solution through a capillary tube against a high potential electric field and high temperature. The charged droplets generated gradually evaporate and the volume contraction increases the charge density on the drops. When the balance between repulsive and cohesive forces is lost in favor of repulsions, the droplets fall apart releasing charged analytes in the gas phase ready to enter the mass-spectrometer. ESI ionization is usually coupled to liquid chromatography [31, 32].

MALDI is based on the laser excitation of a solid matrix containing the sample dispersed within it. There is still much debate about the exact mechanism of ionization. During laser ablation, the matrix absorbs laser light converting it to heat in a small area, causing a plume of hot gas containing ions in an excited state to be ejected from the dried spot. In a secondary process these molecules ionize the analytes [33, 34].

3.3.5 Peptide Fragmentation

In addition to peptide ionization, another important step in peptide-based mass spectrometry is the peptide fragmentation [35]. The peptide ions generated in the ion source of the mass spectrometer are usually referred to as parent ions. The peptide sequence is determined by fragmentation of parent ions by collision with an inert gas as in Collision Induced Dissociation (CID) and in Higher energy Collision induced Dissociation (HCD) [36, 37] or by direct or indirect interaction with electrons as in Electron Capture Dissociation (ECD) and Electron Transfer Dissociation (ETD), respectively [38, 39].

Both CID and HCD methods are based on the collision between accelerated parent ions and inert gas molecules in a dedicated collision cell. The energy released by collision is absorbed by the parent ion increasing its internal energy and inducing the rupture of the most labile bonds. The amount of energy required for fragmentation depends on the size and charge of the parent ions.

Experimental evidence supports the involvement of a proton at the cleavage site for the generation of fragment ions and are summarized in the mobile proton model [40, 41]. Protons responsible for cleavage are supposed to be bound to basic side-chains of parent ions amino acids (*e.g.*, arginine and lysine) and to be mobilized by the collision energy towards the backbone heteroatoms where they catalyze the cleavage. The amount of collisional energy necessary to "mobilize" the proton is proportional to the basicity of the protonated side chains and its calibration is extremely important for the generation of an appropriate number of meaningful fragments avoiding over-fragmentation that generates the less informative internal fragments.

The length of parent ions is extremely important for the generation of informative MS/MS spectra. As a rule of thumb, multiple charged ions of peptides containing 6–20 amino acids generate MS/MS spectra suitable for sequence assignment. Spectra of shorter peptides do not provide enough fragments for a confident assignment of each amino acid and, conversely, the complexity of MS/MS spectra of longer peptides is too high for accurate data interpretation and assignment.

Each of the above-mentioned methods displays preferential patterns of fragmentation. Roepstorff and Fohlman in 1984 and Johnson in 1987 proposed a conventional nomenclature for fragments based on which bond is broken along the peptide backbone and which fragment retains the charge, as summarized in Fig. 3.1c [42, 43]. When the charge is on the N terminus, ions are defined as *a*, *b* or *c*. Otherwise, if the charge is retained on the C terminus, the ions are named *x*, *y* or *z*. Fragments *a* and *x* are generated by the rupture of the bond between the α-carbon and the carbonyl group, *b* and *y* when the peptide bond is broken, and *c* and *z* are generated by the cleavage of the bond between amide nitrogen and α-carbon.

If more than one backbone cleavage happens on the same peptide an internal fragment of the parent ions is generated. These are usually formed by a combination of *b* and *y* cleavages that generates amino-acylium ions and less frequently of *a* and *y* providing amino-immonium ions. The shortest internal fragment generated by *a* and *y* cleavage of a single side-chain are immonium ions.

CID fragmentation of positively charged ions usually generates *b* and *y* fragments and if the molecular ion contains R, N, Q or K residues in its sequence, loss of ammonia (−17 Da) is detectable. In case the sequence includes S, T, E or D residues, loss of water (−18 Da) is detected [44]. HCD fragmentation generates a pool of fragment ions similar to those generated by CID with the exception of ammonia and water losses. ETD and ECD fragmentation methods generate MS/MS spectra mainly populated by *c*, *y*, $z + 1$, and $z + 2$ ions.

3.4 Data Analysis

The sequence of a peptide is determined by analysis of the acquired MS and MS/MS spectra. Two main analytical approaches and a number of algorithms are available for this purpose.

In one case, the analysis follows a *de-novo* sequence-independent interpretation of spectral data. Briefly, the *de-novo* sequencing algorithms interpret the MS/MS spectra and evaluate the mass difference between peaks predicting the whole sequence whose mass is expected to match the mass of the parent ion (*i.e.*, MS *m/z* value). The applicability of these methods is limited by the generation of non-predictable fragments (*e.g.*, excision of side-chains, neutral losses, and internal fragmentation) as well as uneven fragmentation along the polypeptide chain due to preferential cleavages (*e.g.*, in presence of proline residues). These additional peaks lead to an increase of MS/MS spectra complexity and ambiguity reducing the *de-novo* sequencing efficiency [45].

The second approach is based on an uninterpreted search of MS and MS/MS data from a previously generated peptide sequence database. This method is the most commonly applied to wide proteomics experiments. Each experimental MS/MS spectra is compared to a library of predicted spectra generated from sequences of *in-silico* digested proteins [46]. A number of licensed and free on-line and off-line search engines are nowadays available, *e.g.*, Sequest [47], Mascot [48], X!Tandem [49, 50]. In addition to the *m/z* value of parent ions, the MS/MS data, and the database of sequences, additional parameters are required for searching, *e.g.*, the MS and MS/MS match tolerances and the side-chain modifications to be accounted for.

The databases of sequences used can be proteome wide (*e.g.*, Uniprot or SwissProt databases) or a smaller subset of selected sequences tailored to the experimental purpose. The sequences included in the database are *in-silico* digested mimicking the activity of proteases used for the sample preparation (*i.e.*, cleavage site specificity, missed cleavages, C-terminal or N-terminal cleavage). The generated peptides are then modified according to the list of fixed or variable modifications to be searched (*e.g.*, Cys-carbamidomethyl and Met oxidation). Each generated peptide is indexed according to its molecular weight and the MS/MS fragmentation is predicted.

The searching process is based on the cross-correlation of experimental parent ion *m/z* value and fragments peaks with the theoretical values included in the database. The matching of both experimental and theoretical MS and MS/MS values is allowed within a defined tolerance. Based on the number of matched peaks and deviation between theoretical and experimental values the match is scored, and the assigned peptides ranked.

Protein identification is based on the matching and assignment of identified peptides to the sequences collected in the protein database. An identified peptide can be assigned to a single protein (*i.e.*, unique or exclusive peptides) or more proteins in case of sequence homology (*i.e.*, shared or not-exclusive peptides). The assignment of exclusive peptides to a protein permits its identification and the assignment of not-exclusive ones contributes to increase its sequence coverage, thus increasing the confidence of identification.

References

1. Aebersold R, Mann M (2003) Mass spectrometry-based proteomics. Nature 422(6928):198–207. https://doi.org/10.1038/nature01511
2. Gingras AC, Gstaiger M, Raught B, Aebersold R (2007) Analysis of protein complexes using mass spectrometry. Nat Rev Mol Cell Bio 8(8):645–654. https://doi.org/10.1038/nrm2208
3. Granvogl B, Ploscher M, Eichacker LA (2007) Sample preparation by in-gel digestion for mass spectrometry-based proteomics. Anal Bioanal Chem 389(4): 991–1002. https://doi.org/10.1007/s00216-007-1451-4
4. Suttapitugsakul S, Xiao H, Smeekens J, Wu R (2017) Evaluation and optimization of reduction and alkylation methods to maximize peptide identification with MS-based proteomics. Mol BioSyst 13(12): 2574–2582. https://doi.org/10.1039/c7mb00393e
5. Masaki T, Fujihashi T, Nakamura K, Soejima M (1981) Studies on a new proteolytic enzyme from Achromobacter lyticus M497-1. II. Specificity and inhibition studies of Achromobacter protease I. Biochim Biophys Acta 660(1):51–55
6. Masaki T, Tanabe M, Nakamura K, Soejima M (1981) Studies on a new proteolytic enzyme from A chromobacter lyticus M497-1. I. Purification and some enzymatic properties. Biochim Biophys Acta 660(1):44–50
7. Vandermarliere E, Mueller M, Martens L (2013) Getting intimate with trypsin, the leading protease in proteomics. Mass Spectrom Rev 32(6):453–465. https://doi.org/10.1002/mas.21376
8. Glatter T, Ludwig C, Ahrne E, Aebersold R, Heck AJR, Schmidt A (2012) Large-scale quantitative assessment of different in-solution protein digestion protocols reveals superior cleavage efficiency of tandem Lys-C/trypsin proteolysis over trypsin digestion. J

Proteome Res 11(11):5145–5156. https://doi.org/10.1021/pr300273g
9. Birktoft JJ, Breddam K (1994) Glutamyl endopeptidases. Method Enzymol 244:114–126
10. Demidyuk IV, Chukhontseva KN, Kostrov SV (2017) Glutamyl endopeptidases: the puzzle of substrate specificity. Acta Nat 9(2):17–33
11. Mil'gotina EI, Voyushina TL, Chestukhina GG (2003) Glutamyl endopeptidases: structure, function, and practical application. Russ J Bioorg Chem 29(6): 511–522. https://doi.org/10.1023/B:RUBI.0000008891.51576.d0
12. Ono T, Ohara-Nemoto Y, Shimoyama Y, Okawara H, Kobayakawa T, Baba TT, Kimura S, Nemoto TK (2010) Amino acid residues modulating the activities of staphylococcal glutamyl endopeptidases. Biol Chem 391(10):1221–1232. https://doi.org/10.1515/Bc.2010.116
13. Tetaz T, Morrison JR, Andreou J, Fidge NH (1990) Relaxed specificity of endoproteinase-asp-N – this enzyme cleaves at peptide-bonds N-terminal to glutamate as well as aspartate and cysteic acid residues. Biochem Int 22(3):561–566
14. Ingrosso D, Fowler AV, Bleibaum J, Clarke S (1989) Specificity of endoproteinase Asp-N (pseudomonas-fragi) – cleavage at glutamyl residues in 2 proteins. Biochem Biophys Res Commun 162(3):1528–1534. https://doi.org/10.1016/0006-291x(89)90848-6
15. Hagmann ML, Geuss U, Fischer S, Kresse GB (1995) Peptidyl-Asp metalloendopeptidase. In: Proteolytic enzymes: aspartic and metallo peptidases, vol 248. Elsevier, pp 782–787
16. Lopez-Ferrer D, Petritis K, Robinson EW, Hixson KK, Tian ZX, Lee JH, Lee SW, Tolic N, Weitz KK, Belov ME, Smith RD, Pasa-Tolic L (2011) Pressurized pepsin digestion in proteomics. Mol Cell Proteomics 10(2). https://doi.org/10.1074/mcp.M110.001479
17. Fruton JS (1976) The mechanism of the catalytic action of pepsin and related acid proteinases. Adv Enzymol Relat Areas Mol Biol 44:1–36
18. Hamuro Y, Coales SJ, Molnar KS, Tuske SJ, Morrow JA (2008) Specificity of immobilized porcine pepsin in H/D exchange compatible conditions. Rapid Commun Mass Spectrom 22(7):1041–1046. https://doi.org/10.1002/rcm.3467
19. Zhang Z, Smith DL (1993) Determination of amide hydrogen exchange by mass spectrometry: a new tool for protein structure elucidation. Protein Sci 2(4): 522–531. https://doi.org/10.1002/pro.5560020404
20. Swaney DL, Wenger CD, Coon JJ (2010) Value of using multiple proteases for large-scale mass spectrometry-based proteomics. J Proteome Res 9(3): 1323–1329. https://doi.org/10.1021/pr900863u
21. Choudhary G, Wu SL, Shieh P, Hancock WS (2003) Multiple enzymatic digestion for enhanced sequence coverage of proteins in complex proteomic mixtures using capillary LC with ion trap MS/MS. J Proteome Res 2(1):59–67
22. Wisniewski JR, Mann M (2012) Consecutive proteolytic digestion in an enzyme reactor increases depth of proteomic and phosphoproteomic analysis. Anal Chem 84(6):2631–2637. https://doi.org/10.1021/ac300006b
23. Crimmins DL, Mische SM, Denslow ND (2005) Chemical cleavage of proteins in solution. Curr Protoc Protein Sci Chapter 11:Unit 11–14. https://doi.org/10.1002/0471140864.ps1104s40
24. Vit O, Petrak J (2017) Integral membrane proteins in proteomics. How to break open the black box? J Proteome 153:8–20. https://doi.org/10.1016/j.jprot.2016.08.006
25. Karpievitch YV, Polpitiya AD, Anderson GA, Smith RD, Dabney AR (2010) Liquid chromatography mass spectrometry-based proteomics: biological and technological aspects. Ann Appl Stat 4(4):1797–1823. https://doi.org/10.1214/10-Aoas341
26. Suckau D, Resemann A, Schuerenberg M, Hufnagel P, Franzen J, Holle A (2003) A novel MALDI LIFT-TOF/TOF mass spectrometer for proteomics. Anal Bioanal Chem 376(7):952–965. https://doi.org/10.1007/s00216-003-2057-0
27. Susnea I, Bernevic B, Wicke M, Ma L, Liu SY, Schellander K, Przybylski M (2013) Application of MALDI-TOF-mass spectrometry to proteome analysis using stain-free gel electrophoresis. Top Curr Chem 331:37–54. https://doi.org/10.1007/128_2012_321
28. Fenn JB (2002) Electrospray ionization mass spectrometry: how it all began. J Biomol Tech 13(3): 101–118
29. Fenn JB, Mann M, Meng CK, Wong SF, Whitehouse CM (1989) Electrospray ionization for mass spectrometry of large biomolecules. Science 246(4926):64–71
30. Tanaka K, Waki H, Ido Y, Akita S, Yoshida Y, Yoshida T, Matsuo T (1988) Protein and polymer analyses up to m/z 100 000 by laser ionization time-of-flight mass spectrometry. Rapid Commun Mass Spectrom 2(8):151–153. https://doi.org/10.1002/rcm.1290020802
31. Fenn JB, Mann M, Meng CK, Wong SF, Whitehouse CM (1990) Electrospray ionization-principles and practice. Mass Spectrom Rev 9(1):37–70. https://doi.org/10.1002/mas.1280090103
32. Bruins AP (1998) Mechanistic aspects of electrospray ionization. J Chromatogr A 794(1–2):345–357. https://doi.org/10.1016/S0021-9673(97)01110-2
33. Molin L, Seraglia R, Czarnocki Z, Maurin JK, Plucinski FA, Traldi P (2012) On the primary ionization mechanism(s) in matrix-assisted laser desorption ionization. J Anal Methods Chem 2012(14):161865. https://doi.org/10.1155/2012/161865
34. Karas M, Kruger R (2003) Ion formation in MALDI: the cluster ionization mechanism. Chem Rev 103(2): 427–439. https://doi.org/10.1021/cr010376a
35. Brodbelt JS (2016) Ion activation methods for peptides and proteins. Anal Chem 88(1):30–51. https://doi.org/10.1021/acs.analchem.5b04563

36. Olsen JV, Macek B, Lange O, Makarov A, Horning S, Mann M (2007) Higher-energy C-trap dissociation for peptide modification analysis. Nat Methods 4(9): 709–712. https://doi.org/10.1038/Nmeth1060
37. Wells JM, McLuckey SA (2005) Collision-induced dissociation (CID) of peptides and proteins. Biol Mass Spectrom 402:148–185. https://doi.org/10.1016/S0076-6879(05)02005-7
38. Zubarev RA, Kelleher NL, McLafferty FW (1998) Electron capture dissociation of multiply charged protein cations. A nonergodic process. J Am Chem Soc 120(13):3265–3266. https://doi.org/10.1021/ja973478k
39. Syka JEP, Coon JJ, Schroeder MJ, Shabanowitz J, Hunt DF (2004) Peptide and protein sequence analysis by electron transfer dissociation mass spectrometry. Proc Natl Acad Sci U S A 101(26):9528–9533. https://doi.org/10.1073/pnas.0402700101
40. Wysocki VH, Tsaprailis G, Smith LL, Breci LA (2000) Mobile and localized protons: a framework for understanding peptide dissociation. J Mass Spectrom 35(12): 1399–1406. https://doi.org/10.1002/1096-9888(200012)35:12<1399::Aid-jms86>3.0.Co;2-r
41. Boyd R, Somogyi A (2010) The mobile proton hypothesis in fragmentation of protonated peptides: a perspective. J Am Soc Mass Spectrom 21(8):1275–1278. https://doi.org/10.1016/j.jasms.2010.04.017
42. Roepstorff P, Fohlman J (1984) Proposal for a common nomenclature for sequence ions in mass spectra of peptides. Biomed Mass Spectrom 11(11):601. https://doi.org/10.1002/bms.1200111109
43. Johnson RS, Martin SA, Biemann K, Stults JT, Watson JT (1987) Novel fragmentation process of peptides by collision-induced decomposition in a tandem mass spectrometer: differentiation of leucine and isoleucine. Anal Chem 59(21):2621–2625. https://doi.org/10.1021/ac00148a019
44. Papayannopoulos IA (1995) The interpretation of collision-induced dissociation tandem mass-spectra of peptides. Mass Spectrom Rev 14(1):49–73. https://doi.org/10.1002/mas.1280140104
45. Medzihradszky KF, Chalkley RJ (2015) Lessons in de novo peptide sequencing by tandem mass spectrometry. Mass Spectrom Rev 34(1):43–63
46. Cottrell JS (2011) Protein identification using MS/MS data. J Proteome 74(10):1842–1851. https://doi.org/10.1016/j.jprot.2011.05.014
47. Eng JK, McCormack AL, Yates JR (1994) An approach to correlate tandem mass spectral data of peptides with amino acid sequences in a protein database. J Am Soc Mass Spectrom 5(11):976–989. https://doi.org/10.1016/1044-0305(94)80016-2
48. Perkins DN, Pappin DJ, Creasy DM, Cottrell JS (1999) Probability-based protein identification by searching sequence databases using mass spectrometry data. Electrophoresis 20(18):3551–3567. https://doi.org/10.1002/(SICI)1522-2683(19991201)20:18<3551::AID-ELPS3551>3.0.CO;2-2
49. Craig R, Beavis RC (2004) TANDEM: matching proteins with tandem mass spectra. Bioinformatics 20(9):1466–1467. https://doi.org/10.1093/bioinformatics/bth092
50. Craig R, Beavis RC (2003) A method for reducing the time required to match protein sequences with tandem mass spectra. Rapid Commun Mass Spectrom 17(20): 2310–2316. https://doi.org/10.1002/rcm.1198

Probing Protein Complexes Composition, Stoichiometry, and Interactions by Peptide-Based Mass Spectrometry

4

Gianluca Degliesposti

Abstract

The characterization of a protein complex by mass spectrometry can be conducted at different levels. Initial steps regard the qualitative composition of the complex and subunit identification. After that, quantitative information such as stoichiometric ratios and copy numbers for each subunit in a complex or super-complex is acquired. Peptide-based LC-MS/MS offers a wide number of methods and protocols for the characterization of protein complexes. This chapter concentrates on the applications of peptide-based LC-MS/MS for the qualitative, quantitative, and structural characterization of protein complexes focusing on subunit identification, determination of stoichiometric ratio and number of subunits per complex as well as on cross-linking mass spectrometry and hydrogen/deuterium exchange as methods for the structural investigation of the biological assemblies.

Keywords

Proteomics · Protein identification · Stoichiometry · Quantitative proteomics · Cross-linking and mass spectrometry · Hydrogen/deuterium exchange

G. Degliesposti (✉)
MRC Laboratory of Molecular Biology, Cambridge, UK
e-mail: gdeglies@mrc-lmb.cam.ac.uk

4.1 Introduction

Peptide-based LC-MS/MS methods are effective tools for the acquisition of information about protein composition and protein-protein interactions in multisubunit complexes through different experimental protocols. As summarized in Fig. 4.1a, protein identification provides the qualitative composition of the complex. Information about stoichiometric ratios among subunits and their copy number in the active complex requires relative and absolute quantitative investigation, respectively (Fig. 4.1b, c). In addition to qualitative and quantitative information, peptide-based MS investigations allow the collection of structural insights about protein complexes such as protein-protein interactions, structurally ordered regions, and solvent accessible surfaces. As examples of structural methods that rely on peptide-based mass spectrometry, chemical cross-linking (Fig. 4.1d) and hydrogen deuterium exchange will be discussed [1–8].

4.2 Subunit Identification and Characterization

Subunit identification is the starting point for the characterization of a protein complex. After purification, protein complexes are usually checked for the retained biochemical activity. However, the first step for characterization is the identification

M. C. Vega, F. J. Fernández (eds.), *Advanced Technologies for Protein Complex Production and Characterization*, Advances in Experimental Medicine and Biology 1453,
https://doi.org/10.1007/978-3-031-52193-5_4

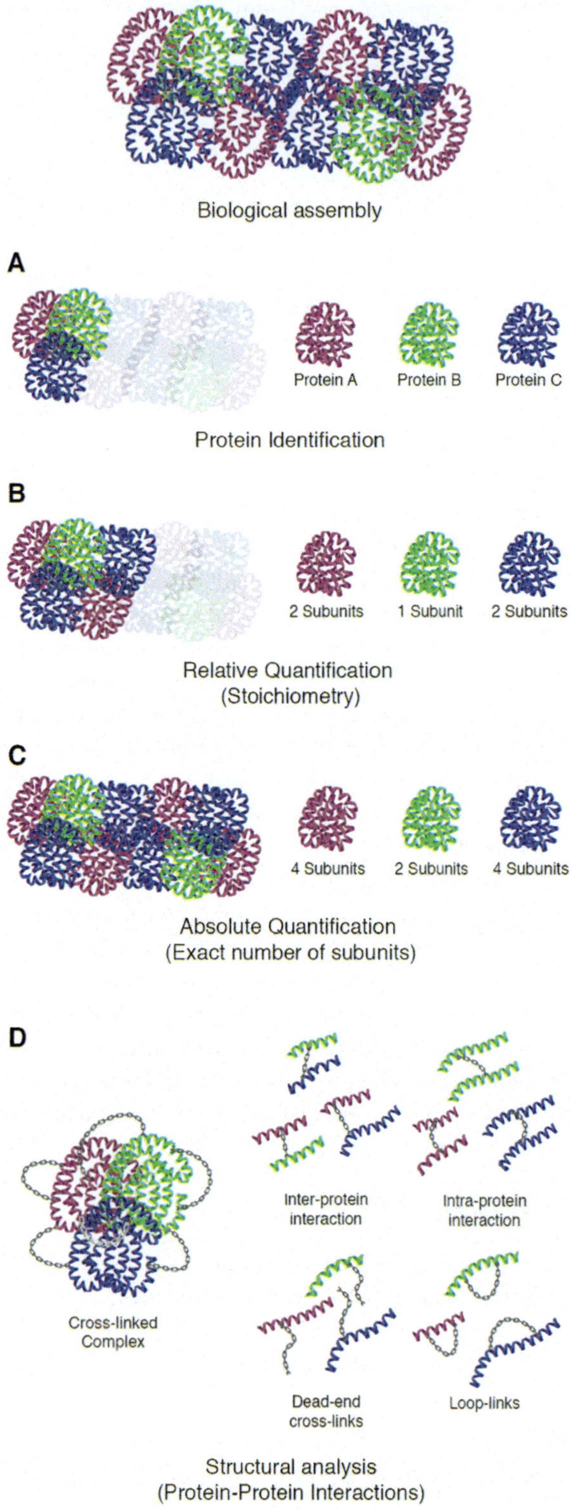

Fig. 4.1 Hypothetical assembly of a protein complex composed by multiple copies of different subunits. Through peptide-based mass spectrometry it is possible (**a**) to identify the proteins that form the complex, (**b**) their stoichiometric ratio by relative quantification and (**c**) the absolute number of copies of each protein by absolute quantification. (**d**) Structural analysis by chemical cross-linking and mass spectrometry allows the mapping of protein-protein interactions

of their exact protein composition. The available information about subunit and protein sequences depends on the way the sample is prepared (*e.g.*, by extraction of native complexes from tissue or expression of recombinant proteins and *in-vitro* assembly of the complex). In case of extraction and purification from cells or tissues the exact composition of a protein complex is unknown and strictly depends on the original source. Non-homogeneous populations of purified complexes are generated by different isoforms of subunits, post-translational modifications, or labile interactions between subunits. In these cases, an accurate identification of each protein by sequencing is extremely important. Where the complex is *in-vitro* assembled starting from purified recombinant proteins, information about protein sequences and post-translational modifications (PTM) of subunits is already available. However, the quantitative information, *e.g.*, stoichiometry and number of copies, of the whole assembly is still unknown.

Protein identification follows the peptide-based MS protocol described in Chap. 3, which is summarized in Fig. 4.2a. Denaturation of the complex and its subsequent fractionation are aimed at improving the identification of all the proteins and is particularly useful when complexes are composed of many subunits. Before protein digestion, the samples are incubated with reducing agents (*e.g.*, DTT or TCEP) to disrupt disulfide bonds. The reverse-reaction is prevented by alkylation of sulfhydryl groups (*e.g.*, with iodoacetamide) to irreversibly block the reactive thiols, thereby impeding the formation of new disulfide bonds. The reduction/alkylation step is omitted in cases where the determination of disulfide bonds is the goal of the characterization. Finally, after sample preparation, proteins are hydrolyzed, usually by a selective enzymatic digestion (*e.g.*, with trypsin), to obtain peptides with suitable features for LC-MS/MS analysis.

The MS and MS/MS data acquired allow peptide identification and assignment to protein sequences. These data also permit qualitative investigation of potential PTMs even though dedicated protocols for sample preparation are necessary for a reliable detection of PTMs [9].

4.3 Quantitative Composition: Stoichiometry and Number of Subunits

For the characterization of the quaternary structure of a protein complex or in supramolecular assemblies, stoichiometric ratios, and number of copies of each subunit have to be determined.

Native mass spectrometry is the gold standard method for the stoichiometry inference of purified and homogeneous complexes [10]. Native MS provides experimental masses of whole complexes and subcomplexes. The comparison and combination of experimental data with theoretical masses allow the determination of the stoichiometric ratio for each subunit. Unfortunately, native-MS experiments require instrumentation and experimental conditions that limit its application only to specialized laboratories preventing its widespread use as a routine method.

A more accessible alternative to native mass spectrometry is offered by protocols of quantitative peptide-based MS that can be applied in any proteomics laboratory. Using bottom-up approaches, the stoichiometry of a protein complex is identified by relative quantification comparing the concentrations of each protein in the sample. On the other hand, absolute quantification is required for the identification of the number of copies of each subunit in the whole biological assembly. Two different methods for quantitative investigation of protein complexes are commonly used. One is based on classical quantitative MS methods (*e.g.*, targeted mass spectrometric analysis using labeled peptides) while the second method is a label-free proteomics approach, based on data dependent MS acquisition [7].

A way to determine the stoichiometric ratios in protein complexes is by comparing the concentration of each subunit. Mass spectrometry does not allow a direct comparison of protein concentrations. The peptide concentration is proportional to MS intensity which depends on the amino acid sequence that determines its ionization efficiency. For this reason, MS intensities of different peptides cannot be compared without a preliminary normalization against those of

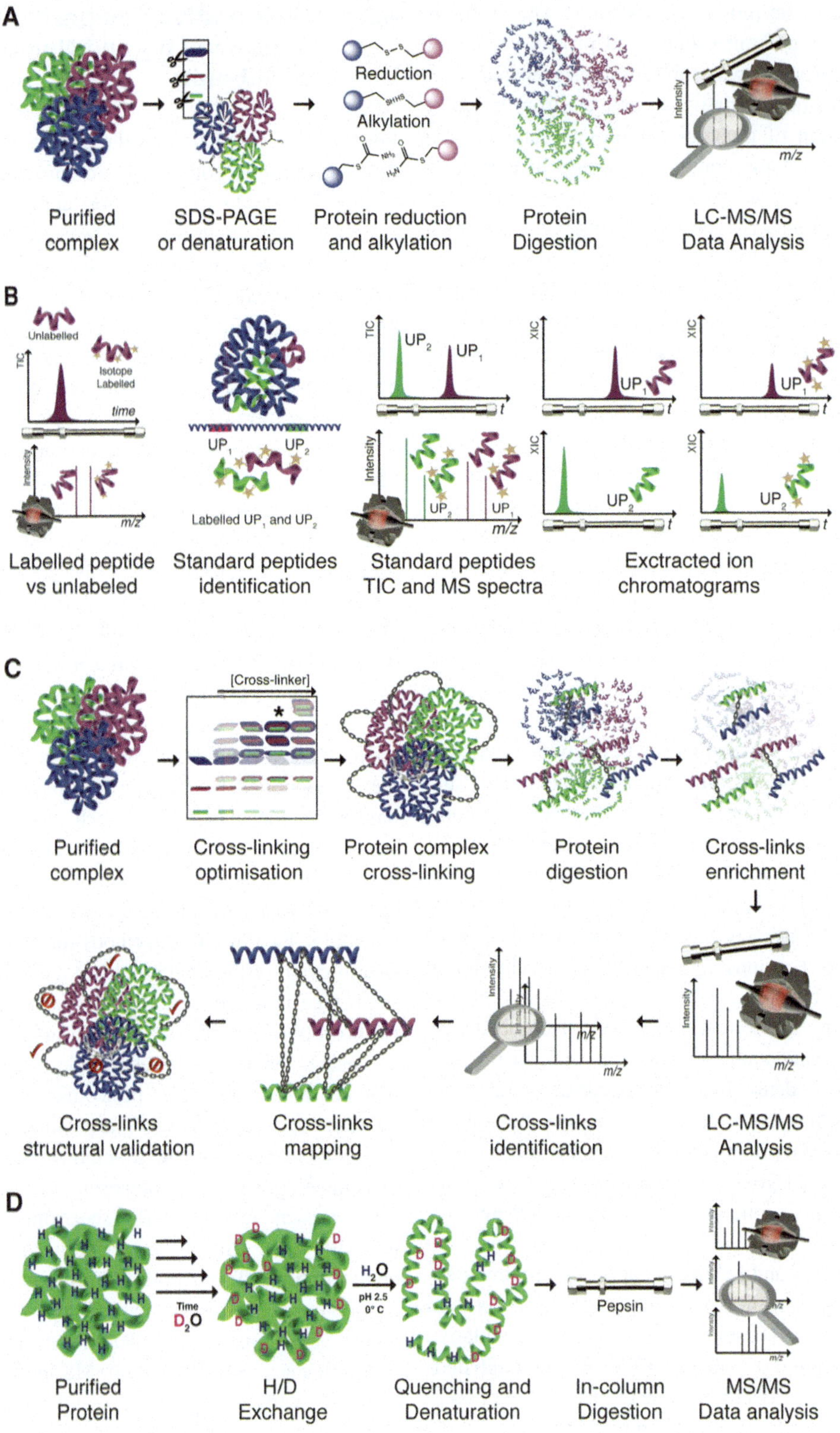

Fig. 4.2 Summary of peptide-based mass-spectrometry methods. (**a**) protein identification, (**b**) protein quantitation, and structural investigation by (**c**) chemical cross-linking, or (**d**) hydrogen deuterium exchange

standard peptides. Therefore, for quantitative purposes the definition and preparation of a quantified set of reference peptides is necessary. The ionization efficiency is not affected by isotopic composition, which makes isotope labeled peptides the best reference compounds for quantitative purpose (Fig. 4.2b).

The standard peptides can be produced in different ways and experimental conditions. Methods like SILAC (stable isotope labeling with amino acids in cell culture) or ^{15}N labeling allow the expression of whole isotopically labeled proteins [11, 12]. In this way each peptide of the complex is produced as a labeled isoform defining the best condition for quantitative investigation. The inclusion of specifically labeled amino acids in the protein limits these methods only to *in-vitro* expressed complexes. For complexes extracted from tissues, or if the expression of whole labeled proteins is unachievable, a not comprehensive set of reference peptides for each protein must be defined, designed, and produced, *e.g.*, by chemical synthesis or expression.

Uniqueness and high ionization efficiency are key requirements for reference peptides. The selection of the best candidates as standards follows a preliminary qualitative MS screening. At least two unique peptides displaying high ionization efficiency for each protein must be identified (Fig. 4.2b). For each peptide three MS/MS transitions are recorded in a Selected Reaction Monitoring (SRM) data acquisition method [13–15].

The addition of standards in quantitative analysis requires accuracy and precision. To optimize this step, several methods for standard preparation and addition have been developed and applied to real experimental conditions [5, 16].

In 2003 Gerber and colleagues reported an absolute quantification protocol named AQUA [17]. This is based on spiking the sample with known amounts of isotopically labelled reference peptides (termed AQUA peptides). As mentioned above, isotopically labelled standards allow a direct comparison for both endogenous and standard peptides. By normalizing the intensities of endogenous peptides to those of the standards it is possible to measure the relative concentration of each subunit of a complex. This approach has two main requirements for a reliable quantification based on AQUA peptides. First, the precise addition of standards for each protein at a known equimolar concentration. Second, the complete digestion of all the proteins. These requirements in turn represent the limitations and pitfalls of this quantitative methods.

QconCAT (Quantitative conCatamers) [18] is an improved approach for the production and addition of equimolar amounts of standards aimed at overcoming the mentioned limitations of AQUA peptides. This is based on the generation of a concatemer of reference peptides separated by a specific proteolytic site (*e.g.*, trypsin). The *in-vitro* or *in-vivo* expression of concatenated peptides allows the generation of an artificial isotopically labeled *QconCAT* polypeptide that after quantitation is added to the undigested complex. The enzymatic digestion of the sample generates both endogenous (unlabeled) and artificial (labeled) peptides. If the digestion proceeds to completion, an equal concentration of all standard peptides in the mixture is achieved. The accuracy of the analysis depends on the digestion efficiency of both sample and QconCAT peptide. The lower the number of missed cleavage sites in both the sample and QconCAT peptides the higher is the accuracy of the analysis. However, different hydrolytic rates for endogenous peptides and the artificial QconCAT protein are expected due to different sequence and secondary/tertiary structure that can interfere with the reaction kinetics.

In order to guarantee equal amounts of free reference peptides after digestion, a simplified version of QconCAT based on tandem peptides was developed [19]. In this method two representative peptides are fused together preserving a proteolytic site and are added to the protein complex before digestion. The enzymatic digestion generates an equal amount of both peptides regardless of the hydrolytic efficiency.

An alternative quantitative approach aimed to improve the accuracy by assuring equimolar amounts of reference peptides is the EtEP (Equimolarity through equalizer peptide) [20]. In this method, reference labeled peptides are synthesized fused to an equalizer peptide including a proteolytic site between the two. Before the addition to the sample, the fused peptides are digested with trypsin and quantified by peptide-based MS using the equalizer peptide to normalize the MS intensities. After adjusting the concentration of each peptide, a standard solution at equimolar concentrations can be prepared and used for tandem mass tag (TMT) [21], isobaric tag for relative and absolute quantitation (iTRAQ), and mass-differential tag for relative and absolute quantitation (mTRAQ) investigations of the protein concentrations in the complex sample.

Label-free methods are an alternative tool for quantitative targeted MS approaches which are based on LC-MS/MS data dependent acquisition. Intensity-based absolute quantification (iBAQ) [22] and exponentially modified protein abundance index (emPAI) [23] are widely applied methods for label-free proteomics and have been used effectively to determine the stoichiometry of complex subunits.

The intensity-based absolute quantification method (iBAQ) relies on the correlation between the MS intensity of detected peptides and the protein concentration. A standard solution composed by an accurate selection of quantified proteins covering a wide range of concentrations is added to the sample (e.g., the commercially available Universal Proteomics Standard). The iBAQ index for each protein identified by LC-MS/MS analysis is then calculated as the ratio of the sum of the mass intensities of all the peptides assigned to an identified protein ($I_{Observed}$) and its theoretical number of peptides generated by *in-silico* digestion ($N_{Observable}$), Eq. (4.1).

$$\mathrm{iBAQ} = \frac{\sum \mathrm{I}_{Observed}}{\mathrm{N}_{Observable}} \tag{4.1}$$

The linear regression model obtained by correlating the logarithmic transformed iBAQ indexes *versus* the molar amount of the standard proteins allows the extrapolation of the absolute quantification of each protein in the sample.

The exponentially modified protein abundance index (emPAI) is based on a similar concept to iBAQ but, instead of the MS intensities, it is based on the number of peptides identified and assigned. The protein abundance index (PAI, Eq. 4.2) is calculated for each protein identified as the ratio between the number of peptides observed ($N_{observed}$) for an identified protein and the total number of theoretical observable peptides ($N_{observable}$) for the same protein.

$$PAI = \frac{N_{observed}}{N_{observable}} \tag{4.2}$$

$$emPAI = 10^{PAI} - 1 \tag{4.3}$$

$$Protein\ content\ (mol\%) = \frac{emPAI}{\sum emPAI} \tag{4.4}$$

The emPAI values (Eq. 4.3) are calculated as an index of the abundance of a protein. The sum of all the emPAI values can be considered as the overall amount of proteins in the sample. For this reason, by dividing the emPAI value of a protein for the sum of the emPAI values of all the proteins in the sample, the relative amount of a given protein in the sample can be determined, Eq. 4.4.

4.4 Structural Investigation of Protein Complexes

In addition to the qualitative and quantitative analysis for subunit identification and quantification, the contribution of mass spectrometry to the characterization of protein complexes extends to structural biology. In recent years, cryo-electron microscopy (cryo-EM) has become, together with X-ray crystallography or NMR, one of the most widely applied methods for structural characterization of protein complexes. The requirement of a monodisperse protein-protein complex in vitreous ice instead of a crystal has made cryo-EM a very attractive method in comparison to X-ray crystallography for the investigation of complexes difficult to crystallize (*e.g.*, large membrane complexes). Mass spectrometry, through different approaches, is a supporting tool providing information about protein-protein interactions and protein folding (*e.g.*, secondary and tertiary structure or flexible regions) useful for the structural modeling of protein complexes and the better understanding of their dynamics [24–26].

4.4.1 Cross-Linking and Mass Spectrometry

For a long time, protein cross-linking was used to stabilize labile protein complexes during extraction and purification. In the last years, the peptide-based mass spectrometric analysis of cross-linked proteins (XL-MS) has become a very popular support technique to the structural analysis of protein complexes, providing useful constraints for the structural modeling of cryo-EM, NMR or X-ray crystallography data [4, 27, 28].

Cross-linking experiments are based on the formation of covalent bonds between the cross-linking reagents and the amino acids side chains of the proteins. Cross-linking reagents are composed by two reactive groups spaced by a linker chain. The reactive groups define the selectivity toward a certain group of amino acids and the spacer length sets the distance cut-off between the reaction centers on the surface of cross-linkable proteins.

Cross-linking reactions generate different types of products (*i.e.*, intra-protein, inter-protein, and dead-end cross-links) as depicted in Fig. 4.1d. Intra-protein or inter-protein cross-links are formed when both the reactive sites of a cross-linker react with residues of the same protein or of two different proteins, respectively. On the other hand, dead-end cross-links (or mono-links) are generated when only a single reactive site of the cross-linker reacts with the protein and the other is hydrolyzed by buffer or reversible secondary reactions.

The digestion of a cross-linked complex generates a mixture of peptides: free or unmodified peptides, modified or dead-end peptides as well as intra- and inter-protein cross-linked peptides. Among the intra-protein peptides, cyclic intra-protein cross-linked peptides (*i.e.*, loop-peptides) are observed when the cross-linker bridges two amino acid that after digestion are on the same peptide.

The intra-protein and inter-protein cross-links are the most informative products provided by XL-MS experiments. However, their molar fraction is lower than that of free peptides. This is due to a number of factors such as incomplete cross-linking reaction, multiple peptide-peptide combinations, and cross-linking by-products (*e.g.*, dead-end cross-links) generated during the reaction. The MS detection of cross-linked peptides is therefore complicated by their comparatively lower concentration in the digests. To overcome this limitation, a strategy for their enrichment is encouraged. Two main differences between cross-linked and linear peptides are the higher molecular weight and the net charge. Therefore, these features can be exploited for enrichment protocols based on size exclusion (SEC) and strong cation exchange (SCX) chromatography [29, 30]. A typical workflow for cross-linking and mass spectrometry investigation is summarized in Fig. 4.2c.

A number of cross-linking reagents are nowadays commercially available, and the choice depends on the experimental questions and technical features required. They have been extensively revised and classified according to their features but their list is constantly updated [31].

The choice of the best cross-linking reagent for structural investigations is based on amino-acid selectivity, solubility and membrane permeability, length, and chemical features of linker chain important for enrichment, MS detection, and identification of cross-linked peptides.

Amino-acid selectivity is required to restrict the number of cross-linked peptides with same sequence but different connectivity, which translates in smaller molar fractions of each isoform, higher complexity of mass spectrometric data, and wider database of potential cross-links to search, all leading to exponentially longer time for data analysis and less efficient identification.

Most of the cross-linkers commercially available and commonly used for XL-MS investigations react with amino-acid side chains containing primary amine, hydroxyl groups, or carboxylic acids (Fig. 4.3). The hydrophilicity and the wide

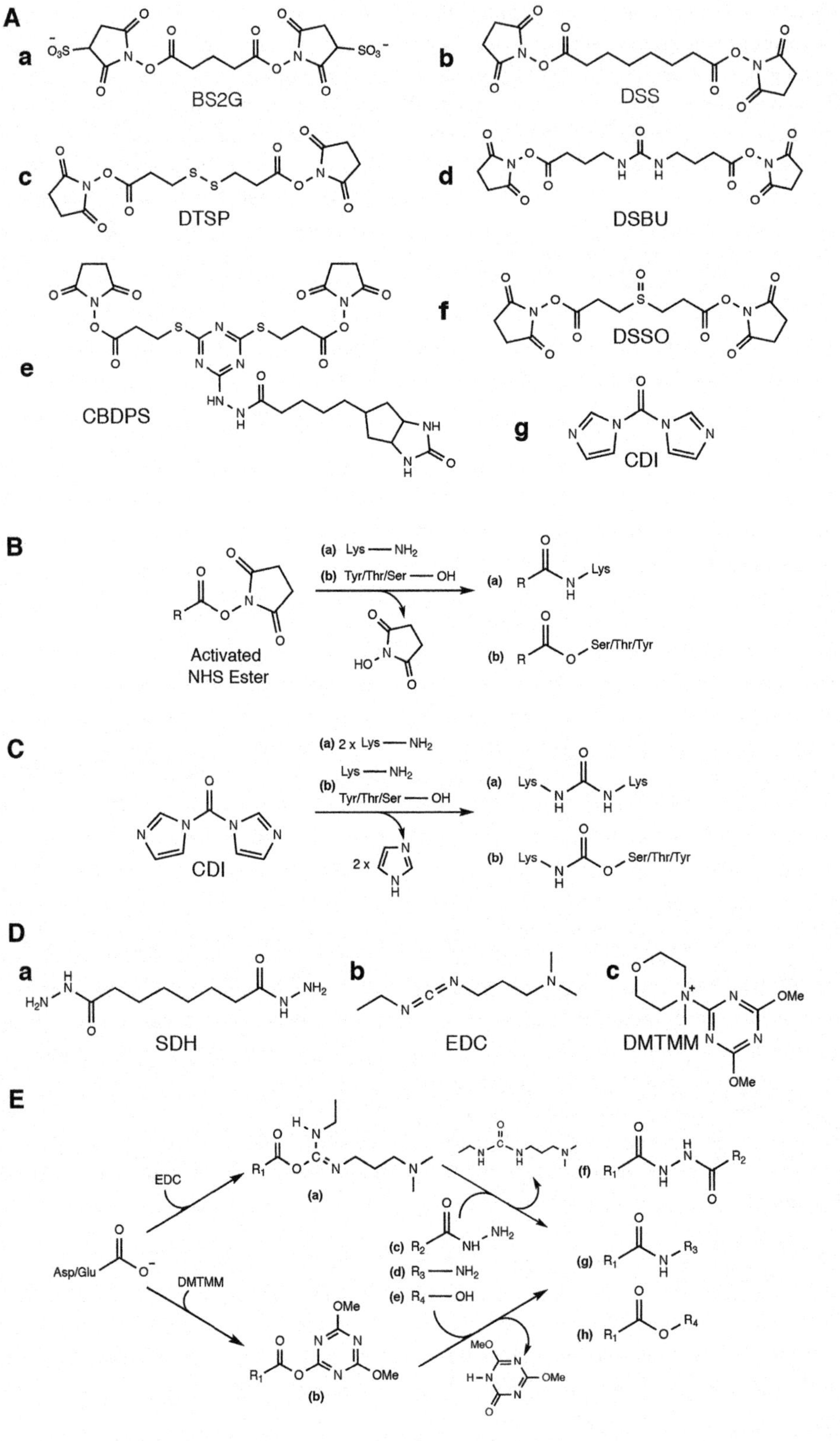

Fig. 4.3 Cross-linkers for mass spectrometry research. (**a**) Primary amine and hydroxyl specific cross-linking reagents based on NHS ester (*a–f*) and carbonyldiimidazole (*g*) reactive group. (**b**) Reaction products of activated NHS esters. (**c**) Reaction products of carbonyldiimidazole (CDI). (**d**) Example of

distribution of these amino acids on protein surfaces make them good targets for studying protein-protein interactions.

The most commonly reported reagents for structural cross-linking experiments are the homobifunctional N-hydroxysuccinimide (NHS) esters (Fig. 4.3a, *a–f*). Introduced as cross-linking reagents in the mid-1970s, this class of compounds reacts with nucleophiles, preferentially primary amines (*e.g.*, lysine side chains and amino termini of proteins) to form stable amides (Fig. 4.3b, *a*) [32]. However, a secondary reactivity towards the hydroxyl group (*e.g.*, serine, threonine, and tyrosine side chains) allows the formation of stable esters detectable by MS analysis (Fig. 4.3b, *b*) [33, 34]. In addition, NHS esters promptly react with both sulfhydryl and imidazolyl groups of cysteine and histidine residues, forming reversible covalent bonds not detectable by XL-MS workflows.

NHS esters react in aqueous buffers free of primary amines, ammonia salts, and other reactive nucleophiles (*e.g.*, thiols and imidazole). A basic pH between 7 and 8 increases the reactivity with the protein as well as the rate of hydrolysis catalyzed by hydroxide ions. The lower the temperature the slower the reaction kinetics is, and a balance between time and temperature is required based on the stability of the protein complex. Two (2) h at 4 °C or 30 min at 37 °C are common conditions for unstable and very stable complexes, respectively.

Similar selectivity to NHS-esters has recently been reported for 1,1-carbonyldiimidazole (CDI) (Fig. 4.3b, *g*) [35]. This reagent, previously employed for the synthesis of amides, urea and carbamates [36–38], reacts with primary amines and hydroxyl groups of amino acid side chains (Fig. 4.3c). CDI generates stable urea or carbamate by reacting with two primary amines or an amine and a hydroxyl group, respectively. In contrast with NHS-esters, CDI mono-links are unstable and promptly regenerate the free amino acid side chain. CDI cross-linking reaction is efficiently incubated in aqueous buffers at slightly basic pH (between 7.2 and 8) at temperatures above 10 °C.

An alternative to lysine-specific reagents is offered by the carboxy-selective reagents. Their application to chemical XL-MS investigation was reported for the first time in 2008 by Petr Novak [39] based on the reaction of dihydrazide cross-linkers with pre-activated carboxylic groups to form stable bonds (Fig. 4.3d, *a*). The activation of carboxylic acid was initially reported using EDC (1-Ethyl-3-(3-dimethylaminopropyl)-carbodiimide hydrochloride) (Fig. 4.3d, *b*) [38, 40], and more recently achieved using DMTMM (4-(4,6-dimethoxy-(1,3,5)-triazin-2-yl)-4-methyl-morpholinium-chloride) (Fig. 4.3d, *c*) [41, 42]. Both EDC and DMTMM allow a single-pot reaction with dihydrazide-based cross-linkers but in substantially different reaction conditions (Fig. 4.3e). EDC activation was described as a 2-h reaction in pyridine/hydrochloride buffer (pH 5.2) at room temperature [39]. On the other hand, DMTMM activation is allowed in aqueous buffer (*e.g.*, HEPES or PBS) at pH 7.4 at 37 °C for 45 min. The milder conditions required by DMTMM extend its application to complexes unstable under non-physiological conditions [41].

Solubility of cross-linking reagents is another important feature to take into consideration. Hydrophobicity is an advantage for the analysis of purified membrane proteins or for *in-vivo*

Fig. 4.3 (continued) dihydrazides (*a*) and EDC and DMTMM activators (*b–c*) required for cross-linking acidic residues. (**e**) activation of carboxyl side chain by DMTMM and EDC. The DMTMM activation of carboxylic acids goes through a nucleophilic aromatic substitution of the N-methyl-morpholine with the carboxylic acid generating an activated ester intermediate (*a*). EDC activation of carboxylic acids forms a reactive O-acylisourea intermediate (*b*). Both the activated intermediates react with dihydrazide nitrogen (*c*) forming a stable bond (*f*). In addition, reactions with either primary amines (*d*), *e.g.*, lysine side chains and N termini, or hydroxyl groups (*e*) from serine, threonine, and tyrosine side chains, occur forming stable amide bonds (*g*) or esters (*h*) cross-linking. Hydrophobic cross-linkers are membrane permeable and can react with regions of proteins on both the external and the internal side of membranes or micelles. Furthermore, they can cross the membrane of living cells and fix transient complexes *in-vivo*. However, the low water solubility of hydrophobic cross-linkers requires their pre-solubilization in organic solvents like DMSO). To limit the chaotropic activity of organic solvents that can interefere with the native folding of the protein complex, their concentration in the sample solution should be kept as low as possible.

The water solubility of NHS-esters is low and stock solutions are usually prepared in DMSO. Sulfo-NHS are soluble analogues of NHS esters (Fig. 4.3a, *a*). Their negatively charged sulfo-substituent on the succinimide ring increases the solubility in aqueous buffers preventing membrane permeability. The solubility of dihydrazide cross-linkers is promoted by the polarity of dihydrazide functional groups and limited by the length of the hydrophobic chain spacer.

The main role of linker chains in cross-linking reagents is to set the distance cut-off between two amino acids. The longer the distance the wider the probed surface is. Consequently, the higher the chance of cross-link formation and the lower the spatial resolution of the structural information acquired are. The highest resolution is reached by *zero-length* cross-linkers, which promote the condensation of amino acid side chains without introducing any additional atom, *e.g.*, EDC and DMTMM (Fig. 4.3d, *b*, *c*), or inserting a single carbonyl group, *e.g.*, CDI (Fig. 4.3a, *g*) [35, 39, 41]. Despite the high-resolution potential of zero-length cross-linkers, the chain length of most reported reagents for XL-MS studies is around 12–14 Å (*e.g.*, DSS, BS3, DSSO, and DSBU).

In addition to setting the distance between reactive groups, additional features of spacer chain such as biotinylation, isotopic composition, and incorporation of a MS cleavable group play primary roles for the enrichment or identification of cross-linked peptides.

Biotinylated linkers were designed to simplify the enrichment of cross-linked peptides by pull-down with avidin or streptavidin coated beads. An example of a biotinylated crosslinker is the homo-bifunctional NHS-ester CBDPS (Cyanur-Biotin-Dimercapto-Propionyl-Succinimide) (Fig. 4.3a, *e*) [43, 44].

Isotopic composition and inclusion of MS cleavable groups are chemical features of the linker chain aimed at improving the discrimination between cross-linked and free peptides during the MS data analysis.

Isotopically labeled cross-linkers are isoforms of the same reagents with a different isotopic composition of the spacer chains. By mixing the reagents at a defined ratio of heavy and light isoforms, cross-links with mass and MS intensity proportional to the number of isotopes and the isoform concentration are generated. After digestion and MS analysis, the search of cross-linked peptides is first focused on the detection of parent ion pairs corresponding to heavy and light isoforms of the same cross-link. After that, the MS/MS spectra of the shortlisted pairs are searched against the sequence database restricting the search space and simplifying the assignment procedure [45]. Isotope-labeled cross-linker selective for both primary amine and carboxylic acid are commercially available and reported in a number of studies [29, 41, 46–48].

An additional strategy aimed at simplifying and driving the MS data analysis toward the identification of cross-linked peptides is based on cleavable linkers. The spacer chain of these reagents contains a chemical bond that is fragmented by Collision Induced Dissociation (CID or HCD) or Electron Transfer Dissociation (ETD) during the MS analysis. The inclusion of a MS-cleavable linker allows the fragmentation of cross-linked peptides into two modified linear peptides improving their detection in very complex samples and in proteome-wide experiments [49–55]. Disuccinimidyl Dibutyric Urea (DSBU, also known as BuUrBu; Fig. 4.3d, *d*) [56], and Disuccinimidyl Sulfoxide (DSSO; Fig. 4.3d, *g*) [57] are examples of cleavable reagents that are commercially available. The wide range of investigation offered by cleavable crosslinkers has led to the fast development of new compounds with different reactivity such as photoactivatable reagents [35, 58–60].

A further class of cleavable linkers is represented by thiol-cleavable reagents, *e.g.*, di-thio-succinimidyl-propionate DSP or DTSP (Fig. 4.3a, *c*) [61]. Characterized by linker chains containing a disulfide bond, these reagents are both MS and chemically cleavable. These features are attractive for the transient stabilization of labile complexes during their extraction and purification. The reduction of the disulfide bond after digestion releases tagged linear peptides whose identification is easier than that of cross-linked peptides improving the protein sequence coverage and the disclosure of networks of protein interactions.

Cross-linked peptides are analyzed by reversed phase liquid chromatography and mass spectrometry. The peptides are separated on a gradient of acetonitrile acidified with formic acid and tailored to the sample complexity. The data acquisition protocols vary according to the cross-linker and the instrument used. MS/MS methods based on data dependent acquisition and CID or HCD fragmentation are used for both isotope-labelled and cleavable cross-linkers (*e.g.*, DSBU) [53, 56, 62]. Furthermore, ETD fragmentation and MS^3 methods have been reported for the cleavable cross-linker DSSO [63].

The MS data analysis and the searching strategy depend on the cross-linker features and the data acquisition protocol used. Several software packages are nowadays available for the analysis of MS data acquired from cross-linking experiments: XQuest, StavroX/MeroX, and XlinkX form a non-exhaustive list of the most frequently reported software packages [41, 50, 64, 65]. Cross-links identification rely on the *in-silico* generation of a database of cross-linked peptides used for the search and assignment of MS data. Databases are usually seeded with decoy proteins (*e.g.*, reversed protein sequences) to provide a statistical validation of the assignment by calculating the false discovery rate (FDR). As a final validation step, the spectral data assigned to cross-links are manually interrogated.

The list of validated cross-linked peptides is used to generate topological maps of interactions between the proteins in the complexes and several software and applications are available for this purpose, *e.g.*, Xvis, xlnet [66, 67]. In addition to the topological mapping, the identified cross-links can be mapped and checked on available structural data. For this purpose, a dedicated software program (Xlinks) has been developed and integrated in 3D viewer software [68].

4.4.2 Hydrogen Deuterium Exchange (HDX)

Methods for mass spectrometric data acquisition are designed and tailored to improve the identification of unknown proteins or the quantitative determination of known proteins. Proteins have a flexible and dynamic nature necessary to exert their chemical and biological activity. Structural investigation methods such as X-ray crystallography and cryo-EM provide mostly static representations of the most stable conformations in specific experimental conditions. Insights into the conformational changes and flexibility of the proteins provide a deeper understanding of protein function and mechanism of activity. As summarized in Fig. 4.2d, Hydrogen Deuterium Exchange (HDX) is a method for the study of conformational dynamics of proteins and protein complexes. It is based on monitoring the rate of proton exchange between protein backbone and deuterated solvent. The rate of proton exchange of a protein region is proportional to both solvent exposure and structural organization. Defined secondary structures such as α-helices and β-sheets display a slower exchange rate than unstructured regions (*e.g.*, loops or disordered segments). Therefore, the comparison of deuterium incorporation rates allows the identification of structured and unstructured regions of the proteins. Peptidyl hydrogens, due to their low exchange rate (ranging from fractions of a second to days), can be effectively monitored by mass spectrometry. The exchange rates of other hydrogens (*e.g.*, those bound to N, O, and S of amino acid side-chains) are too high to be monitored without back-exchange interferences.

Hydrogen/Deuterium exchange is based on acid-base reactions. The exchange rate is influenced by pH, temperature and, as mentioned above, structural factors (*e.g.*, secondary, tertiary, and quaternary structures of proteins and protein complexes). Ice cold buffer at pH 2.5 minimizes the exchange rate of peptidyl hydrogens for an unstructured peptide leading to a half-time of exchange longer than 1 h. Conversely, the exchange rate is increased by higher temperature and higher or lower pH values.

Once the reaction conditions are defined (temperature, buffer, and pH) and kept constant, the intrinsic exchange rate (k_i) for an unstructured peptide can be determined providing a measure of the influence of the primary sequence on the exchange rate [69].

The kinetic constants for hydrogen/deuterium exchange of folded proteins (k_{ex}) differ by several order of magnitude from those of unstructured peptides (Fig. 4.4a). This difference is due to secondary structure stabilization by hydrogen bonding that requires a transient rupture and reformation of the bonds to exchange the hydrogens with solvent deuterium. The dynamic equilibrium of protein unfolding/refolding permits amide hydrogen exchange even in the most structured regions. Figure 4.4b summarise the hydrogen deuterium exchange in a structured protein. Constants k_1 *and* k_{-1} are the unfolding and refolding rate constants, respectively, and k_2 is the hydrogen deuterium exchange rate constant in the unfolded protein (approximately equal to the intrinsic exchange rate $k_2 \approx k_i$). The overall exchange rate can be split into two contributions: the unfolding/refolding equilibrium (k_1/k_{-1}) and the intrinsic exchange rate of the unfolded sequence k_2. Based on the ratio between k_2 and k_{-1} two main hydrogen exchange kinetic profiles are identifiable. If the exchange constant k_2 is much larger than the refolding constant k_{-1} ($k_2 >> k_{-1}$) the exchange is very rapid and takes place every time a hydrogen bond is broken. In these conditions the overall exchange kinetic constant k_{ex} depends only on the unfolding rate kinetic constant k_1 ($k_{ex} = k_1$). This condition, termed *EX_1 kinetics* or monomolecular exchange, is not very common and is mostly observed when unfolded states are favored by sequence or induced by denaturants. A more common situation displays a refolding rate constant (k_{-1}) much larger than the exchange rate constant (k_2). In these conditions when $k_{-1} >> k_2$, the rupture and regeneration of a hydrogen-bond is faster than the hydrogen exchange. Known as *EX_2 kinetics* or bimolecular exchange, this exchange kinetics (see equation below) allows the determination of the unfolding equilibrium constant K_{unf} and the kinetic constant k_2 (Eq. 4.1) [128]. Eq. 4.5 [70].

$$k_{ex} = {}^{k_1}/_{k_{-1}} \times k_2 = K_{unf} \times k_2 \qquad (4.5)$$

For certain proteins or complexes, a combination of EX_1 and EX_2 profiles coexist and can be discriminated by the spectral data [71, 72].

HDX experiments start with the protein incubation in deuterated water for defined amounts of time allowing the hydrogen to freely exchange with deuterium. The number of time-points depends on the experimental design and on the structure investigated, and appropriate incubation times can range from seconds to days. The exchange reaction is quenched by adding an ice-cold buffer solution at pH 2.5. These conditions limit the back-exchange of the backbone amide incorporated deuterium and are kept

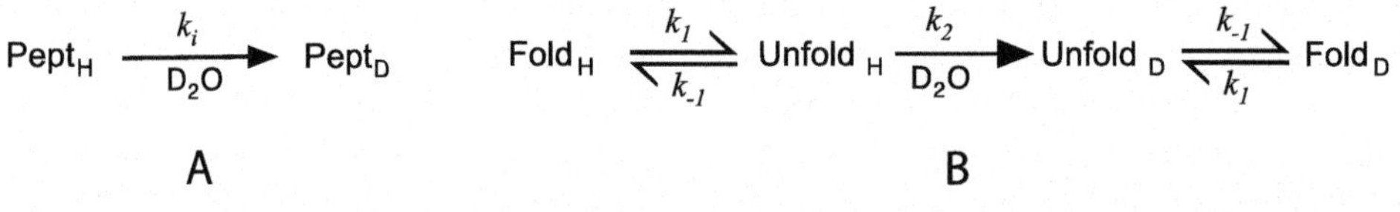

Fig. 4.4 Hydrogen deuterium exchange reaction. (**a**) Single step chemical reaction of Hydrogen/Deuterium Exchange for a generic unstructured peptide (Pept) occurring at an intrinsic exchange rate (k_i), and allowed in deuterated water (D_2O). (**b**) Representation of the multiple steps reaction of Hydrogen/Deuterium Exchange for a hypothetical protein region. The overall exchange rate depends on two contributions: the unfolding (k_1) and refolding (k_{-1}) rate constants, and the intrinsic exchange rate of the unfolded region (k_2)

constant for all the following analytic steps. In a peptide-based HDX-MS approach, the protein is digested using an acid functional protease (*e.g.*, pepsin) before LC-MS/MS analysis. To prevent deuterium back-exchange, the chromatographic step is required to be rapid and performed at low temperature. Pepsin functionalized LC columns are available for rapid digestion and direct data acquisition during LC-MS/MS. This approach generates overlapping short peptides allowing a high sequence coverage and determination of exchange rates at peptide resolution.

The data analysis starts with the verification of deuterium incorporation, then continue with the peptide identification and the evaluation of the amount of hydrogens exchanged. The successful incorporation of deuterium is checked by comparing the isotopic profile of unlabeled and deuterated samples. For both unlabeled and completely deuterated peptides the isotopic distribution is expected to appear as a Cauchy distribution with shifted *m/z* values. However, regardless of the relative amount of incorporated deuterium, the isotope distribution of labeled peptides displays a symmetrical Gaussian distribution. This is due to deuterium back-exchange and the maximum deuterium incorporation is therefore identified by the maximum shift of *m/z* values [73].

Once the incorporation of deuterium is confirmed, data analysis proceeds with the peptide identification. Pepsin, in contrast with trypsin, has a non-specific activity with a preference for hydrophobic amino acids making the hydrolysis unpredictable and peptide identification more complicated. Therefore, different approaches from those described for protein identification in Chap. 3 are required for data interpretation.

Peptide identification is based on the generation of a peptide list from the LC-MS/MS analysis of a non-deuterated sample. The *m/z* value of each MS ion detected is screened against a database containing all the possible peptide sequences generated by sample digestion and pre-assigned to a peptide on an exact mass-based approach. The MS/MS data are then used to confirm the assignment. The peptide list generated will include the mono-isotopic mass, the charge state, and the retention time of the identified ions together with the amino acid sequence assigned. The aim of this list is the generation of an indexed database of assigned peptides to be used for the assignment of deuterated isoforms.

The deuterium incorporation is measured by relative quantification between the deuterated samples and the non-deuterated ones. If the exchange of a given peptide follows an EX_2 kinetic profile, deuterium incorporation increases proportionally with exposure time. The identification of deuterated samples is based on the precomputed peptide list, considering parameters such as charge state, shift of *m/z* values, and increase in mass of the isotope distribution. Once the deuterated peptide is assigned, a relative quantification of the incorporated deuterium is possible. Therefore, for each peptide it is possible to evaluate the relative amount of deuterium as a function of time. A visual summary of the exchange is achieved by plotting the amount of deuterium incorporation over time. For similar purposes, many other representations that consider the whole sequence of the protein can be generated. For example, by rendering the deuterium levels on peptide maps or on the ribbon of 3D models or using difference plots to compare different experimental conditions for the same proteins [74–77].

The interest on HDX-MS for the investigation of proteins and protein complexes from both structural and functional point of view is constantly increasing. The development of new methods as well as the availability of automated systems and instrumental techniques has allowed the application of this technique to a wider list of targets, ranging from simple ligand-protein interactions to protein-protein interactions. The clear advantage of HDX-MS in comparison to other methods for peptide-based mass spectrometry relies on the time course investigation leading to a better observation of dynamic of structural changes [78–81].

References

1. Aebersold R, Mann M (2003) Mass spectrometry-based proteomics. Nature 422(6928):198–207. https://doi.org/10.1038/nature01511
2. Gingras AC, Gstaiger M, Raught B, Aebersold R (2007) Analysis of protein complexes using mass

spectrometry. Nat Rev Mol Cell Biol 8(8):645–654. https://doi.org/10.1038/nrm2208
3. Domon B, Aebersold R (2006) Review – mass spectrometry and protein analysis. Science 312(5771): 212–217. https://doi.org/10.1126/science.1124619
4. Yu C, Huang L (2018) Cross-linking mass spectrometry: an emerging technology for interactomics and structural biology. Anal Chem 90(1):144–165. https://doi.org/10.1021/acs.analchem.7b04431
5. Wohlgemuth I, Lenz C, Urlaub H (2015) Studying macromolecular complex stoichiometries by peptide-based mass spectrometry. Proteomics 15(5–6): 862–879. https://doi.org/10.1002/pmic.201400466
6. Konermann L, Pan J, Liu Y-H (2011) Hydrogen exchange mass spectrometry for studying protein structure and dynamics introduction: protein folding, dynamics, and function. Chem Soc Rev 40(40): 1224–1234. https://doi.org/10.1039/c0cs00113a
7. Ankney JA, Muneer A, Chen X (2018) Relative and absolute quantitation in mass spectrometry-based proteomics. Annu Rev Anal Chem 11:49–77. https://doi.org/10.1146/annurev-anchem-061516-045357
8. Kaur U, Johnson DT, Chea EE, Deredge DJ, Espino JA, Jones LM (2019) Evolution of structural biology through the lens of mass spectrometry. Anal Chem 91(1):142–155. https://doi.org/10.1021/acs.analchem.8b05014
9. Silva AMN, Vitorino R, Domingues MRM, Spickett CM, Domingues P (2013) Post-translational modifications and mass spectrometry detection. Free Radic Biol Med 65:925–941. https://doi.org/10.1016/j.freeradbiomed.2013.08.184
10. Marcoux J, Robinson CV (2013) Twenty years of gas phase structural biology. Structure 21(9):1541–1550. https://doi.org/10.1016/j.str.2013.08.002
11. Conrads TP, Alving K, Veenstra TD, Belov ME, Anderson GA, Anderson DJ, Lipton MS, Pasa-Tolic L, Udseth HR, Chrisler WB, Thrall BD, Smith RD (2001) Quantitative analysis of bacterial and mammalian proteomes using a combination of cysteine affinity tags and N-15-metabolic labeling. Anal Chem 73(9): 2132–2139. https://doi.org/10.1021/ac001487x
12. Ong SE, Blagoev B, Kratchmarova I, Kristensen DB, Steen H, Pandey A, Mann M (2002) Stable isotope labeling by amino acids in cell culture, SILAC, as a simple and accurate approach to expression proteomics. Mol Cell Proteomics 1(5):376–386. https://doi.org/10.1074/mcp.M200025-MCP200
13. Picotti P, Aebersold R (2012) Selected reaction monitoring-based proteomics: workflows, potential, pitfalls and future directions. Nat Methods 9(6): 555–566. https://doi.org/10.1038/Nmeth.2015
14. Bereman MS, MacLean B, Tomazela DM, Liebler DC, MacCoss MJ (2012) The development of selected reaction monitoring methods for targeted proteomics via empirical refinement. Proteomics 12(8): 1134–1141. https://doi.org/10.1002/pmic.201200042
15. Lange V, Picotti P, Domon B, Aebersold R (2008) Selected reaction monitoring for quantitative proteomics: a tutorial. Mol Syst Biol 4:222. https://doi.org/10.1038/msb.2008.61
16. Schmidt C, Urlaub H (2012) Absolute quantification of proteins using standard peptides and multiple reaction monitoring. Methods Mol Biol 893:249–265. https://doi.org/10.1007/978-1-61779-885-6_17
17. Gerber SA, Rush J, Stemman O, Kirschner MW, Gygi SP (2003) Absolute quantification of proteins and phosphoproteins from cell lysates by tandem MS. Proc Natl Acad Sci U S A 100(12):6940–6945. https://doi.org/10.1073/pnas.0832254100
18. Beynon RJ, Doherty MK, Pratt JM, Gaskell SJ (2005) Multiplexed absolute quantification in proteomics using artificial QCAT proteins of concatenated signature peptides. Nat Methods 2(8):587–589. https://doi.org/10.1038/nmeth774
19. Winter D, Hung CW, Jaskolla TW, Karas M, Lehmann WD (2012) Enzyme-cleavable tandem peptides for quantitative studies in MS-based proteomics. Proteomics 12(23–24):3470–3474. https://doi.org/10.1002/pmic.201200290
20. Holzmann J, Pichler P, Madalinski M, Kurzbauer R, Mechtler K (2009) Stoichiometry determination of the MP1-p14 complex using a novel and cost-efficient method to produce an equimolar mixture of standard peptides. Anal Chem 81(24):10254–10261. https://doi.org/10.1021/ac902286m
21. Thompson A, Schafer J, Kuhn K, Kienle S, Schwarz J, Schmidt G, Neumann T, Hamon C (2003) Tandem mass tags: a novel quantification strategy for comparative analysis of complex protein mixtures by MS/MS. Anal Chem 75(8):1895–1904. https://doi.org/10.1021/ac0262560
22. Schwanhausser B, Busse D, Li N, Dittmar G, Schuchhardt J, Wolf J, Chen W, Selbach M (2011) Global quantification of mammalian gene expression control. Nature 473(7347):337–342. https://doi.org/10.1038/nature10098
23. Ishihama Y, Oda Y, Tabata T, Sato T, Nagasu T, Rappsilber J, Mann M (2005) Exponentially modified protein abundance index (emPAI) for estimation of absolute protein amount in proteomics by the number of sequenced peptides per protein. Mol Cell Proteomics 4(9):1265–1272. https://doi.org/10.1074/mcp.M500061-MCP200
24. Walzthoeni T, Leitner A, Stengel F, Aebersold R (2013) Mass spectrometry supported determination of protein complex structure. Curr Opin Struct Biol 23(2):252–260. https://doi.org/10.1016/j.sbi.2013.02.008
25. Sinz A (2018) Cross-linking/mass spectrometry for studying protein structures and protein-protein interactions: where are we now and where should we go from here? Angew Chem Int Ed Engl 57(22): 6390–6396. https://doi.org/10.1002/anie.201709559
26. Konermann L, Pan J, Liu YH (2011) Hydrogen exchange mass spectrometry for studying protein structure and dynamics. Chem Soc Rev 40(3): 1224–1234. https://doi.org/10.1039/c0cs00113a

27. Sinz A (2006) Chemical cross-linking and mass spectrometry to map three-dimensional protein structures and protein-protein interactions. Mass Spectrom Rev 25(4):663–682. https://doi.org/10.1002/mas.20082
28. Rappsilber J (2011) The beginning of a beautiful friendship: cross-linking/mass spectrometry and modelling of proteins and multi-protein complexes. J Struct Biol 173(3):530–540. https://doi.org/10.1016/j.jsb.2010.10.014
29. Leitner A, Reischl R, Walzthoeni T, Herzog F, Bohn S, Forster F, Aebersold R (2012) Expanding the chemical cross-linking toolbox by the use of multiple proteases and enrichment by size exclusion chromatography. 11. https://doi.org/10.1074/mcp.M111.014126
30. Schmidt R, Sinz A (2017) Improved single-step enrichment methods of cross-linked products for protein structure analysis and protein interaction mapping. Anal Bioanal Chem 409(9):2393–2400. https://doi.org/10.1007/s00216-017-0185-1
31. Petrotchenko EV, Borchers CH (2010) Crosslinking combined with mass spectrometry for structural proteomics. Mass Spectrom Rev 29(6):862–876. https://doi.org/10.1002/mas.20293
32. Hermanson GT (2013) Homobifunctional crosslinkers. In: Hermanson GT (ed) Bioconjugate techniques. Academic, Boston, pp 275–298. https://doi.org/10.1016/b978-0-12-382239-0.00005-4
33. Mädler S, Bich C, Touboul D, Zenobi R (2009) Chemical cross-linking with NHS esters: a systematic study on amino acid reactivities. J Mass Spectrom 44(5): 694–706. https://doi.org/10.1002/jms.1544
34. Kalkhof S, Sinz A (2008) Chances and pitfalls of chemical cross-linking with amine-reactive N-hydroxysuccinimide esters. Anal Bioanal Chem 392(1–2):1–8. https://doi.org/10.1007/s00216-008-2231-5
35. Hage C, Iacobucci C, Rehkamp A, Arlt C, Sinz A (2017) The first zero-length mass spectrometry-cleavable cross-linker for protein structure analysis. Angew Chem Int Ed 56(46):14551–14555. https://doi.org/10.1002/anie.201708273
36. Montalbetti CAGN, Falque V (2005) Amide bond formation and peptide coupling. Tetrahedron 61(46): 10827–10852. https://doi.org/10.1016/j.tet.2005.08.031
37. Padiya KJ, Gavade S, Kardile B, Tiwari M, Bajare S, Mane M, Gaware V, Varghese S, Harel D, Kurhade S (2012) Unprecedented "in water" imidazole carbonylation: paradigm shift for preparation of urea and carbamate. Org Lett 14(11):2814–2817. https://doi.org/10.1021/ol301009d
38. Hermanson GT (2013) Zero-length crosslinkers. In: Hermanson GT (ed) Bioconjugate techniques. Academic, Boston, pp 259–273. https://doi.org/10.1016/b978-0-12-382239-0.00004-2
39. Novak P, Kruppa GH (2008) Intra-molecular cross-linking of acidic residues for protein structure studies. Eur J Mass Spectrom (Chichester) 14(6):355–365. https://doi.org/10.1255/ejms.963
40. Yamada H, Imoto T, Fujita K, Okazaki K, Motomura M (1981) Selective modification of aspartic acid-101 in lysozyme by carbodiimide reaction. Biochemistry 20(17):4836–4842. https://doi.org/10.1021/bi00520a005
41. Leitner A, Joachimiak LA, Unverdorben P, Walzthoeni T, Frydman J, Förster F, Aebersold R (2014) Chemical cross-linking/mass spectrometry targeting acidic residues in proteins and protein complexes. Proc Natl Acad Sci U S A 111(26): 9455–9460. https://doi.org/10.1073/pnas.1320298111
42. Kunishima M, Kawachi C, Morita J, Terao K, Iwasaki F, Tani S (1999) 4-(4,6-dimethoxy-1,3,5-triazin-2-yl)-4-methyl-morpholinium chloride: an efficient condensing agent leading to the formation of amides and esters. Tetrahedron 55(46):13159–13170. https://doi.org/10.1016/S0040-4020(99)00809-1
43. Petrotchenko EV, Serpa JJ, Cabecinha AN, Lesperance M, Borchers CH (2014) "Out-gel" tryptic digestion procedure for chemical cross-linking studies with mass spectrometric detection. J Proteome Res 13(2):527–535. https://doi.org/10.1021/pr400710q
44. Petrotchenko EV, Serpa JJ, Borchers CH (2011) An isotopically coded CID-cleavable biotinylated cross-linker for structural proteomics. Mol Cell Proteomics 10(2):M110.001420. https://doi.org/10.1074/mcp.M110.001420
45. Seebacher J, Mallick P, Zhang N, Eddes JS, Aebersold R, Gelb MH (2006) Protein cross-linking analysis using mass spectrometry, isotope-coded cross-linkers, and integrated computational data processing. J Proteome Res 5(9):2270–2282. https://doi.org/10.1021/pr060154z
46. Fiedorczuk K, Letts JA, Degliesposti G, Kaszuba K, Skehel M, Sazanov LA (2016) Atomic structure of the entire mammalian mitochondrial complex I. Nature 538(7625):406–410. https://doi.org/10.1038/nature19794
47. Casanal A, Kumar A, Hill CH, Easter AD, Emsley P, Degliesposti G, Gordiyenko Y, Santhanam B, Wolf J, Wiederhold K, Dornan GL, Skehel M, Robinson CV, Passmore LA (2017) Architecture of eukaryotic mRNA 3′-end processing machinery. Science 358(6366):1056–1059. https://doi.org/10.1126/science.aao6535
48. Martino F, Pal M, Munoz-Hernandez H, Rodriguez CF, Nunez-Ramirez R, Gil-Carton D, Degliesposti G, Skehel JM, Roe SM, Prodromou C, Pearl LH, Llorca O (2018) RPAP3 provides a flexible scaffold for coupling HSP90 to the human R2TP co-chaperone complex. Nat Commun 9(1):1501. https://doi.org/10.1038/s41467-018-03942-1
49. Liu F, Lossl P, Rabbitts BM, Balaban RS, Heck AJR (2018) The interactome of intact mitochondria by cross-linking mass spectrometry provides evidence for coexisting respiratory supercomplexes. Mol Cell Proteomics 17(2):216–232. https://doi.org/10.1074/mcp.RA117.000470
50. Liu F, Rijkers DT, Post H, Heck AJ (2015) Proteome-wide profiling of protein assemblies by cross-linking

mass spectrometry. Nat Methods 12(12):1179–1184. https://doi.org/10.1038/nmeth.3603

51. Chavez JD, Lee CF, Caudal A, Keller A, Tian R, Bruce JE (2018) Chemical crosslinking mass spectrometry analysis of protein conformations and Supercomplexes in heart tissue. Cell Syst 6(1):136. https://doi.org/10.1016/j.cels.2017.10.017
52. Sinz A (2018) Crosslinking mass spectrometry goes in-tissue. Cell Syst 6(1):10–12. https://doi.org/10.1016/j.cels.2018.01.005
53. Arlt C, Gotze M, Ihling CH, Hage C, Schafer M, Sinz A (2016) Integrated workflow for structural proteomics studies based on cross-linking/mass spectrometry with an MS/MS cleavable cross-linker. Anal Chem 88(16):7930–7937. https://doi.org/10.1021/acs.analchem.5b04853
54. Haupl B, Ihling CH, Sinz A (2016) Protein interaction network of human protein kinase D2 revealed by chemical cross-linking/mass spectrometry. J Proteome Res 15(10):3686–3699. https://doi.org/10.1021/acs.jproteome.6b00513
55. Iacobucci C, Gotze M, Ihling CH, Piotrowski C, Arlt C, Schafer M, Hage C, Schmidt R, Sinz A (2018) A cross-linking/mass spectrometry workflow based on MS-cleavable cross-linkers and the MeroX software for studying protein structures and protein-protein interactions. Nat Protoc 13(12):2864–2889. https://doi.org/10.1038/s41596-018-0068-8
56. Muller MQ, Dreiocker F, Ihling CH, Schafer M, Sinz A (2010) Cleavable cross-linker for protein structure analysis: reliable identification of cross-linking products by tandem MS. Anal Chem 82(16):6958–6968. https://doi.org/10.1021/ac101241t
57. Kao AH, Chiu CL, Vellucci D, Yang YY, Patel VR, Guan SH, Randall A, Baldi P, Rychnovsky SD, Huang L (2011) Development of a novel cross-linking strategy for fast and accurate identification of cross-linked peptides of protein complexes. Mol Cell Proteomics 10(1):M110.002212. https://doi.org/10.1074/mcp.M110.002212
58. Iacobucci C, Hage C, Schafer M, Sinz A (2017) A novel MS-cleavable azo cross-linker for peptide structure analysis by free radical initiated peptide sequencing (FRIPS). J Am Soc Mass Spectrom 28(10):2039–2053. https://doi.org/10.1007/s13361-017-1744-6
59. Iacobucci C, Piotrowski C, Rehkamp A, Ihling CH, Sinz A (2018) The first MS-cleavable, photo-thiol-reactive cross-linker for protein structural studies. J Am Soc Mass Spectrom. https://doi.org/10.1007/s13361-018-1952-8
60. Hage C, Ihling CH, Gotze M, Schafer M, Sinz A (2017) Dissociation behavior of a TEMPO-active Ester cross-linker for peptide structure analysis by free radical initiated peptide sequencing (FRIPS) in negative ESI-MS. J Am Soc Mass Spectrom 28(1):56–68. https://doi.org/10.1007/s13361-016-1426-9
61. Bennett KL, Kussmann M, Bjork P, Godzwon M, Mikkelsen M, Sorensen P, Roepstorff P (2000) Chemical cross-linking with thiol-cleavable reagents combined with differential mass spectrometric peptide mapping – a novel approach to assess intermolecular protein contacts. Protein Sci 9(8):1503–1518. https://doi.org/10.1110/ps.9.8.1503
62. Muller MQ, Dreiocker F, Ihling CH, Schafer M, Sinz A (2010) Fragmentation behavior of a thiourea-based reagent for protein structure analysis by collision-induced dissociative chemical cross-linking. J Mass Spectrom 45(8):880–891. https://doi.org/10.1002/jms.1775
63. Liu F, Lossl P, Scheltema R, Viner R, Heck AJR (2017) Optimized fragmentation schemes and data analysis strategies for proteome-wide cross-link identification. Nat Commun 8:15473. https://doi.org/10.1038/ncomms15473
64. Gotze M, Pettelkau J, Schaks S, Bosse K, Ihling CH, Krauth F, Fritzsche R, Kuhn U, Sinz A (2012) StavroX-A software for analyzing crosslinked products in protein interaction studies. J Am Soc Mass Spectrom 23(1):76–87. https://doi.org/10.1007/s13361-011-0261-2
65. Gotze M, Pettelkau J, Fritzsche R, Ihling CH, Schafer M, Sinz A (2015) Automated assignment of MS/MS cleavable cross-links in protein 3D-structure analysis. J Am Soc Mass Spectrom 26(1):83–97. https://doi.org/10.1007/s13361-014-1001-1
66. Grimm M, Zimniak T, Kahraman A, Herzog F (2015) xVis: a web server for the schematic visualization and interpretation of crosslink-derived spatial restraints. Nucleic Acids Res 43(W1):W362–W369. https://doi.org/10.1093/nar/gkv463
67. Combe CW, Fischer L, Rappsilber J (2015) xiNET: cross-link network maps with residue resolution. Mol Cell Proteomics 14(4):1137–1147. https://doi.org/10.1074/mcp.O114.042259
68. Kosinski J, von Appen A, Ori A, Karius K, Muller CW, Beck M (2015) Xlink analyzer: software for analysis and visualization of cross-linking data in the context of three-dimensional structures. J Struct Biol 189(3):177–183. https://doi.org/10.1016/j.jsb.2015.01.014
69. Bai YW, Milne JS, Mayne L, Englander SW (1993) Primary structure effects on peptide group hydrogen-exchange. Proteins 17(1):75–86. https://doi.org/10.1002/prot.340170110
70. Wales TE, Engen JR (2006) Hydrogen exchange mass spectrometry for the analysis of protein dynamics. Mass Spectrom Rev 25(1):158–170. https://doi.org/10.1002/mas.20064
71. Weis DD, Wales TE, Engen JR, Hotchko M, Ten Eyck LF (2006) Identification and characterization of EX1 kinetics in H/D exchange mass spectrometry by peak width analysis. J Am Soc Mass Spectrom 17(11):1498–1509. https://doi.org/10.1016/j.jasms.2006.05.014
72. Miranker A, Robinson CV, Radford SE, Aplin RT, Dobson CM (1993) Detection of transient protein-folding populations by mass-spectrometry. Science 262(5135):896–900. https://doi.org/10.1126/science.8235611

73. Wales TE, Eggertson MJ, Engen JR (2013) Considerations in the analysis of hydrogen exchange mass spectrometry data. In: Matthiesen R (ed) Mass spectrometry data analysis in proteomics. Humana Press, Totowa, pp 263–288. https://doi.org/10.1007/978-1-62703-392-3_11
74. Houde D, Berkowitz SA, Engen JR (2011) The utility of hydrogen/deuterium exchange mass spectrometry in biopharmaceutical comparability studies. J Pharm Sci 100(6):2071–2086. https://doi.org/10.1002/jps.22432
75. Tiyanont K, Wales TE, Aste-Amezaga M, Aster JC, Engen JR, Blacklow SC (2011) Evidence for increased exposure of the Notch1 metalloprotease cleavage site upon conversion to an activated conformation. Structure 19(4):546–554. https://doi.org/10.1016/j.str.2011.01.016
76. Tsutsumi S, Mollapour M, Prodromou C, Lee CT, Panaretou B, Yoshida S, Mayer MP, Neckers LM (2012) Charged linker sequence modulates eukaryotic heat shock protein 90 (Hsp90) chaperone activity. Proc Natl Acad Sci U S A 109(8):2937–2942. https://doi.org/10.1073/pnas.1114414109
77. Street TO, Lavery LA, Verba KA, Lee CT, Mayer MP, Agard DA (2012) Cross-monomer substrate contacts reposition the Hsp90 N-terminal domain and prime the chaperone activity. J Mol Biol 415(1):3–15. https://doi.org/10.1016/j.jmb.2011.10.038
78. Gladkova C, Maslen SL, Skehel JM, Komander D, Masson GR, Maslen SL, Williams RL, Mevissen TET, Kulathu Y, Mulder MPC, Geurink PP, Maslen SL, Gersch M, Elliott PR, Burke JE, van Tol BDM, Akutsu M, Oualid FE, Kawasaki M, Freund SMV, Ovaa H, Komander D, van Maldegem F, Maslen S, Johnson CM, Chandra A, Ganesh K, Skehel M, Rada C (2018) Mechanism of parkin activation by PINK1. Nature 559(7714):410–414. https://doi.org/10.1038/s41586-018-0224-x
79. Masson GR, Maslen SL, Williams RL (2017) Analysis of phosphoinositide 3-kinase inhibitors by bottom-up electron-transfer dissociation hydrogen/deuterium exchange mass spectrometry. Biochem J 474(11):1867–1877. https://doi.org/10.1042/Bcj20170127
80. Mevissen TET, Kulathu Y, Mulder MPC, Geurink PP, Maslen SL, Gersch M, Elliott PR, Burke JE, van Tol BDM, Akutsu M, El Oualid F, Kawasaki M, Freund SMV, Ovaa H, Komander D (2016) Molecular basis of Lys11-polyubiquitin specificity in the deubiquitinase Cezanne. Nature 538(7625):402. https://doi.org/10.1038/nature19836
81. van Maldegem F, Maslen S, Johnson CM, Chandra A, Ganesh K, Skehel M, Rada C (2015) CTNNBL1 facilitates the association of CWC15 with CDC5L and is required to maintain the abundance of the Prp19 spliceosomal complex. Nucleic Acids Res 43(14):7058–7069. https://doi.org/10.1093/nar/gkv643

Discovery and Characterization of Linear Motif Mediated Protein-Protein Complexes

5

András Zeke, Anita Alexa, and Attila Reményi

Abstract

There are myriads of protein-protein complexes that form within the cell. In addition to classical binding events between globular domains, many protein-protein interactions involve short disordered protein regions. The latter contain so-called linear motifs binding specifically to ordered protein domain surfaces. Linear binding motifs are classified based on their consensus sequence, where only a few amino acids are conserved. In this chapter we will review experimental and *in silico* techniques that can be used for the discovery and characterization of linear motif mediated protein-protein complexes involved in cellular signaling, protein level and gene expression regulation.

Keywords

Linear motif · Protein-peptide binding · Interactome analysis · Phosphoswitches

A. Zeke · A. Alexa · A. Reményi (✉)
Institute of Organic Chemistry, HUN-REN Research Center for Natural Sciences, Budapest, Hungary
e-mail: remenyi.attila@ttk.hu

5.1 Introduction

Two decades of exploration on the disordered part of the proteome highlighted the role of unstructured protein segments in protein-protein complex formation. Characterization of the disordered proteome of various organisms revealed that there might be thousands of functional sites within these poorly characterized protein regions, possibly governing millions of protein-protein interactions (PPI) within the cell [1]. Protein-protein complexes involving some disordered fragments are thus abundant and are different from classical PPIs forming between two globular domains. The fundamental difference is that disordered protein regions bind to their interacting domains via amino acids that are sequential, thus linearly organized, in the interacting protein, hence they are also referred to as short linear motifs (SLiMs) [2]. Linear motifs are physically defined by their ability to undergo a disorder-to-order transition upon binding to their partners (folding when binding). There is a lot of confusion about SLiMs – especially in the older literature (>10 years ago). Although various protein stretches were often designated by biologists as motifs, only true linear motifs (lacking a fixed, intrinsic structure of their own) can act as autonomous binding elements.

In contrast to protein interfaces formed between globular domains, where interacting residues come from different parts of the 3D

M. C. Vega, F. J. Fernández (eds.), *Advanced Technologies for Protein Complex Production and Characterization*, Advances in Experimental Medicine and Biology 1453,
https://doi.org/10.1007/978-3-031-52193-5_5

structure, the contact residues of linear motifs are contained within a relatively short (3–25 amino acids) stretch. In principle, this should make the identification of linear motifs binding to the same protein domain straightforward, based on their sequence similarity. Unfortunately, the low information content of SLiM sequences renders them elusive for simple consensus motif based *in silico* searches [3]. This is because only a few fixed positions (as low as 2–3 amino acid residues, albeit more typically 4–5) define a particular type of linear motifs, called classes. The Eukaryotic Linear Motif (ELM) database currently contains more than 200 motif classes with more than 2000 instances [4]. Linear motifs generally bind to shallow, but dedicated surfaces located on well-structured protein domains (catalytic and adaptor domains alike) and can adopt diverse conformations upon binding (Fig. 5.1a). As these binding elements are functionally modular, SLiMs can also display an island-like evolutionary conservation, having the characteristics of isolated peptides [5]. Linear motifs tend to remain highly exposed to the solvent, even at their bound, folded state. This allows a great deal of flexibility in their amino acid sequence as residues not contacting the partner protein surface can often be arbitrary [5].

The binding affinity of linear motifs varies widely, from the nanomolar to the high micromolar range – depending on the number of contact residues, the entropic cost of becoming ordered, and the nature of the secondary interactions between motif and the domain surface (Fig. 5.1b). In addition, SLiMs can be subject to secondary modifications: thus they can function as dynamic switches [6]. Phosphorylation of residues in key positions often alters the binding affinity of linear motifs, which forms the molecular basis of feedback regulation and dynamic responses in cellular signaling [7]. This chapter will include a short review of experimental PPI techniques that can be used for linear binding motif discovery. We shall also discuss how *in silico* predictions can be used to find putative motif classes and motif instances at the level of proteomes, as well as the importance and pitfalls of cell-based validation tools. Finally, we highlight how linear motifs are used as building blocks in nature, to make functional protein assemblies and highly dynamic complexes.

5.2 Experimental PPI Techniques for Linear Binding Motif Discovery

Protein-protein interaction compendia were first assembled by researchers interested in the molecular basis of specific physiological processes. This was a slow-paced approach and binding between two proteins were primarily probed by pull-down and co-immunoprecipitation (co-IP) experiments. Each experiment required the careful preparation of the bait and its binders (called preys) for *in vitro* binding studies, while prey identification for a given bait in cell-based assays depended on specific antibodies or required sophisticated analytical know-how (*e.g.*, mass spectrometry-based sequencing of tiny amounts of in-gel digested protein material). In these low-scale experiments identification of new interaction partners required a great effort and thus progress was slow. Eventually, the technology for handling entire open reading frame (ORF) collections was developed. In addition, new techniques for the *in vitro* and cell-based identification of binary interaction pairs emerged (*e.g.*, yeast two hybrid, Y2H, and protein-fragment complementation assay, PCA) that turned out to be suitable for large-scale experiments [8]. Furthermore, affinity purification (AP) tags attached to bait proteins allowed the identification of intact complexes whose components could be identified by mass spectrometry (AP-MS) [9]. These developments enabled high-throughput techniques to explore protein-protein connectivity maps at the level of proteomes, giving hope that understanding of cellular functions at systems level will shortly be within reach. Thus, specific cellular functions (*e.g.*, cellular signaling, metabolism and transcription) could be put into a bigger context by piecing smaller PPI networks together through common nodes [10]. As it turned out, these generic high-throughput PPI discovery tools still have fundamental limitations and need

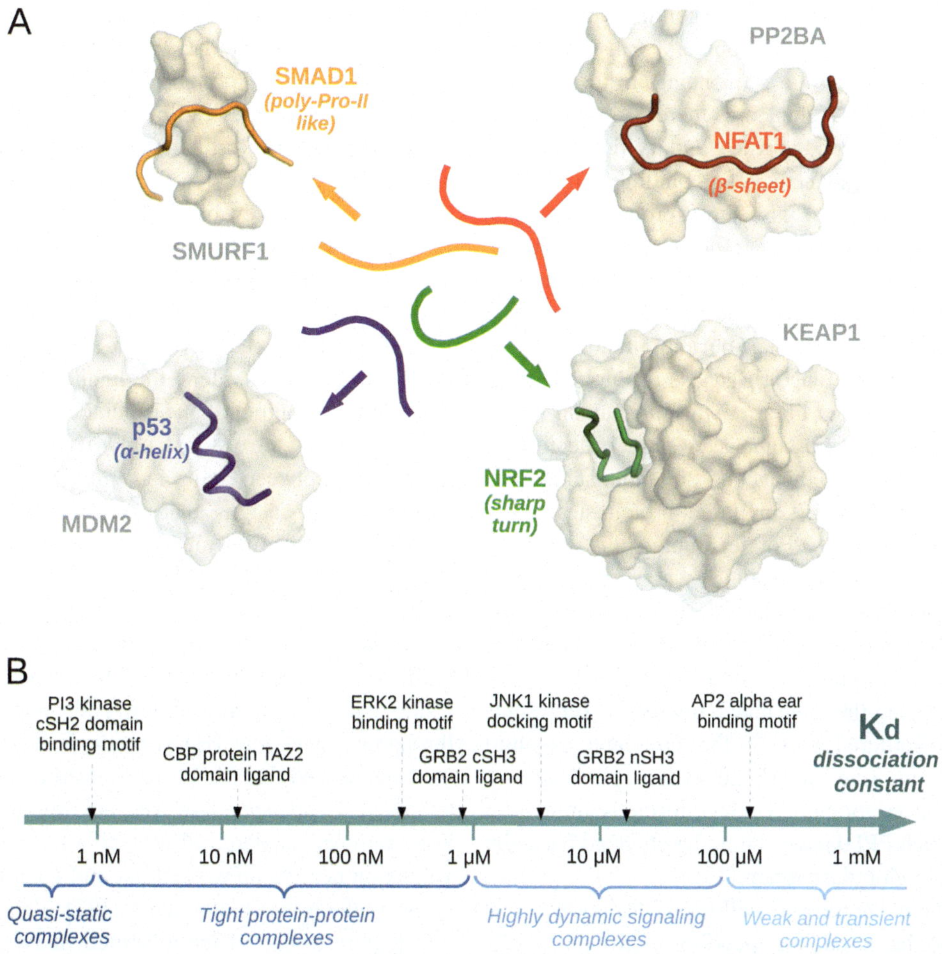

Fig. 5.1 Properties of linear motif mediated protein-protein complexes. (**a**) Linear motifs undergo a disorder-to-order transition upon binding and become folded. They can adopt an enormous variety of conformations in their bound form, including α-helices (MDM2-p53 motif), β-sheets (PP2BA-NFAT1 motif), type II polyproline helices (SMURF1-SMAD1 motif) and turns (KEAP1-NRF2 motif). (**b**) These interactions vary in their binding strength, across almost six orders of magnitude. The strongest linear motifs might show a K_D dissociation constant below the nanomolar range, while the weakest are nearly millimolar. However, most known examples have a binding affinity between $K_D = 100$ nM and 10 μM

to be complemented by low-scale PPI validation studies.

About two decades ago it already became apparent through classical low-scale deletion mapping (where the regions responsible for a binary interaction could be identified by systematic deletions of segments from full-length proteins) that many protein-protein interactions are mediated by surprisingly short fragments. These SLiMs are autonomous and bind to structured domains of their partners. Currently, there are several high-throughput PPI technologies available for researchers interested in exploring linear motif mediated interactions [11]. Chemical synthesis on a solid support might also be used to generate a library size of 100–1000 unique short peptides, which can then be probed for binding to globular proteins [12]. Phage display is a more biological approach that can be applied to select short protein fragments binding to a functionalized surface [13]. The successful binders can be easily identified by DNA sequencing from complex libraries ($>10^{10}$ clones) as there is a clonal relationship between the phage and its displayed protein sequence that enables its binding to the target protein surface. Only short protein fragments (*e.g.*, SLiMs) can be efficiently displayed by phages, therefore this technique is a great tool to identify linear motifs binding with high affinity (in the low micromolar, but more realistically in the nanomolar range). Because of the high complexity of phage libraries, the full human proteome can be represented as a large collection of short protein fragments (10–20 amino acids), and then probed against any target protein surface for interaction [14]. For example, this type of proteomic phage display was used to find peptides that can bind to PDZ (postsynaptic density protein 95/discs large/zona occludens 1) domains found in the human proteome [15]. Using the same phage library that represents the disordered part of the proteome as short fragments (*e.g.*, as 15–20 amino-acid-long partially overlapping stretches), formerly unimaginably rich PPI datasets could be obtained basically for any globular protein [13].

There are several cell-based techniques that could be used for peptide motif discovery, although they were initially applied as general PPI discovery tools. These include the most widely used yeast two hybrid (Y2H), bacterial two-hybrid (B2H) and its mammalian equivalent (M2H), where a bait-prey interaction will activate the transcription of a reporter gene or enzyme [16, 17]. A yeast surface display is now also available, that could only be paralleled by phage display [18]. In contrast to these, affinity purification mass spectrometry (AP-MS) is more adequate to gain information about protein-protein complexes present in a native cellular environment, while it is less suitable to detect low affinity interactions. Conversely, Y2H and M2H are far better suited to detect low affinity and transient PPIs, which are often the hallmarks of SLiM mediated interactions. There are other cell-based techniques that might also be used for proteome level analysis, albeit they are better suited to validate the hits of primary screens (see later). PCA based techniques exploit that some reporter proteins (*e.g.*, GFP or luciferase) can be split up and fold only if their two fragments are held together by bait-prey binding [19–21]. In addition, the mammalian protein-protein interaction trap (MAPPIT) analysis is an interesting two-hybrid based assay, designed to ameliorate the high false negative rate of classical two-hybrid methods. Here, the ligand-dependent transcriptional output of an artificial cytokine signaling pathway is restored only if there is an interaction between bait and prey [22]. The various assay designs that can be potentially used to explore SLiM-mediated interactions and complexes are listed in Table 5.1.

Early estimates on peptide motif mediated PPIs reckoned that 15–40% of protein associations within the cell might be mediated by linear motifs. Using the above-described low-scale and high-throughput screening (HTS) methods it is now well established that SLiMs are indeed extremely abundant. Challenges, however, are still abound. For example, standardized protocols for HTS hit validation need to be worked out and gain acceptance by the research community [23]. Validation will require testing of the identified SLiM in the context of the full-length protein and in cell-based assays in order to eliminate the high false positive discovery rate of some current HTS techniques. On the other hand, as most SLiM mediated interactions are weak (> micromolar in affinity), experimental methods involving washing steps to eliminate nonspecific binding are plagued by false negatives. Overall, results of generic HTS studies need to be better aligned with that of more classical low-scale studies on SLiMs, or at least the discrepancies need to be understood. Finally, the importance of posttranslational modifications (PTM) also needs to be addressed, as it is now established that short linear motif mediated interactions are also greatly affected by various PTMs (*e.g.*, phosphorylation, acetylation, sumoylation) [6].

Table 5.1 Summary of PPI techniques with pros and cons

Name of main method	Sub-method or variant	Recommended for	Potential disadvantages	References
Peptide or protein arrays	Solid-phase peptide array binding (with radioactive detection)	Strong binding events known to rely on a small epitope	Not suitable for "naive" screening, due to limited array sizes. Hydrophobic epitopes might not be exposed correctly.	[12]
Biological surface display selection	Classical phage display (M13 bacteriophage)	Any linear motif dependent interaction (with unknown motifs)	Will not capture transient interactions. Library bias can be a problem for longer motifs	[13]
	Phage display with modifying enzyme (e.g. phosphorylation)	Enzyme-substrate type interactions (transient binding)	Only useful for certain enzymes. Capturing of modified eptitopes needs to be robust.	[43]
	Yeast surface display	Glycosylated, or other large extracellular epitopes	Relatively limited library size. More suitable for folded domains than motifs.	[18]
Affinity purification – Mass spectrometry (AP-MS)	AP-MS with tandem affinity purification (TAP-tagging)	Well-defined, constitutive, soluble complexes (regardless of size)	Has difficulty in capturing dynamic and low-affinity ineractions. Results need to be adjusted for cellular protein abundance.	[9]
Protein fragment complementation assays (PCA)	Split-luciferase based assays (e.g. NanoLuc)	Interactions already suspected, based on other methods.	Not well suited for large-scale screening. Fusion points need to be optimized for each protein.	[40]
	Bimolecular fluorescent complementation (BiFC)	Interactions already suspected, based on other methods.	Not well suited for large-scale screening. Fusion points need to be optimized for each protein.	[20, 21]
Two-hybrid screening (2H)	Yeast two-hybrid, classical Gal4 (Y2H)	Strong, preferably nuclear protein complexes	Comes with a considerable false positive rate. Self-activating clones are a problem.	[8]
	Yeast two-hybrid (Y2H) with modifying enzyme	Conditional interactions (positive switches), such as phosphotyrosines	The modifying enzyme system needs to be operational in yeast and orthogonal to endogenous systems (rarely satisfied).	[42]
	Bacterial two-hybrid (B2H), adenyl cyclase based	Large libraries, well suitable for non-nuclear proteins	Due to copy number variation of plasmids, the read-out can be quite noisy.	[16]
	Mammalian two-hybrid (M2H), Gal4 based	Complexes reliant on organism-specific secondary modifications	Limited library size.	[17]
Signaling pathway reconstitution	Mammalian protein-protein interaction trap (MAPPIT)	Protein complexes only formed correctly in mammalian cells	Short-lived interactions will not be detected. Nuclear proteins can be problematic.	[22]
Proximity-based protein labelling	In vivo biotinylation assays (streptavidin capture / MS)	Complicated cellular complexes (including weak interactions)	Labelling does not necessarily imply a direct interaction. Labelling efficiency varies among proteins.	[34]
	Solid-phase enzyme activity arrays (e.g. phosphorylation)	Direct or indirect enzyme-substrate interactions (including dymanic ones)	Needs to be optimized for each particular enzyme. Scale-up can be a problem.	[5]

(continued)

Table 5.1 (continued)

Name of main method	Sub-method or variant	Recommended for	Potential disadvantages	References
	Enzyme-fused protein activity arrays (e.g. SAMDI)	Can detect weak and transient interactions in vitro	Not broadly used. Fusion protein engineering can be problematic.	[33]
In vivo protein crosslinking	Crosslinking with mass spectrometry detection (XL-MS)	Weak interactions, only forming under certain conditions	Very prone to artifacts. Not particularly suitable for large-scale interactomics.	[35, 36]

5.3 Prediction of Linear Motifs *In Silico*

The simplest goal one can have in mind with linear motifs is to find new instances of otherwise well-characterized motif classes (which also means discovering new partnerships). Establishing a motif requires a detailed structural knowledge about the interaction, to clarify which amino acids (positions) are critical for binding, and which substitutions are tolerated without the loss of interaction. The resulting consensus motifs might contain wildcards ("." or x), where any amino acid is allowed, positions with multiple ambiguities [abc], or those with particular amino acids excluded [^d]. With the regular expression (RegEx) formalism, it is particularly easy to search for novel motif candidates in any protein sequence [24]. This can be conveniently done on online servers, such as the Eukaryotic Linear Motifs (ELM) suite. The ELM suite also features an extensive collection of manually curated, experimentally validated linear motifs as well as details for selected examples. Thus, it is an excellent tool to scan our protein of interest against a complete collection of established linear motifs [4].

A similar, but more challenging task is the identification of arbitrary motifs in complete proteomes. ScanProsite offers an easy opportunity to do just that: after defining a linear motif, the user is free to search for potential occurrences in any proteome of interest [25]. However, it should be noted that the output of sequence-based searches should always be filtered before interpreting the results. To remove spurious and structurally inaccessible motifs, which may match to the query sequence but are buried inside a folded domain or are located in a cellular compartment that is inaccessible for the binding partner, some filters are routinely used. These structural or topological filters are often combined with evolutionary filters. This second step removes newly-evolved sequences that might not have any important biological function in the given organism. The SlimSearch4 suite incorporates all these filtering options to find user-defined motifs in many proteomes, and it is the suggested method of choice when working with proteomes of model organisms [26].

In silico methods may also be applied to discover linear motifs directly from interactomes (Fig. 5.2a). *De novo* identification of linear motifs, with no structural knowledge on the interaction, is probably the most difficult task. However, in addition to heuristic identification of shared motifs in PPI partner sets, it is also possible to identify motifs with fully automated methods [27]. First, one needs to establish a highly reliable set of interactor proteins. Then (after filtering) the remaining sequences should be subjected to suitable software, such as MEME [28]. These *de novo* methods most commonly apply an alphabet building approach to come up with a regular expression that covers as many instances of partner proteins as possible. While powerful, such approaches have many potential pitfalls: they only allow exact characters (with no chemical similarity handling) and have difficulties with non-standard amino acids (such as secondary modification sites). Because of the general degeneracy of linear motifs, they also require a rather high number of valid interactors (which is often unavailable). Therefore, newer methods have also been developed, using hidden Markov chain models to make them more sensitive, although these are still somewhat error-prone [29, 30].

For *in silico* methods the caveat is the underlying assumption that all protein-protein

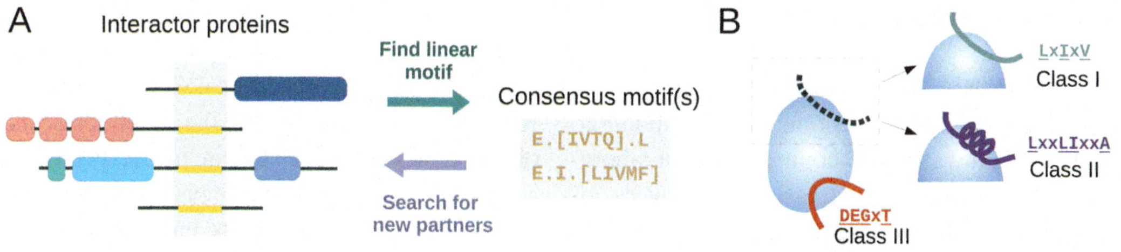

Fig. 5.2 Approaches and challenges of *in silico* motif discovery. (**a**) *In silico* methods can be used in two ways. One might aim to extract the linear motif from a reliable dataset of interactor proteins, in the form of consensus motif(s). Once a motif has been found and characterized, it can also be used to predict new members of the interactome, previously missed by experiments. (**b**) Binary interactions between two proteins may be based on multiple consensus motifs: there could be multiple valid motif conformations for a single binding site (classes I–II in the hypothetical example) or there are multiple binding sites (class III) per protein domain

interactions use the same surface, the same geometry, and the same motif type. While this might be true for a few selected examples, protein-protein interactions may be vastly more complex (Fig. 5.2b). The fact that the same partner protein, the same domain, and even the same surface can admit multiple SLiMs complicates their detection considerably [5]. Therefore, it is generally advised to assume multiple consensus motifs for each interaction. Another related problem is the inherent degeneracy of linear motifs. Moreover, natural SLiMs are often not optimized for binding strength. Consequently, they exist as multiple, relaxed versions of a hypothetical optimal consensus [31]. Because of these shortcomings, purely sequence-based *in silico* approaches should be complemented by structure-based scoring schemes whenever possible [32]. Naturally, this requires structural templates (protein-peptide complexes) that represent all viable SLiM binding modes [5]. Testing three-dimensional complementarity of peptides to the partner domain surface will then help to filter hits before starting their experimental validation.

5.4 Capture and Validation of Protein-Linear Motif Based Interactions

Generic HTS PPI and *in silico*-based predictions described above are likely to have many false positive hits. Ideally, each new linear motif instance should be validated by focused low-scale or other experimental screens, orthogonal to the primary screen. There is no universal solution for this validation step, albeit there are some good practices that need to be followed. Putative SLiM mediated interactions need to be validated *in vitro* if the primary screen was cellular. Conversely, cell-based assays should be performed for confirmation if the primary screen was *in vitro*. Furthermore, if an interaction was detected between full-length proteins and it is suspected that a short region from one of the partners mediates binding, then this should be confirmed using an *in vitro* fragment-based experiment. Similarly, if short protein fragments were used in the primary screen, then testing the SLiM *in situ*, in the context of the full-length protein, will be needed. Finally, the biological functionality (not just the binding ability) of a SLiM should be confirmed by mutational analysis in a cellular environment, using a specialized *in vivo* setup.

For weak and transient binary interactions, two-hybrid based (or similar) techniques might be good "first hint" screening methods that can also be automated. However, the initial dataset usually contains a high percentage of spurious hits (false positives) and misses most of the real binders (false negatives). Thus, it is imperative to confirm the results by other cell-based protein-protein interaction (PPI) assays to remove false positives. Experimental confirmation of all hits, similarly to *in silico* generated lists, may still be a demanding task requiring the testing of hundreds of motifs.

Sometimes, special methods need to be developed for motif detection and/or validation. This problem was also encountered in our laboratory, during the proteome-wide discovery of mitogen-activated protein kinase (MAPKs) recruiting motifs [5]. Namely, we found that PPI techniques applying washing steps are unsuitable for the detection of these typically low affinity (1–20 μM) interactions. These motifs (including a benchmarking set of well-known true positives) also consistently gave negative results in phage display. Thus, the kinase recruiting motifs could only be reliably identified with a custom assay, detecting the phosphorylation enhancement for an artificial substrate. Similar enzymatic labeling-based methods have since been effectively used both *in vitro* (*e.g.*, chemical modification of peptide arrays by lysine deacetylase fused proteins) and cells (*e.g.*, biotinylation of partners by biotinyl transferase fused proteins) [33, 34]. Enzymatic modification of binding partners by a "perpetrator" protein in the custom assay can thus often be used to explore or confirm protein-protein interactions. The drawback is that such interactions might not always be direct.

For highly transient interactions, in principle, crosslinkers might also be used to "freeze" short-lived interactions [35]. Crosslinking between proteins can even be controlled by a photochemical process and then mapped to surface sites with high precision using mass spectrometry [36]. However, crosslinking can also be highly unspecific. A considerably milder version of trapping weak and transient interactions is implemented in protein-fragment complementation assays (PCA), where the bait and its prey are covalently bound to the fragments of a third (reporter) protein [37]. Upon interaction, the reporter protein assembles and recovers its own enzymatic activity or fluorescence. When non-functional fragments of fluorescent proteins (*e.g.*, GFP, YFP) are reconstituted after interaction, the method is called bimolecular fluorescence complementation assay (BiFC). BiFC has the advantage that it can also help to visualize the place of the interaction (nucleus, cytoplasm, cell membrane, or cytoskeleton) by fluorescence microscopy. Unfortunately, for weaker interactions (>10 μM) or low protein expression levels, the signal-to-noise ratio of BiFC is often poor, due to the autofluorescence of cells and standard cell culture media. In this case a split luciferase complementation assay tends to give better results, as its enzymatic activity multiplies the signal [37]. Moreover, cells without active luciferase do not give any background luminescence.

For a good PCA experiment the following details need to be considered:

1. Choose an appropriate reporter system (enzyme or fluorescent protein) and reporter protein (Fig. 5.3a). For example, in the case of a simple split luciferase complementation assay, several differently split constructs are available, with variable stability of the complex based on fragment lengths [38].
2. Try to place reporter fragments optimally (to the N-terminus or C-terminus of the protein of interest, POI). At least eight different fusion proteins are possible for one interacting pair of proteins [38, 39]. The best practice is to test every possibility; however, there are ways to predict which combinations are going to be highly functional:
   (a) If the 3D structure of a POI is known, then the optimal orientation of the complementing fragments can be guessed at.
   (b) The reporter fragments need to be as close to the interaction site as possible (in both proteins), otherwise the signal will be low.
   (c) Fusing a reporter fragment to the N-terminus might enhance the expression of our POI, so it is advantageous to choose an N-terminal fragment for a POI with low expression, while a C-terminal fragment for a POI with high expression.
   (d) The creation of the fusion protein must not significantly alter the localization, stability, and biological function of POIs.
3. Determine the length and sequence of the linker between the protein and its reporter

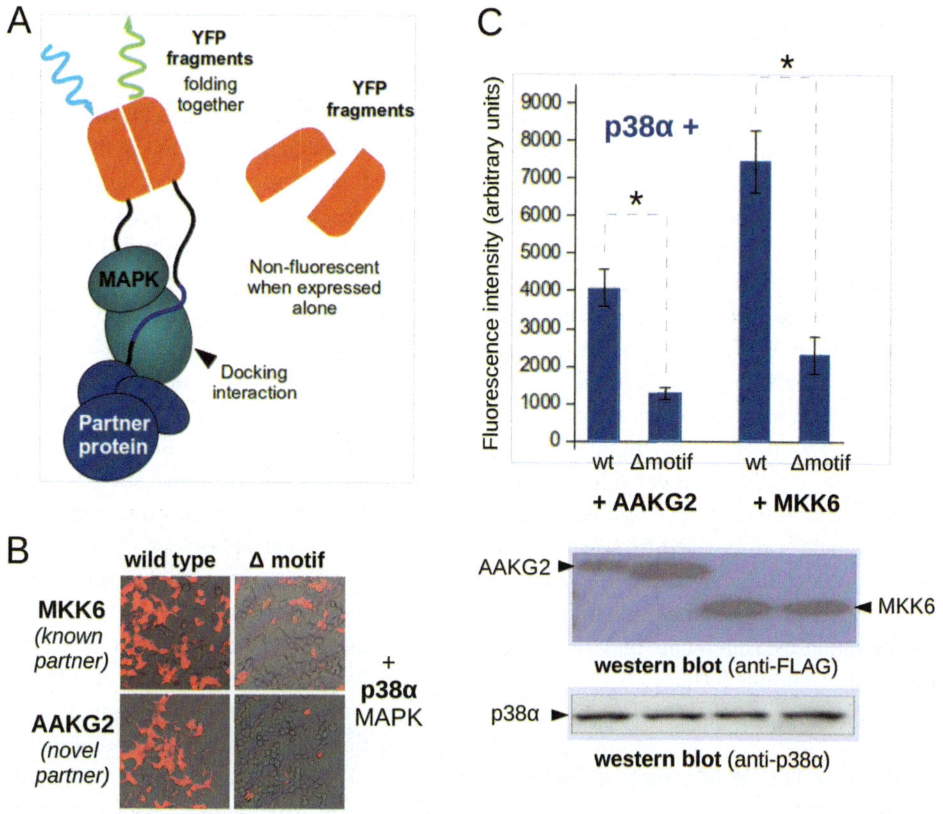

Fig. 5.3 Cellular PCA based techniques in practice. (**a**) Interactions of p38α MAP kinase with docking motif (D-motif) containing partners were examined with N-terminal YFP fragment fusions in a fluorescent protein based PCA experiment. (**b**) Fluorescence microscopy already shows a profound difference between wild-type and mutant proteins lacking the linear motif. (**c**) Differences in fluorescence are quantifiable in experiments if all fragments are expressed at a similar extent, as shown by the Western-blots of samples derived from HEK293 cells transiently transfected with the YFP fragments [5]

fragment. The linkers should be conformationally neutral, hydrophilic, and flexible enough to ensure the proper fit of fusion partners upon protein-protein complex formation and reconstitution of the reporter. Normally, a $(GGSGGS)_n$ sequence with $n = 1$–3 is sufficient.

4. Choosing the right negative controls is possibly the most important aspect of PCA based validation experiments. This is because fluorescent protein fragments can associate and give false positive signal in the absence of bait-prey interaction, particularly under high expression levels. (Sometimes reporter fragments without the POIs give higher signal than the fusion protein, since reporter fragments have higher expression level in the cell than the fusion construct containing a large POI.) Therefore, it is obligatory to include controls to ensure that the enzymatic activity or fluorescence is not due to unspecific effects. The best control is the same globular POI with its linear motif binding surface mutated. Alternatively, key residues of the motif can also be mutated in the linear motif containing partner, or the entire motif can be removed (Fig. 5.3b).
5. Test the expression levels of the fusion proteins (in all cases). This is especially important when using the PCA signal to estimate relative

interaction strengths of different preys binding to the same bait. However, one should also be aware that interactions also depend on the localization and expression of POI and all these affect the reconstitution of the reporter. Including the same epitope tag (*e.g.*, FLAG, V5) into all the expression constructs can help to quantify protein expression levels (Fig. 5.3c).

6. A major drawback of PCA based methods is that they are not well-suited to observe PPI dynamics, as most fusion tags permanently hold bait and prey together once the reporter is folded (which is nevertheless a great advantage to detect weak and transient interactions). Recently, however, fully reversible luciferase complementation assays have also been developed. The latter can be used to detect the PPI dynamics of a cellular signaling relevant transient interaction, for example, enabling complex formation measurements under basal and stimulated conditions [40].

5.5 Linear Motif Mediated PPI Dynamics

From a functional perspective, it is important to note that linear binding motif mediated PPIs are often affected by post-translational modifications [6, 41]. There are several ways protein phosphorylation mechanistically affects PPI dynamics. The best-known case is when phosphorylation creates a binding site for a globular domain (*e.g.*, SH2 domains for tyrosine-containing linear motifs). This can be referred to as a clear ON switch. However, if the affinity constants change only to a lesser degree (less than about an order of magnitude), then a positive phosphoswitch may be regarded as an ON dimmer, where phosphorylation has a more graded effect on linear motif mediated binding. Conversely, there are many examples for OFF switches (loss of binding due to full steric clash) or OFF dimmers (decrease of affinity), where phosphorylation interferes with protein-peptide type interactions (Fig. 5.4a).

De novo identification of linear motifs, whose activity strictly depends on covalent modifications (canonical ON switches) often poses a great technical difficulty. These interactions are unlikely to form in heterologous biological systems (*e.g.*, the "usual" Y2H assays or phage displays) and are invisible to peptide arrays unless they were synthesized with modified amino acids. To circumvent such challenges, specialized screens are necessary. Introduction of mammalian tyrosine kinases into yeast strains (phospho-tyrosine Y2H) or direct phosphorylation of phage surface epitopes with kinase enzymes prior to their selection are just a few solutions that can be applied for this difficult scenario [42, 43]. For *in vitro* screens, phosphorylated linear motifs can be generated conventionally, using either specific kinases or chemical synthesis; but certain modified amino acids (*e.g.*, phospho-Ser) might also be incorporated as an unnatural amino acid during translation of the bait [44].

In cells, the dynamics of linear motif mediated protein complex formation may also be followed by conventional methods. In our example, the dissociation of ribosomal S6 subunit kinase 1 (RSK1) from its PDZ domain-containing scaffold protein (membrane associated guanylate kinase inverted 1, MAGI1) upon epidermal growth factor (EGF) treatment was monitored by a luciferase based PCA experiment. EGF treatment results in the activation of RSK1. Activated RSK1 then autophosphorylates itself in its C-terminal region at Thr733, inside its PDZ binding linear motif. This phosphorylation event acts as an OFF switch, as it sterically disrupts the complex (Fig. 5.4b). RSK1-MAGI-1 association dips at maximal RSK1 activity and dephosphorylation at this site by phosphatases allows re-assembly of the complex [7].

5.6 Conclusion

In recent years, the robustness and reliability of interactomics discovery tools have improved dramatically. Despite this evolution, there is no one-size-fits-all experimental method for all problems. The golden rule is thus to combine different methods (including generic and specific assays) to obtain the most reliable results

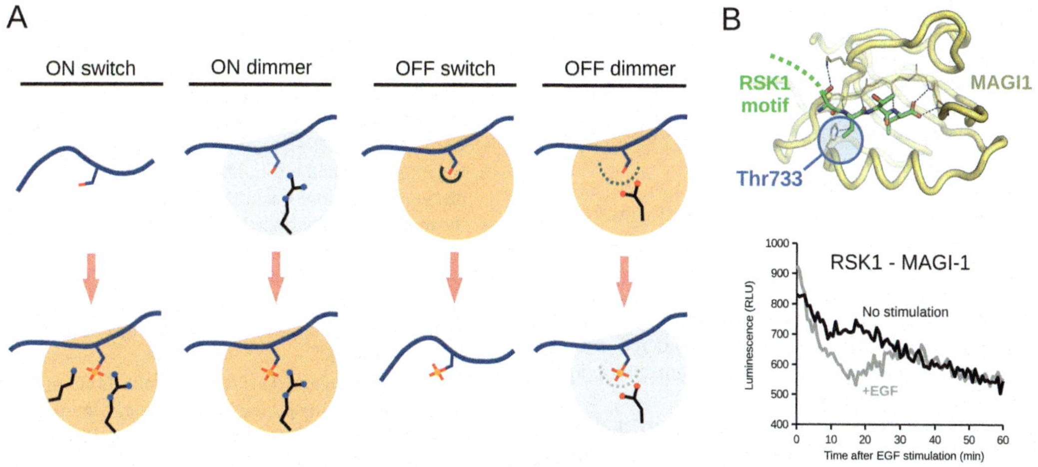

Fig. 5.4 Linear motif-based switches modulated by phosphorylation. (**a**) Classification of phosphoswitches. Phosphorylation might have a profound effect on protein-peptide type interactions, either by promoting it (ON switch), disrupting it (OFF switch) or by altering the binding affinity in a more gradual fashion (ON or OFF dimer). (**b**) As an example of dynamic regulation, the formation of RSK1-MAGI1 complex is controlled by PDZ domain binding and PDZ ligand phosphorylation (on Thr733). The crystal structure of the RSK1 C-terminus (green) in complex with the MAGI1 PDZ domain (brown) is shown on the upper panel. The lower panel shows the results of a luciferase complementation assay, where the complex formation between RSK1 and MAGI1 was monitored in unstimulated (black) and EGF stimulated (gray) HEK293 cells [7]

possible. The same is true for purely *in silico* methods: they work best when coupled with experimental testing, and a sequential approach can be very powerful. Thus, one can use predictions to obtain a putative interactor set, which would then be tested experimentally. The results can be utilized to improve predictions, and the process can be repeated until it converges to a set of well-defined linear motifs.

There are possibly millions of PPIs in the cell that orchestrate cellular events at the molecular level. How can we obtain a biological understanding of these complex PPI networks? What is the role of a given linear binding motif in cellular regulation? Linear motif discovery tools clearly highlighted the abundance of SLiMs in the human proteome. These motifs seem to be particularly abundant in regulatory systems, such as in proteins involved in cellular signaling, in the control of protein degradation and gene expression. Mutations of linear motifs are also seen in diseases such as cancer [45, 46]. At the same time, domain surfaces accommodating SLiMs are attractive targets for novel pharmaceutical agents [47]. However, before SLiMs and their partners could be utilized for biotechnology, medical diagnostics or therapy, their behavior needs to be untangled first. In many proteins, linear motifs form complicated networks, where secondary modifications might trigger each other in a sequential order [48, 49]. By now it has become well-established that in dynamic regulatory networks – particularly involved in cellular signaling, protein level, and gene expression regulation – phosphoswitches, phosphodegrons, or phosphorylation-dependent SUMOylation sites are used as simple building blocks. Considering the great progress made in the field in the last two decades, a more complete exploration of linear motif-controlled protein-protein associations within these networks appears to be a tractable challenge for scientists.

Acknowledgements This work was supported by grants from the National Research, Development and Innovation Office, Hungary (OTKA NN 114309, OTKA PD120973 and KKP_17 126963).

References

1. Tompa P, Davey NE, Gibson TJ, Babu MM (2014) A million peptide motifs for the molecular biologist. Mol Cell 55:161–169. https://doi.org/10.1016/j.molcel.2014.05.032
2. Van Roey K, Uyar B, Weatheritt RJ et al (2014) Short linear motifs: ubiquitous and functionally diverse protein interaction modules directing cell regulation. Chem Rev 114:6733–6778. https://doi.org/10.1021/cr400585q
3. Mészáros B, Simon I, Dosztányi Z (2009) Prediction of protein binding regions in disordered proteins. PLoS Comput Biol 5:e1000376. https://doi.org/10.1371/journal.pcbi.1000376
4. Gouw M, Michael S, Sámano-Sánchez H et al (2018) The eukaryotic linear motif resource – 2018 update. Nucleic Acids Res 46:D428–D434. https://doi.org/10.1093/nar/gkx1077
5. Zeke A, Bastys T, Alexa A et al (2015) Systematic discovery of linear binding motifs targeting an ancient protein interaction surface on MAP kinases. Mol Syst Biol 11:837
6. Van Roey K, Dinkel H, Weatheritt RJ et al (2013) The switches.ELM resource: a compendium of conditional regulatory interaction interfaces. Sci Signal 6:rs7. https://doi.org/10.1126/scisignal.2003345
7. Gógl G, Biri-Kovács B, Póti ÁL et al (2018) Dynamic control of RSK complexes by phosphoswitch-based regulation. FEBS J 285:46–71. https://doi.org/10.1111/febs.14311
8. Vidal M, Fields S (2014) The yeast two-hybrid assay: still finding connections after 25 years. Nat Methods 11:1203–1206
9. Dunham WH, Mullin M, Gingras A-C (2012) Affinity-purification coupled to mass spectrometry: basic principles and strategies. Proteomics 12:1576–1590. https://doi.org/10.1002/pmic.201100523
10. Bonetta L (2010) Interactome under construction. Nature 468:851–852. https://doi.org/10.1038/468851a
11. Blikstad C, Ivarsson Y (2015) High-throughput methods for identification of protein-protein interactions involving short linear motifs. Cell Commun Signal 13:38. https://doi.org/10.1186/s12964-015-0116-8
12. Volkmer R, Tapia V, Landgraf C (2012) Synthetic peptide arrays for investigating protein interaction domains. FEBS Lett 586:2780–2786. https://doi.org/10.1016/j.febslet.2012.04.028
13. Davey NE, Seo M-H, Yadav VK et al (2017) Discovery of short linear motif-mediated interactions through phage display of intrinsically disordered regions of the human proteome. FEBS J 284:485–498. https://doi.org/10.1111/febs.13995
14. Larman HB, Zhao Z, Laserson U et al (2011) Autoantigen discovery with a synthetic human peptidome. Nat Biotechnol 29:535–541. https://doi.org/10.1038/nbt.1856
15. Ivarsson Y, Arnold R, McLaughlin M et al (2014) Large-scale interaction profiling of PDZ domains through proteomic peptide-phage display using human and viral phage peptidomes. Proc Natl Acad Sci 111:2542–2547. https://doi.org/10.1073/pnas.1312296111
16. Karimova G, Gauliard E, Davi M et al (2017) Protein–protein interaction: bacterial two-hybrid. In: Methods in molecular biology. Humana Press, Clifton, pp 159–176
17. Riegel E, Heimbucher T, Höfer T, Czerny T (2017) A sensitive, semi-quantitative mammalian two-hybrid assay. BioTechniques 62:206–214. https://doi.org/10.2144/000114544
18. Cherf GM, Cochran JR (2015) Applications of yeast surface display for protein engineering. Methods Mol Biol 1319:155–175. https://doi.org/10.1007/978-1-4939-2748-7_8
19. Michnick SW, Ear PH, Landry C et al (2011) Protein-fragment complementation assays for large-scale analysis, functional dissection and dynamic studies of protein–protein interactions in living cells. In: Methods in molecular biology. Humana Press, Clifton, pp 395–425
20. Cabantous S, Nguyen HB, Pedelacq J-D et al (2013) A new protein-protein interaction sensor based on tripartite split-GFP Association. Sci Rep 3:2854. https://doi.org/10.1038/srep02854
21. To T-L, Zhang Q, Shu X (2016) Structure-guided design of a reversible fluorogenic reporter of protein-protein interactions. Protein Sci 25:748–753. https://doi.org/10.1002/pro.2866
22. Lemmens I, Lievens S, Tavernier J (2015) MAPPIT, a mammalian two-hybrid method for in-cell detection of protein-protein interactions. Methods Mol Biol 1278:447–455. https://doi.org/10.1007/978-1-4939-2425-7_29
23. Gibson TJ, Dinkel H, Van Roey K, Diella F (2015) Experimental detection of short regulatory motifs in eukaryotic proteins: tips for good practice as well as for bad. Cell Commun Signal 13:42. https://doi.org/10.1186/s12964-015-0121-y
24. Edwards RJ, Palopoli N (2015) Computational prediction of short linear motifs from protein sequences. In: Methods in molecular biology. Humana Press, Clifton, pp 89–141
25. de Castro E, Sigrist CJA, Gattiker A et al (2006) ScanProsite: detection of PROSITE signature matches and ProRule-associated functional and structural residues in proteins. Nucleic Acids Res 34:W362–W365. https://doi.org/10.1093/nar/gkl124
26. Krystkowiak I, Davey NE (2017) SLiMSearch: a framework for proteome-wide discovery and annotation of functional modules in intrinsically disordered regions. Nucleic Acids Res 45:W464–W469. https://doi.org/10.1093/nar/gkx238
27. Davey NE, Shields DC, Edwards RJ (2006) SLiMDisc: short, linear motif discovery, correcting

for common evolutionary descent. Nucleic Acids Res 34:3546–3554. https://doi.org/10.1093/nar/gkl486
28. Bailey TL, Boden M, Buske FA et al (2009) MEME SUITE: tools for motif discovery and searching. Nucleic Acids Res 37:W202–W208. https://doi.org/10.1093/nar/gkp335
29. Haslam NJ, Shields DC (2012) Profile-based short linear protein motif discovery. BMC Bioinform 13:104. https://doi.org/10.1186/1471-2105-13-104
30. Prytuliak R, Volkmer M, Meier M, Habermann BH (2017) HH-MOTiF: de novo detection of short linear motifs in proteins by hidden Markov model comparisons. Nucleic Acids Res 45:10921–10921. https://doi.org/10.1093/nar/gkx810
31. Hertz EPT, Kruse T, Davey NE et al (2016) A conserved motif provides binding specificity to the PP2A-B56 phosphatase. Mol Cell 63:686–695. https://doi.org/10.1016/j.molcel.2016.06.024
32. Sánchez IE, Beltrao P, Stricher F et al (2008) Genome-wide prediction of SH2 domain targets using structural information and the FoldX algorithm. PLoS Comput Biol 4:e1000052. https://doi.org/10.1371/journal.pcbi.1000052
33. O'Kane PT, Mrksich M (2017) An assay based on SAMDI mass spectrometry for profiling protein interaction domains. J Am Chem Soc 139:10320–10327. https://doi.org/10.1021/jacs.7b03805
34. Beck DB, Narendra V, Drury WJ et al (2014) In vivo proximity labeling for the detection of protein–protein and protein–RNA interactions. J Proteome Res 13:6135–6143. https://doi.org/10.1021/pr500196b
35. Yang B, Tang S, Ma C et al (2017) Spontaneous and specific chemical cross-linking in live cells to capture and identify protein interactions. Nat Commun 8:2240. https://doi.org/10.1038/s41467-017-02409-z
36. Pham ND, Parker RB, Kohler JJ (2013) Photocrosslinking approaches to interactome mapping. Curr Opin Chem Biol 17:90–101. https://doi.org/10.1016/j.cbpa.2012.10.034
37. Morell M, Ventura S, Avilés FX (2009) Protein complementation assays: approaches for the in vivo analysis of protein interactions. FEBS Lett 583:1684–1691. https://doi.org/10.1016/j.febslet.2009.03.002
38. Kudla J, Bock R (2016) Lighting the way to protein-protein interactions: recommendations on best practices for bimolecular fluorescence complementation analyses. Plant Cell 28:1002–1008. https://doi.org/10.1105/tpc.16.00043
39. Kerppola TK (2006) Design and implementation of bimolecular fluorescence complementation (BiFC) assays for the visualization of protein interactions in living cells. Nat Protoc 1:1278–1286. https://doi.org/10.1038/nprot.2006.201
40. Dixon AS, Schwinn MK, Hall MP et al (2016) NanoLuc complementation reporter optimized for accurate measurement of protein interactions in cells. ACS Chem Biol 11:400–408. https://doi.org/10.1021/acschembio.5b00753
41. Wu C-G, Chen H, Guo F et al (2017) PP2A-B' holoenzyme substrate recognition, regulation and role in cytokinesis. Cell Discov 3:17027. https://doi.org/10.1038/celldisc.2017.27
42. Grossmann A, Benlasfer N, Birth P et al (2015) Phospho-tyrosine dependent protein-protein interaction network. Mol Syst Biol 11:794
43. Shah NH, Löbel M, Weiss A, Kuriyan J (2018) Fine-tuning of substrate preferences of the Src-family kinase Lck revealed through a high-throughput specificity screen. elife 7. https://doi.org/10.7554/eLife.35190
44. Rogerson DT, Sachdeva A, Wang K et al (2015) Efficient genetic encoding of phosphoserine and its nonhydrolyzable analog. Nat Chem Biol 11:496–503. https://doi.org/10.1038/nchembio.1823
45. Uyar B, Weatheritt RJ, Dinkel H et al (2014) Proteome-wide analysis of human disease mutations in short linear motifs: neglected players in cancer? Mol BioSyst 10:2626–2642. https://doi.org/10.1039/C4MB00290C
46. Mészáros B, Zeke A, Reményi A et al (2016) Systematic analysis of somatic mutations driving cancer: uncovering functional protein regions in disease development. Biol Direct 11:23. https://doi.org/10.1186/s13062-016-0125-6
47. Corbi-Verge C, Kim PM (2016) Motif mediated protein-protein interactions as drug targets. Cell Commun Signal 14:8. https://doi.org/10.1186/s12964-016-0131-4
48. Yu S, Wang F, Tan X et al (2018) FBW7 targets KLF10 for ubiquitin-dependent degradation. Biochem Biophys Res Commun 495:2092–2097. https://doi.org/10.1016/j.bbrc.2017.11.187
49. Hietakangas V, Anckar J, Blomster HA et al (2006) PDSM, a motif for phosphorylation-dependent SUMO modification. Proc Natl Acad Sci 103:45–50. https://doi.org/10.1073/pnas.0503698102

Protein-Protein Binding Kinetics by Biolayer Interferometry

6

Jorge Santos-López, Sara Gómez, Francisco J. Fernández, and M. Cristina Vega

Abstract

The specific kinetics and thermodynamics of protein-protein interactions underlie the molecular mechanisms of cellular functions; hence the characterization of these interaction parameters is central to the quantitative understanding of physiological and pathological processes. Many methods have been developed to study protein-protein interactions, which differ in various features including the interaction detection principle, the sensitivity, whether the method operates *in vivo*, *in vitro*, or *in silico*, the temperature control, the use of labels, immobilization, the amount of sample required, the number of measurements that can be accomplished simultaneously, or the cost. Bio-Layer Interferometry (BLI) is a label-free biophysical method to measure the kinetics of protein-protein interactions. Label-free interaction assays are a broad family of methods that do not require protein modifications (other than immobilization) or labels such as fusions with fluorescent proteins or transactivating domains or chemical modifications like biotinylation or reaction with radionuclides. Besides BLI, other label-free techniques that are widely used for determining protein-protein interactions include surface plasmon resonance (SPR), thermophoresis, and isothermal titration calorimetry (ITC), among others.

Keywords

Biolayer interferometry · Label-free techniques · Protein-protein interactions · Binding kinetics · Complement system · C5a anaphylatoxin · Antibody · Streptavidin biosensor

6.1 Biolayer Interferometry

BLI is an *in-vitro* optical technique that can register molecular association and disassociation kinetics onto the tip of a biosensor by white light interferometry [1–4]. In detail, the binding and detaching of a protein analyte is monitored by measuring changes in the interference pattern from white light (obtained from a tungsten lamp) reflected by two layers at the biosensor tip: the surface of the biosensor tip

J. Santos-López · M. C. Vega (✉)
Centro de Investigaciones Biológicas Margarita Salas (CIB-CSIC), Madrid, Spain
e-mail: cvega@cib.csic.es

S. Gómez
Centro de Investigaciones Biológicas Margarita Salas (CIB-CSIC), Madrid, Spain

Universidad Europea de Madrid, Madrid, Spain

F. J. Fernández
Abvance Biotech SL, Madrid, Spain

M. C. Vega, F. J. Fernández (eds.), *Advanced Technologies for Protein Complex Production and Characterization*, Advances in Experimental Medicine and Biology 1453,
https://doi.org/10.1007/978-3-031-52193-5_6

where molecules attach (variable) and an internal reference layer (constant). Any changes in the thickness of the molecular layer on the surface of the sensor tip, by molecular coupling or disengaging, is associated to a variation in the optical path length of this layer and so it causes a shift in the interference pattern of the reflected white light from the two surfaces. This modification of the reflected light can be measured in real-time by a detector and is reported as a wavelength shift ($\Delta\lambda$) in nm, which is the magnitude of binding response plotted versus time in BLI sensorgrams (Fig. 6.1). Importantly, BLI signal is sensitive to molecules that associate or dissociate directly to the biosensor tip surface or to other molecules previously bound to this surface. In accordance with this, qualitative and quantitative information about molecular interactions and sample concentration can be obtained in BLI experiments.

Several BLI platforms have been developed that display slightly different specifications and capabilities (Fig. 6.2) [3–5]. Overall, low sample quantities are required for an assay, from about 200 μL down to only 4 μL. Moreover, samples can be recovered after the assay and used in further experiments. High sensitivity can be reached by these platforms, with a lower limit for the molecular size that can be detected that varies from 150 Da to 10,000 Da, and an affinity range from 0.1 mM to 10 pM (Octet systems) or from 1 mM to 0.1 nM (BLItz system). Regarding the number of channels, BLI platforms offer flexibility. While some platforms can only read one or two samples at the same time using one or two channels, some Octet systems have 8 or 16 channels, thereby enabling high-throughput approaches.

BLI biosensors are disposable fiber-optic devices whose tips are made from a biocompatible matrix that is uniform, non-denaturing, and minimizes non-specific binding, and are conveniently derivatized to be able to bind biomolecules such as proteins or nucleic acids covalently or non-covalently. An important requisite for BLI assays is preserving the molecular activity of the biomolecules while attached to the biosensor. Multiple commercial biosensors are available according to the tip derivatization, binding features, and intended or potential use. These include biosensors coated with specific antibodies, Ni-NTA, streptavidin, or proteins A, G, or L, as well as some amino-reactive biosensors [3, 4].

Generally, BLI biosensors are divided into three categories depending on their application: binding assays, quantitation assays, or both. The suitable application for a given biosensor is established by several sensor properties: molecular binding specificity, stability, and affinity; assay reproducibility; and the quantitation dynamic range. Although biosensor disposal is recommended after one use, some biosensors may be regenerated by removing the molecules captured on their tips. Regeneration procedures aim to detach the highest quantity possible of the attached molecules while preserving the quality and the proper function of the biosensor tip (biocompatible matrix, its derivatization, and, if any, other captured molecules that should remain retained). The number of use-regeneration cycles compatible with a biosensor depends on the biosensor type and the application, and should be determined experimentally [3, 4].

BLI assays are carried out employing a dip-and-read approach, in which the tip of a biosensor is sequentially submerged in solutions with or without the molecules assayed for binding. The solution support is in constant rotation to avoid the formation of concentration gradients and to prevent rebinding events [6]. The interactions analyzed in BLI experiments can involve different types of biomolecules, including proteins, nucleic acids, lipids, peptides, and small molecules.

According to the type of information obtained, two types of BLI experiments can be distinguished [1, 4]: binding and quantitation experiments.

6.1.1 Binding Experiments

In binding experiments, qualitative and quantitative information about kinetic and equilibrium properties of molecular interactions is sought. Typically, the interaction studied occurs between

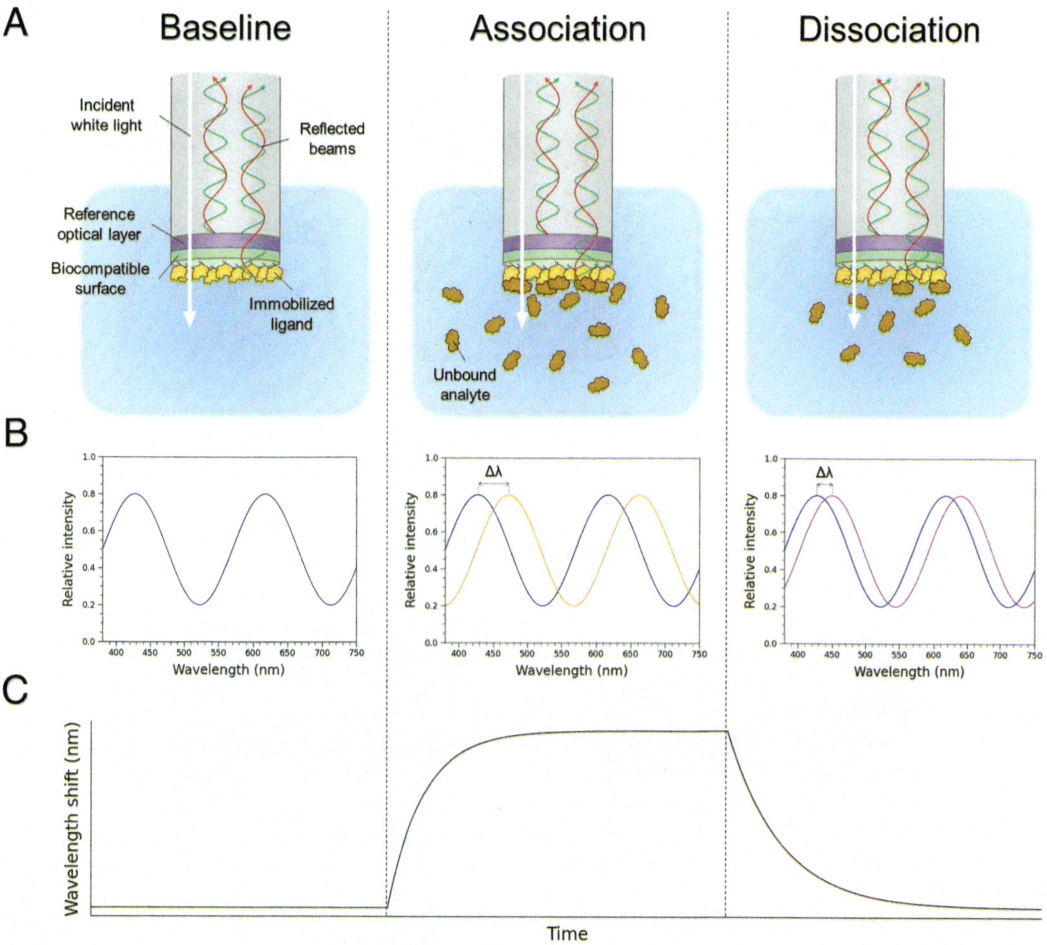

Fig. 6.1 Principle of the BLI assay. (**a**) The sensor tip coated with an interacting partner (ligand) is sequentially dipped into assay buffer, a solution of the interacting partner (analyte), and assay buffer in the baseline, association and dissociation phases, respectively. (**b**) Throughout the assay, the interference pattern of reflected light from the reference layer and the molecular layer onto the biocompatible surface is continuously monitored. During the association and dissociation phases, the molecular layer's thickness changes. These changes are tracked by measuring shifts in the interference pattern of the reflected light compared to the pattern observed during the baseline phase (denoted as Δλ). (**c**) The binding curve for the BLI assay is derived by plotting the wavelength shift in reflected light against time

two molecules that are sequentially exposed to the biosensor. The first interacting molecule, referred to as ligand or bait, is immobilized onto the biosensor tip surface with high affinity. Next, binding to and detaching from the sensor tip of the other molecule of the pair, referred to as the analyte, is monitored. A set of association and dissociation curves (sensorgrams) is acquired by repeating the experiment with different analyte concentrations. From this set of sensorgrams, the association rate constant (k_a) and the dissociation rate constant (k_d) of the interaction can be obtained by fitting the data with an adequate binding model, and then the binding affinity constants (K_D and K_A) can be calculated [7–9-]. Alternatively to the standard kinetic strategy, approaches have been developed to derive the equilibrium affinity constants without measuring the kinetic parameters. These approaches can be helpful to determine equilibrium constants of

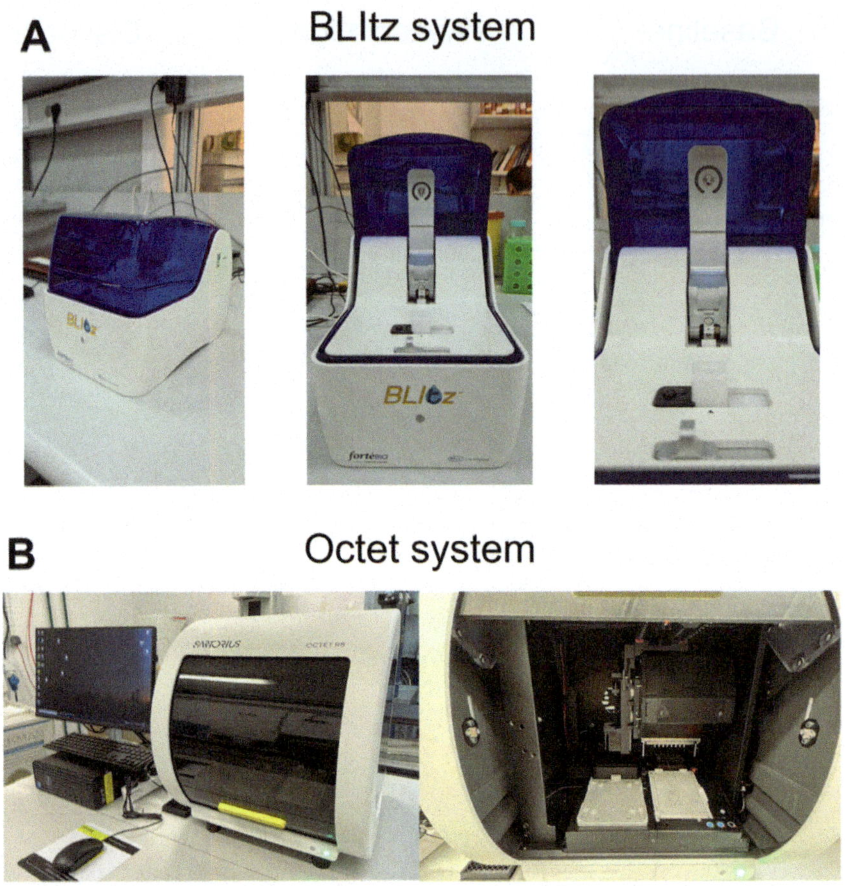

Fig. 6.2. Two commercially available BLI platforms. (**a**) FortéBio BLItz system. Photographs taken by Dr. Sergio Navas-Yuste (CIB-CSIC). (**b**) FortéBio Octet R8 system. Images were kindly provided by Dr. Alberto Marina (IBV-CSIC)

interactions with complex kinetics or that are disrupted by the immobilization step [7, 10]. Using either kinetic or equilibrium strategies, BLI binding experiments have been applied to elucidate the structural basis of macromolecular interactions and to the identification, comparison, and validation of binders and interaction modulators targeting specific proteins and protein complexes (see examples in Sect. 6.3).

Examples of biosensors suitable for binding assays include streptavidin (SA), aminopropylsilane (APS), amine reactive 2nd generation (AR2G), or anti-human Fc capture sensors [11].

6.1.2 Quantitation Experiments

In this type of experiments, we seek to determine the concentration of a macromolecule in a sample. These assays require a BLI experimental design sensitive to a molecular interaction involving the molecule to quantify (analyte) and a standard solution series of the analyte. For example, the analyte could be known to bind the unmodified surface of a certain biosensor or a specific ligand susceptible to be captured on the biosensor's tip, in a dose-dependent (quantitative) fashion. Once a suitable experimental design is selected, BLI binding curves are measured for all available analyte concentrations. From these binding curves, a standard curve is generated by plotting the BLI signal at a specific time point or the initial association rate against the known analyte concentration; to determine the analyte concentration in the unknown samples, only the association kinetic phase is necessary. Next, the analyte concentrations of the unknown samples are determined by interpolating their corresponding measurements in the standard curve previously prepared [12–16]. Remarkably, the high specificity of the technique allows to use samples that are not completely purified and differentiate between active and inactive proteins. Anti-human IgG Fc, anti-murine IgG Fc, or protein A are examples of some of the biosensors available for quantitative purposes [17].

6.1.3 BLI in Comparison with Other Techniques

Compared to other techniques for the study of molecular interactions and quantitation [6, 18, 19], BLI displays some remarkable features.

BLI can be used to probe macromolecular interactions in rather crude or particulate samples, such as cell lysates, serum, periplasmic extracts, or solutions containing viral-like particles; this is possible because the BLI signal hardly depends on unbound molecules or the refractive index [20–23]. Besides, the dip-and-read approach facilitates these measurements because it does not require microfluidics or other costly instrumentation. Surface plasmon resonance (SPR), another label-free technique, records changes in the refractive index as the analyte binds the immobilized ligand and it does depend on microfluidics [1]; therefore, the SPR instrument can easily become clogged with crude samples.

The simple dip-and-read method characteristic of BLI measurements has been easily parallelized for high-throughput applications, thereby accelerating the collection of binding data faster and more efficiently than other techniques [8].

Compared to equilibrium binding assays like ELISA, BLI has the advantage of directly measuring the association and dissociation kinetic constants, thus affording a deeper understanding of the interaction. The kinetic dissociation constant, k_d, is regarded as one the most important parameter to assess the efficacy of a therapeutic. Interactions with similar affinity can exhibit markedly different association and dissociation rates, therefore experimentally determining the kinetic constants provides more insight than the K_D value [24]. While the initial investment in BLI instruments and biosensors can be substantial, the overall maintenance costs for BLI platforms are relatively low, rendering them cost-effective in the long run [25, 26]. Furthermore, development of BLI assays is straightforward, and comprehensive kinetic data can be rapidly acquired.

BLI presents several limitations that merit attention. A significant concern is that the binding event must take place on the biocompatible surface of the sensor, where one of the interaction partners is immobilized. As a result, it is imperative to validate that the observed interaction accurately reflects native binding events and is not affected by factors such as mass transfer limitations, preferential orientations, or conformational changes [27].

Another issue arises from analyte rebinding during the dissociation phase, an undesirable event exacerbated by the absence of a flow system to remove dissociated analyte. This phenomenon distorts the dissociation curve and complicates kinetic analysis, especially for slow dissociating interactions [8]. To mitigate this artifact, researchers have used "sink strategies" by introducing a specific competitor for the analyte in the dissociation buffer, thereby preventing the analyte from rebinding the ligand [7].

A problem with the basic BLI platforms (*e.g.*, BLItz) is lack of temperature control, which is an important drawback compared to the SPR instruments and the more advanced BLI systems such as the Octet RED96. The latter can control the assay temperature between 15 and 40 °C. Likewise, absence of humidity control in the BLI platforms may be an issue with sensitive samples or when the maximum assay time can cause substantial evaporation [8, 28, 29].

With respect to the reproducibility of the results obtained by BLI, there are sources of variability that must be borne in mind. Firstly, the physical nature of the readout method and features of the platform and biosensor impose limitations in sensitivity for some applications; for example, limited lower-size sensitivity may restrict binding studies with small molecules [7]. However, this lower-size limit could be advantageous for the study of effects of small allosteric modulators in the interaction between two bigger molecules [28]. Furthermore, certain experimental variation among biosensor batches can be observed and the binding parameters determined can show a limited correlation with those obtained by other techniques, such as SPR [8, 29].

6.2 BLI Assay Development

Assay development for BLI starts with a basic experimental design that takes into consideration the purpose (kinetics, quantitation, or both), the proteins involved in the interaction (or the protein and the small molecule or the nucleic acid binding partner), and the choice of sensor.

To choose which binding partner will be immobilized (the ligand) and which one will be in solution (the analyte), we use some or all of the following criteria [2, 3, 30]:

1. A protein partner should be the ligand if it has a limited availability (as the sample expenditure is less for the ligand than the analyte), it is already tagged (and there are BLI sensors for the fusion/tag), it is the smallest of the two binding partners (as BLI signal increases with the size of the analyte), and it can be safely and stably immobilized on the sensor.
2. A protein partner should be the analyte when it cannot be successfully purified or is available only in complex mixtures.
3. Lastly, a ligand-analyte layout with the lowest interaction valency might be preferred as this simplifies data analysis. In other words, the partner with the highest valency should be immobilized, other factors being equal.

Several controls should be run when performing a BLI experiment [3]:

1. Buffer compatibility: the BLI signal from a fresh sensor exposed to the assay buffer should be flat and stable (buffer control).
2. Nonspecific binding of the analyte to the sensor tip: record the BLI signal from a fresh sensor (without ligand immobilized) exposed to the highest analyte concentration; the signal should be flat and constant, ideally zero (baseline) or close to zero.
3. Nonspecific binding to the ligand: record the BLI signal from a sensor, with ligand immobilized, exposed to a solution with a known non binder (*e.g.*, an isotype control antibody, a scrambled peptide or nucleic acid)
4. Positive control: perform a complete assay with a known binder, thus ensuring that the ligand remains suitable for binding assays after immobilization.
5. Buffer reference: data analysis requires subtracting the background signal (the blank) from all solution components lest the analyte. To obtain a blank curve, perform a complete assay using a solution without analyte.

A simplified workflow for a BLI experiment contains the following steps, which have been described in greater detail elsewhere [3, 4, 26, 28, 31].

6.2.1 Prepare Buffers, Regeneration Solution, Ligand, and Analyte Samples

A suitable assay buffer guarantees the stability of both ligand and analyte and preserves the interaction. It should be compatible with the sensor's biomatrix coating and capture molecules (if present). Buffer compatibility with the BLI signal should be confirmed experimentally, especially if its composition is complex or unknown (*e.g.*, cell lysates, biological fluids). Nonspecific binding to the sensor tip might contribute significantly to the binding curves of some experiments, particularly when working with weak interactions and high concentrations of analyte [32]. In those cases, blocking agents can be added to reduce nonspecific binding; useful blocking agents are 1–2% (w/v) bovine serum albumin (BSA), polyethylene glycol (PEG), nonfat milk, casein, gelatin, non-ionic detergents like Tween 20/Polysorbate 20 (PS20), and saccharides, especially sucrose [32, 33]. If detergents are used, they should be added at working concentrations below the critical micelle concentration. The compatibility with BLI is known for common biochemicals; *e.g.*, dimethylsulfoxide (DMSO) is compatible [34].

Although the optical dip-and-read format allows the use of crude extracts and particulate material as the analyte in BLI experiments, it's highly recommended that both ligand and analyte samples be as pure and free of aggregates as possible. Furthermore, the ligand and the analyte should be dialyzed or dissolved in assay buffer and their concentration should be known accurately.

Ligand concentration for sensor coating should be adjusted to prevent either unsaturated or saturated biosensors. Poorly saturated sensors might show too low BLI signals in response to incubation with the analyte. Conversely, sensors which have too high ligand concentrations can show anomalous BLI signals due to steric hindrance, binding surface hiding, weaker nonspecific interactions at high enough analyte concentration, or rebinding events at lower analyte concentration. To optimize ligand concentration, at least three different concentrations covering one order of magnitude (*e.g.*, 5–50 μg/mL) should be tested for the association and dissociation kinetics of the analyte at a fixed concentration, and the ligand concentration that yields an acceptable nonsaturated BLI signal and adequate kinetics should be chosen (Fig. 6.3a). If ligand binding is low or poorly reproducible, overnight incubation of the sensors in ligand solutions at 4 °C can greatly increase immobilization results [33, 35].

The analyte concentration range should also be optimized experimentally. For quantitation assays, it's critical to determine the sensor's dynamic range for the analyte before obtaining a reliable standard curve. For kinetic experiments, at least four analyte dilutions must be assayed spanning a range of concentrations from 10-fold above to 10-fold below the estimated K_D. To achieve this, it's convenient to prepare a concentrated analyte stock and then make a serial dilution of the analyte with a dilution factor of 2–3. However, the concentration range of analyte may be limited by analyte solubility or assay sensitivity; in those cases, the actual concentration range of analyte can be adjusted based on practical considerations.

6.2.2 Hydrate the Biosensor in Assay Buffer to Minimize Nonspecific Signal

Biosensors should be hydrated for at least 10 min before starting an assay, and they can be stored in assay buffer overnight at 4 °C. After hydration, their tips must not be let dry out and should only be in contact with the assay solutions to maintain quality and functionality.

6.2.3 Maintain a Stable Assay Temperature

Allow sufficient time to equilibrate all solutions and biosensors to reach a stable temperature before starting the assay. This is particularly important for the BLI systems without temperature control (*e.g.*, BLItz). Samples can be maintained on ice up to the beginning of the assay.

6.2.4 Immobilize the Ligand onto the Biosensor Surface

Monitoring the ligand immobilization is recommended to determine the loading level and reproducibility, and the stability of the ligand attached onto the biosensor surface. The immobilization step starts recording the signal of a fresh sensor in assay buffer until a stable baseline is established. Next, the sensor tip is transferred to the ligand solution at the chosen concentration and the association kinetics is recorded until the desired signal level is reached. Then, the sensor

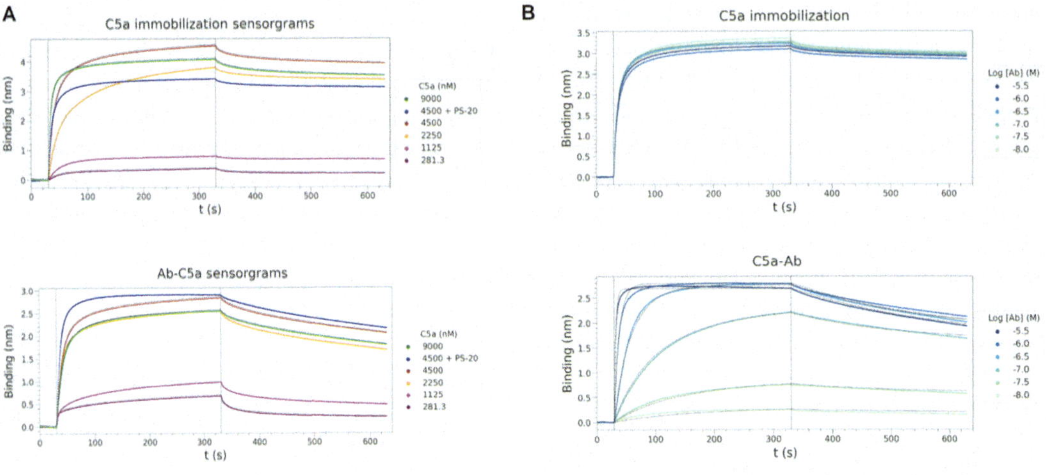

k_a (10^4 M^{-1} s^{-1})	k_d (10^{-3} s^{-1})	K_D (nM)	χ^2	R^2
11.2 (0.1)	1.37 (0.01)	12.2 (0.2)	2.65	0.999

Fig. 6.3 Development of a BLI assay to characterize the interaction between the C5a anaphylatoxin and an antibody (Ab). The experiment was carried on a BLItz system at room temperature, immobilizing biotinylated C5a onto streptavidin (SA) biosensors, and using the Ab as analyte. The assay buffer consisted of 10 mM HEPES (pH 7.4), 150 mM NaCl, 0.34 mM EDTA, supplemented with 0.02% (w/v) PS20 as indicated. Compatibility of the buffer with the BLI assay was previously confirmed (data not shown). The steps of the immobilization and the ligand-analyte interaction were the same: a baseline step in assay buffer for 30 s, an association step in a C5a or Ab solution for 300 s, and a dissociation step in assay buffer for 300 s. (**a**) Optimization of C5a concentration for immobilization. Several concentrations of biotinylated C5a were tested (top) and then the interaction kinetics with an Ab at a constant concentration (1000 nM) was monitored (bottom). PS20 was included in the standard assay buffer since the detergent decreased nonspecific binding. (**b**) Characterization of the interaction between C5a and an Ab. After C5a immobilization at 4500 nM (top), several sensorgrams were obtained using varying Ab concentrations (middle). Corrected data of Ab:C5a interaction were globally fitted to the 1:1 binding model integrated in the BLItz software, and the fitted curves (black) were plotted together with the sensorgrams (middle). The fitting kinetics parameters, with their standard error in brackets, and goodness-of-fit statistics are shown (bottom). In the sensorgrams, the vertical dashed lines separate distinct steps

tip is dipped back into assay buffer to remove nonspecifically bound ligand and monitor the stability of the BLI signal (Fig. 6.3a).

6.2.5 Measure Association and Dissociation Kinetics

After ligand immobilization, the sensor tip is transferred once again to assay buffer and the BLI signal is recorded to establish the binding baseline. In cases where ligand can dissociate from the sensor tip, this step can be foregone. Next, the sensor tip is dipped into the analyte solution at the first assayed concentration, and the association phase of the ligand:analyte interaction can be recorded (Fig. 6.3b). Sufficient time should be allowed for the BLI signal to reach the steady-state plateau, at least for the analyte solution at the highest concentration. Immediately afterward, the sensor tip is moved to assay buffer to record the dissociation phase of the interaction (Fig. 6.3b). Enough time should be allowed for 30–50% of the analyte to dissociate from the sensor; if the interaction is very stable, the dissociation time may have to be extended to 15–30 min. After the first sensorgram is fully recorded, this multi-step procedure must be repeated until all analyte concentrations have been assayed.

6.2.6 Regenerate Sensor Tip Surface (Optional)

If the sensor tips are to be re-used, a suitable regeneration buffer has to be used to remove as much of the remaining bound analyte as possible. Commonly used regeneration buffer formulations include low-pH buffers, detergents (*e.g.*, sodium dodecyl sulfate), and high ionic strength buffers. Typically, several short-time exposures to the regeneration buffer are more successful than a single, longer incubation. As the harsh conditions employed for sensor regeneration may damage the ligand, it's imperative to confirm the sensor's state by dipping it into the highest analyte concentration tested and compare binding capacity and kinetics to those previously recorded. Ideally, a successfully regenerated sensor should maintain >90% of the original binding capacity and equivalent binding kinetics.

6.2.7 Data Processing, Analysis, and Fitting

After recording all sensorgrams and before data analysis can start, two corrections are applied. Firstly, the blank curve at zero analyte (assay buffer only) is subtracted from each sensorgram to correct the background signal. Secondly, the start of the association and dissociation steps are aligned to the end of the immediately previous baseline or association phase, respectively.

Once these corrections have been applied, the corrected sensorgrams are fitted using a mathematical binding model. Fitting can be applied to individual curves (local fitting) or to all curves simultaneously (global fitting). In general, it's better to perform global fitting with the simplest possible binding model, considering the maximal binding signal as a variable, since the level of immobilized ligand can be different for each run. Sensorgrams, fitted curves, kinetic parameters, and fitting statistics must be examined thoroughly, paying especial attention to the curves corresponding to low and high analyte concentrations, which can show weak BLI signal and aggregation, respectively. Besides visual agreement between the experimental and fitting sensorgrams, several criteria have been proposed to evaluate the quality of fitting results: parameter-associated errors should not exceed 10% of the parameter value; the fitting residuals should be confined inside the interval defined by ±10% of the maximum response of the fitted curve; the goodness-of-fit χ^2 statistic should be smaller than 3; and the R^2 statistic greater than 95% [3].

Other recommendations to bear in mind to perform an optimal BLI experiment include:

- Whenever possible, use the same assay buffer for sensor hydration, ligand, analyte, and baseline and dissociation steps; and use the same tube/well with assay buffer for baseline and the corresponding dissociation step.
- Minimize nonspecific binding by optimizing ligand and analyte concentrations and the assay buffer composition.
- Optimize step times to ensure fully equilibrated (flat) baselines and enough dissociation for reliable data fitting.
- Do not reuse biosensors without regenerating them.
- Avoid foam and bubbles in all solutions since they may disturb the BLI signal.

6.3 Examples of Protein-Protein Interactions Studied by BLI

Since its introduction, BLI has grown into an easy-to-perform, versatile, label-free method for the measurement of protein-protein interaction kinetics and affinities in a wide variety of relevant settings. Especially, in the field of structural biology, BLI has made important contributions since the strength of protein-protein interactions governs the stability of multisubunit complexes over the time frame of experiments.

We provide below several examples of the application of BLI for the characterization of protein-protein interactions.

Grela et al. used BLI to characterize the binding kinetics of the interaction between the human ribosomal proteins P1 and P2, which belong to a pentameric complex called P-stalk, and a ricin A chain (RTA) [36]. P-stalk is thought to be involved in cellular processes such as the recruitment of translational GTPases. For BLI, several variants of P1 and P2 including the full-length proteins and deletion mutants lacking the C-terminal domain were expressed in *E. coli*, purified, and assembled into all possible P1-P2 heterodimers. RTA was selected as the ligand, which was captured on an NTA sensor through a hexahistidine tag on the ligand, and the P1-P2 complexes were used as analytes. For each P1-P2 heterodimer, binding isotherms were recorded at various analyte concentration, and the curves were globally fitted to a standard 1:1 binding model. Additionally, the full-length pentameric complex was expressed with a polycistronic expression system, purified, and tested in the same BLI experimental format for RTA binding. Results showed that the pentameric complexes had the highest affinity to RTA, and that the C-terminal domain of P1 (but not P2) is critical for the interaction [36].

In another application of BLI for structural biology, the interaction of the multidomain PfbA adhesin from the Gram-positive pathogen *Streptococcus pneumoniae* with several human proteins was investigated [37]. Here, the full-length, N-terminal, and central helix domains of PfbA were produced separately and used as ligands for BLI assays. Binding assays were performed for all known analytes (plasminogen, fibrinogen, and human serum albumin) sourced from commercial suppliers. Sensorgrams were fitted to a standard 1:1 binding model, with the result that the C-terminal domain of PfbA bound to fibrinogen, while the central region of PfbA bound to human serum albumin and plasminogen. These results were interpreted as evidence for a modular structure of PfbA, whereby the different domains are functionally independent, targeting distinct ligands [37].

The dip-and-read format characteristic of BLI has also allowed weak/slow interactions to be characterized. A case in point is the interaction between the human anaphylatoxin C5a and the moonlighting protein glyceraldehyde 3-phosphate dehydrogenase (GAPDH) from various Gram-positive and Gram-negative pathogens [38–41]. This interaction has been detected and characterized by various methods, but a quantitative measurement of binding kinetics had been challenging owing to the weak or transient nature of the interaction. Recently, our group has determined the crystal structure of GAPDH from the Gram-negative pathogen *Leptospira interrogans* [41], and proved that it can bind C5a using BLI and other complementary techniques. To obtain a direct measurement of the interaction binding parameters, we immobilized biotinylated C5a (sourced from Abvance Biotech) at 34 μg/mL on SA sensors, and dipped them on either assay buffer (blank) or into a concentrated GAPDH solution (0.23 mM, ~9 mg/mL). The binding curve with analyte yielded clear evidence for a slow ($k_a = 37.3\ M^{-1}\ s^{-1}$), yet specific binding. Interestingly, the complex, once assembled, seemed rather strong, as we could observe little or no dissociation over the 300-s dissociation step. This observation may hold relevance for the interaction *in vivo*, as GAPDH has been identified as part of the cell wall and the secretome of many bacterial pathogens. Avidity effects on surfaces, combined with the slow-dissociating interaction, would result in effective sequestration of C5a, precluding macrophages and neutrophiles from being recruited to the site of infection.

One of the main classes of molecular interactions studied by BLI is antigen-antibody binding. The high affinity and specificity of these interactions have propelled antibodies as prominent biologics in basic research and for diagnosing and treating many disorders. Therefore, antibody production and characterization are central endeavors for academic research as well as the pharmaceutical industry. BLI has proven valuable for applications in antibody development, including the screening and characterization stages. Some of these applications include comparing and classifying antibodies based on their kinetic and affinity properties; blocking assays that explore antigen-hindering interactions; antibody isotyping; and epitope binning by antigen binding competition assays [1, 31, 42–46].

BLI has been successfully used to develop antibodies derived from Ig new antigen receptors

(IgNARs) from sharks. One exciting example concerns IgNARs targeting the receptor-binding domain (RBD) of the spike glycoprotein (S) of the severe acute respiratory syndrome coronavirus 2 (SARS-CoV-2), which have potential applications in COVID-19 diagnostics and therapeutics [47]. In this assay, two complementary experimental designs were used: (1) Each of four single domain antibodies derived from IgNARs (VNAR) were immobilized on Protein A biosensors and purified RBD was the analyte; and (2) biotinylated RBD was immobilized on SA biosensors and VNAR-Fc fusions derived from the four initial VNARs to RBD were used as analytes. Interestingly, while the VNAR showed affinities for RBD in the 10^{-6}–10^{-8} M range, each VNAR-Fc exhibited increased affinities (10^{-7}–10^{-9} M) with respect to the corresponding VNAR. Next, to evaluate the capacity of the VNAR-Fc for blocking the RBD binding to the human angiotensin-converting enzyme 2 (ACE2), a critical interaction for the infectivity of SARS-CoV-2, biotinylated RBDs from wild-type (WT), delta and omicron variants of SARS-CoV-2 were immobilized on SA biosensors, incubated with the antibodies, and then exposed to ACE2. Remarkably, the best blocking antibody was JM-2-FC, which had previously shown an intermediate affinity to WT RBD (10^{-8} M); and while the antibody's ability to block WT and delta RBD varied between 12% and 86%, it did not exceed 20% for omicron RBD. Next, using biotinylated RBD immobilized on SA biosensors and VNAR-Fc at saturating concentrations, competition assays revealed that three non-overlapping epitopes were recognized on RBD by this set of antibodies. Lastly, five biparatopic VNAR were produced by combining noncompetitive VNAR and they were characterized with the previous RBD-VNAR binding and blocking assays. Results indicated that nearly all the biparatopic antibodies exhibited higher affinities to RBD in the VNAR (10^{-8}–10^{-9} M) and VNAR-Fc (10^{-10}–10^{-12} M) formats than their corresponding monovalent forms. Furthermore, the biparatopic VNAR-Fc exhibited blocking capacities of the distinct RBD-ACE2 interactions between 28% and 73% [47].

BLI has also been extensively used to characterize protein-protein interactions involved in physiological and pathological processes. Several protein-protein interactions relevant in cell signaling have been studied by BLI, including the binding of arrestin-3 to GPCR [48], the signal regulatory protein α (SIRP α) recognition by integrin macrophage antigen 1 (Mac-1) [49], the caspase 9 interaction with protein phosphatase 2A (PP2A) [50], or the fortilin binding to transforming growth factor beta 1 (TGF-β1) [51]. Similarly, the binding of the transcription factor GrgA to the σ^{66} factor of the RNA polymerase of *Chlamydia* has been addressed by BLI [26]. Furthermore, the high-affinity binding (subnanomolar) between the skeletal muscle myosin (SkM), with procoagulant activity, and the coagulation factor XI (FXI) was analyzed by BLI [52]. In the last years, multiple aspects of SARS-CoV-2 biology have been addressed using BLI, such as the effect of RBD mutations in the RBD-ACE2 interaction [53, 54], the binding of the S protein of SARS-CoV-2 to the C-type lectins DC-SIGN and L-SIGN [55], or the structural relevance of N-glycan sites (N165 and N234) of the S protein for its binding to ACE2 [56].

BLI assays have also been developed for the study of complement system activation, a topic of great interest for academia and industry given the potent effector functions of complement. In one case, BLI was used to address several unresolved questions concerning the interaction of the Fab fragment of IgG/IgM by C1q of the C1 complex, a crucial event in the activation of the classical pathway of complement [57]. In a first approximation, the researchers designed a BLI assay consisting of IgG immobilized on Protein L or SA sensors and using C1q as analyte; this assay was used to measure the binding kinetics and affinity of C1q for IgG1, IgG2, and IgG4 antibodies. The measured affinities for the various IgG isotypes were in agreement with their capacity to elicit complement-dependent cytotoxicity (CDC). Interestingly, in absence of the specific antigen, IgG1 antibodies displayed similar kinetics and nanomolar affinity for C1q, although they produced notably different CDC effects. The most striking antibody was Trastuzumab, which does not cause CDC. However, in presence of antigen the binding capacity of Trastuzumab to C1q was markedly hindered, emphasizing the importance of the antigen-antibody binding in the complement activation by the classical pathway [57]. Afterwards, a BLI experiment for the kinetic and affinity characterization of IgM-C1q interactions was optimized. Similarly to IgG1-C1q interactions, the obtained affinities of C1q for different IgM were in the nanomolar range, although the binding and dissociation kinetics of the IgM-C1q interactions were slightly slower [27]. More recently, a BLI approach in which aminopropylsilane (APS) biosensors loaded with polyethylene glycol (PEG) are exposed to anti-PEG IgM diluted in untreated serum has proved to be useful to evaluate complement activation induced by anti-PEG IgM in presence of antigen [58].

Besides enabling the characterization of numerous protein-protein interactions, BLI has also been used to screen modulators of specific interactions and study their regulatory function [28, 33, 50]. For example, a study on the interaction between the cyclin-dependent kinase 2 (CDK2) and cyclin A (CycA) and some known inhibitors of CDK2 by BLI has revealed interesting aspects of their regulatory activities. The optimized BLI procedure had N-terminally biotinylated CDK2 immobilized on SA sensors and CycA as the analyte. No differences in kinetics or nanomolar affinity were detected when GST-CDK2 was used as ligand (immobilized onto anti-GST sensors) or when biotinylated-CDK2 was phosphorylated on its activation loop. Next, the modulating effects of some CDK2 inhibitors on CDK2-CycA binding properties were analyzed using three methodological variants: 1) CDK2 was incubated with an inhibitor at a fixed concentration before performing the binding assay at several CycA concentrations; 2) CDK2 was exposed to several inhibitor concentrations before the binding assay with a fixed concentration of CycA; and 3) the inhibitor was added after the association of CycA to CDK2. Interestingly, the inhibitors BMS265246 and dinaciclib, which target the ATP binding site of CDK2, increased the affinity of CycA to CDK2 by decreasing the k_d. In contrast, the cyclin competitors AUZ 454 and MC14 reduced the CDK2-CycA affinity by apparently different mechanisms: while MC14 seemed to hinder CycA association to CDK2, AUZ 454 was suspected to form a ternary complex with CDK2-CycA thereby decreasing the affinity by increasing both k_a and k_d [35].

Other novel applications of BLI have been described in the context of protein-protein interactions, such as the study of enzymatic activities and post-translational modifications. For example, a BLI approach to follow the horseradish peroxidase (HRP) activity has been developed using tyramide, an HRP substrate that covalently binds to nearby tyrosine and tryptophan residues by HRP activity in presence of H_2O_2. In the assay, immobilized HRP is exposed to a reaction mixture containing tyramide and the oxidase activity is followed by the BLI signal which is proportional to tyramide binding to the protein immobilized on the biosensor [59]. Another BLI assay has been optimized to study the activity of the protease of human immunodeficiency virus type 1 (HIV-1). This protease assay is based on monitoring the proteolysis of an immobilized substrate by BLI, as the BLI signal diminishes as a consequence of substrate fragment loss on the biosensor [22]. Finally, in a recent article, a BLI procedure to analyze protein ubiquitination has been described that monitors the increase in BLI signal accompanying the post-translational modification of an immobilized ligand substrate carried out by a ubiquitination reaction mixture [60].

6.4 Concluding remarks

Biolayer Interferometry (BLI) is a robust biophysical technique for the quantification of kinetic parameters pertinent to a variety of molecular interactions, encompassing protein-protein, protein-nucleic acid, and protein-small molecule affinities. This technique allows the detection of interferometry wavelength shifts induced by the molecular binding event at the sensor's tip.

Notably, BLI obviates the necessity for fluorescent or radioactive labeling of the interacting entities. Nevertheless, immobilization of one interaction partner, commonly referred to as the ligand, on the functionalized sensor surface is requisite. The other partner, the analyte, is brought into contact with the immobilized ligand via sensor immersion ("dip") into the analyte-containing solution. Subsequent recording ("read") of binding isotherms or sensorgrams at varying analyte concentrations enables the elucidation of key kinetic parameters, including the association, dissociation, and equilibrium binding constant (ka, kd, and K_D). The technique's "dip-and-read" format offers compatibility with complex, unrefined samples.

Acknowledgments This work was funded by the Spanish Ministerio de Ciencia, Innovación y Universidades-FEDER grants RTI2018-102242-BI00 (MCV), the Spanish Ministerio de Ciencia e Innovación-Recovery, Transformation and Resilience Plan (PRTR) grant PDC2022-133713-I00 (MCV), grant S2022/BMD-7278 of the Regional Government of Madrid (MCV), the European Commission – NextGenerationEU through CSIC's Global Health Platform ("PTI Salud Global") (SGL2103020) (MCV), and the CSIC Special Intramural Grant PIE201620E064 (MCV). It was additionally supported by the Research Network on Complement in Health and Disease (RED2022-134750-T). KdlP was supported by an Industrial PhD grant (IND2018-010094) awarded by the Spanish Ministerio de Economía y Competitividad. JSL acknowledges the support of the PhD program in Molecular Biosciences of the Universidad Autónoma de Madrid (UAM) and the Ministry of Education, Culture and Sports of Spain (FPU Grant 17/06090). KdlP acknowledges the support of the PhD program in Biochemistry, Molecular Biology and Biomedicine of the Universidad Complutense de Madrid (UCM).

References

1. Concepcion J, Witte K, Wartchow C, Choo S, Yao D, Persson H, Wei J, Li P, Heidecker B, Ma W, Varma R, Zhao L-S, Perillat D, Carricato G, Recknor M, Du K, Ho H, Ellis T, Gamez J, Howes M, Phi-Wilson J, Lockard S, Zuk R, Tan H (2009) Label-free detection of biomolecular interactions using biolayer interferometry for kinetic characterization. Comb Chem High Throughput Screen 12:791–800. https://doi.org/10.2174/138620709789104915
2. Kumaraswamy S, Tobias R (2015) Label-free kinetic analysis of an antibody–antigen interaction using biolayer interferometry. In: Meyerkord CL, Fu H (eds) Protein-Protein Interactions. Springer, New York, pp 165–182
3. Sultana A, Lee JE (2015) Measuring protein-protein and protein-nucleic acid interactions by biolayer interferometry. Curr Protoc Protein Sci 79. https://doi.org/10.1002/0471140864.ps1925s79
4. Apiyo DO (2017) Biolayer interferometry (Octet) for label-free biomolecular interaction sensing. In: Schasfoort RBM (ed) Handbook of surface plasmon resonance, 2nd edn. The Royal Society of Chemistry, London, pp 356–397
5. Cleaver S, Gardner M, Barlow A, Ferrari E, Soloviev M (2023) Fast protocols for characterizing antibody–peptide binding. In: Cretich M, Gori A (eds) Peptide microarrays. Springer, New York, pp 83–101
6. Nirschl M, Reuter F, Vörös J (2011) Review of transducer principles for label-free biomolecular interaction analysis. Biosensors 1:70–92. https://doi.org/10.3390/bios1030070
7. Abdiche Y, Malashock D, Pinkerton A, Pons J (2008) Determining kinetics and affinities of protein interactions using a parallel real-time label-free biosensor, the Octet. Anal Biochem 377:209–217. https://doi.org/10.1016/j.ab.2008.03.035
8. Yang D, Singh A, Wu H, Kroe-Barrett R (2016) Comparison of biosensor platforms in the evaluation of high affinity antibody-antigen binding kinetics. Anal Biochem 508:78–96. https://doi.org/10.1016/j.ab.2016.06.024
9. Martin SR, Ramos A, Masino L (2021) Biolayer interferometry: protein–RNA interactions. In: Daviter T, Johnson CM, McLaughlin SH, Williams MA (eds) Protein-ligand interactions. Springer, New York, pp 351–368
10. Weeramange CJ, Fairlamb MS, Singh D, Fenton AW, Swint-Kruse L (2020) The strengths and limitations of using biolayer interferometry to monitor equilibrium titrations of biomolecules. Protein Sci 29:1004–1020. https://doi.org/10.1002/pro.3827
11. Ingale J, Wyatt R (2015) Kinetic analysis of monoclonal antibody binding to HIV-1 gp120-derived hyperglycosylated cores. Bio-Protocol 5. https://doi.org/10.21769/BioProtoc.1615
12. Kol S, Kallehauge TB, Adema S, Hermans P (2015) Development of a VHH-based erythropoietin quantification assay. Mol Biotechnol 57:692–700. https://doi.org/10.1007/s12033-015-9860-7
13. Zhang H, Li W, Luo H, Xiong G, Yu Y (2017) Quantitative determination of testosterone levels with biolayer interferometry. Chem Biol Interact 276:141–148. https://doi.org/10.1016/j.cbi.2017.05.013
14. Carvalho SB, Moreira AS, Gomes J, Carrondo MJT, Thornton DJ, Alves PM, Costa J, Peixoto C (2018) A detection and quantification label-free tool to speed up downstream processing of model mucins. PLoS ONE 13:e0190974. https://doi.org/10.1371/journal.pone.0190974
15. Gao S, Zheng X, Wu J (2018) A biolayer interferometry-based enzyme-linked aptamer sorbent

assay for real-time and highly sensitive detection of PDGF-BB. Biosens Bioelectron 102:57–62. https://doi.org/10.1016/j.bios.2017.11.017
16. Gao S, Li Q, Zhang S, Sun X, Zheng X, Qian H, Wu J (2022) One-step high-throughput detection of low-abundance biomarker BDNF using a biolayer interferometry-based 3D aptasensor. Biosens Bioelectron 215:114566. https://doi.org/10.1016/j.bios.2022.114566
17. Dysinger M, King LE (2012) Practical quantitative and kinetic applications of bio-layer interferometry for toxicokinetic analysis of a monoclonal antibody therapeutic. J Immunol Methods 379:30–41. https://doi.org/10.1016/j.jim.2012.02.017
18. Rao VS, Srinivas K, Sujini GN, Kumar GNS (2014) Protein-protein interaction detection: methods and analysis. Int J Proteomics 2014:1–12. https://doi.org/10.1155/2014/147648
19. Biswas P (2018) Modern biophysical approaches to study protein–ligand interactions. Biophys Rev Lett 13:133–155. https://doi.org/10.1142/S1793048018300013
20. Carvalho SB, Moleirinho MG, Wheatley D, Welsh J, Gantier R, Alves PM, Peixoto C, Carrondo MJT (2017) Universal label-free in-process quantification of influenza virus-like particles. Biotechnol J 12:1700031. https://doi.org/10.1002/biot.201700031
21. Overacker RD, Plitzko B, Loesgen S (2021) Biolayer interferometry provides a robust method for detecting DNA binding small molecules in microbial extracts. Anal Bioanal Chem 413:1159–1171. https://doi.org/10.1007/s00216-020-03079-5
22. Miczi M, Diós Á, Bozóki B, Tőzsér J, Mótyán JA (2021) Development of a bio-layer interferometry-based protease assay using HIV-1 protease as a model. Viruses 13:1183. https://doi.org/10.3390/v13061183
23. Li A, Harris RJ, Fry BG, Barnes AC (2021) A single-step, high throughput, and highly reproducible method for measuring IgM quantity and avidity directly from fish serum via biolayer interferometry (BLI). Fish Shellfish Immunol 119:231–237. https://doi.org/10.1016/j.fsi.2021.10.003
24. Wilson JL, Scott IM, McMurry JL (2010) Optical biosensing: kinetics of protein A-IGG binding using biolayer interferometry. Biochem Mol Biol Educ 38:400–407. https://doi.org/10.1002/bmb.20442
25. Petersen R (2017) Strategies using bio-layer interferometry biosensor technology for vaccine research and development. Biosensors 7:49. https://doi.org/10.3390/bios7040049
26. Desai M, Di R, Fan H (2019) Application of biolayer interferometry (BLI) for studying protein-protein interactions in transcription. J Vis Exp 59687. https://doi.org/10.3791/59687
27. Chouquet A, Pinto AJ, Hennicke J, Ling WL, Bally I, Schwaigerlehner L, Thielens NM, Kunert R, Reiser J-B (2022) Biophysical characterization of the oligomeric states of recombinant immunoglobulins type-M and their C1q-binding kinetics by biolayer interferometry. Front Bioeng Biotechnol 10:816275. https://doi.org/10.3389/fbioe.2022.816275
28. Shah NB, Duncan TM (2014) Bio-layer interferometry for measuring kinetics of protein-protein interactions and allosteric ligand effects. J Vis Exp 51383. https://doi.org/10.3791/51383
29. Ullah SF, Moreira G, Datta SPA, McLamore E, Vanegas D (2022) An experimental framework for developing point-of-need biosensors: connecting bio-layer interferometry and electrochemical impedance spectroscopy. Biosensors 12:938. https://doi.org/10.3390/bios12110938
30. Zhao H, Boyd LF, Schuck P (2017) Measuring protein interactions by optical biosensors. Curr Protoc Protein Sci 88. https://doi.org/10.1002/cpps.31
31. Noy-Porat T, Alcalay R, Mechaly A, Peretz E, Makdasi E, Rosenfeld R, Mazor O (2021) Characterization of antibody-antigen interactions using biolayer interferometry. STAR Protoc 2:100836. https://doi.org/10.1016/j.xpro.2021.100836
32. Dubrow A, Zuniga B, Topo E, Cho J-H (2022) Suppressing nonspecific binding in biolayer interferometry experiments for weak ligand–analyte interactions. ACS Omega 7:9206–9211. https://doi.org/10.1021/acsomega.1c05659
33. Müller-Esparza H, Osorio-Valeriano M, Steube N, Thanbichler M, Randau L (2020) Bio-layer interferometry analysis of the target binding activity of CRISPR-Cas effector complexes. Front Mol Biosci 7:98. https://doi.org/10.3389/fmolb.2020.00098
34. Wartchow CA, Podlaski F, Li S, Rowan K, Zhang X, Mark D, Huang K-S (2011) Biosensor-based small molecule fragment screening with biolayer interferometry. J Comput Aided Mol Des 25:669–676. https://doi.org/10.1007/s10822-011-9439-8
35. Tambo CS, Tripathi S, Perera BGK, Maly DJ, Bridges AJ, Kiss G, Rubin SM (2023) Biolayer interferometry assay for cyclin-dependent kinase-cyclin association reveals diverse effects of Cdk2 inhibitors on cyclin binding kinetics. ACS Chem Biol 18:431–440. https://doi.org/10.1021/acschembio.3c00015
36. Grela P, Li X-P, Horbowicz P, Dźwierzyńska M, Tchórzewski M, Tumer NE (2017) Human ribosomal P1-P2 heterodimer represents an optimal docking site for ricin A chain with a prominent role for P1 C-terminus. Sci Rep 7:5608. https://doi.org/10.1038/s41598-017-05675-5
37. Beulin DSJ, Radhakrishnan D, Suresh SC, Sadasivan C, Yamaguchi M, Kawabata S, Ponnuraj K (2017) *Streptococcus pneumoniae* surface protein PfbA is a versatile multidomain and multiligand-binding adhesin employing different binding mechanisms. FEBS J 284:3404–3421. https://doi.org/10.1111/febs.14200
38. Querol-García J, Fernández FJ, Marin AV, Gómez S, Fullà D, Melchor-Tafur C, Franco-Hidalgo V, Albertí S, Juanhuix J, Rodríguez De Córdoba S, Regueiro JR, Vega MC (2017) Crystal structure of

glyceraldehyde-3-phosphate dehydrogenase from the gram-positive bacterial pathogen A. vaginae, an immunoevasive factor that interacts with the human C5a anaphylatoxin. Front Microbiol 8:541. https://doi.org/10.3389/fmicb.2017.00541
39. Fernández FJ, Gómez S, Vega MC (2019) Pathogens' toolbox to manipulate human complement. Semin Cell Dev Biol 85:98–109. https://doi.org/10.1016/j.semcdb.2017.12.001
40. Gómez S, Querol-García J, Sánchez-Barrón G, Subias M, González-Alsina À, Franco-Hidalgo V, Albertí S, Rodríguez De Córdoba S, Fernández FJ, Vega MC (2019) The antimicrobials anacardic acid and curcumin are not-competitive inhibitors of gram-positive bacterial pathogenic glyceraldehyde-3-phosphate dehydrogenase by a mechanism unrelated to human C5a anaphylatoxin binding. Front Microbiol 10:326. https://doi.org/10.3389/fmicb.2019.00326
41. Navas-Yuste S, De La Paz K, Querol-García J, Gómez-Quevedo S, Rodríguez De Córdoba S, Fernández FJ, Vega MC (2023) The structure of Leptospira interrogans GAPDH sheds light into an immunoevasion factor that can target the anaphylatoxin C5a of innate immunity. Front Immunol 14:1190943. https://doi.org/10.3389/fimmu.2023.1190943
42. Lad L, Clancy S, Kovalenko M, Liu C, Hui T, Smith V, Pagratis N (2015) High-throughput kinetic screening of hybridomas to identify high-affinity antibodies using bio-layer interferometry. SLAS Discov 20:498–507. https://doi.org/10.1177/1087057114560123
43. Kamat V, Rafique A (2017) Designing binding kinetic assay on the bio-layer interferometry (BLI) biosensor to characterize antibody-antigen interactions. Anal Biochem 536:16–31. https://doi.org/10.1016/j.ab.2017.08.002
44. Choi JR, Kim MJ, Tae N, Wi TM, Kim S-H, Lee ES, Kim DH (2020) BLI-based functional assay in phage display benefits the development of a PD-L1-targeting therapeutic antibody. Viruses 12:684. https://doi.org/10.3390/v12060684
45. Bell BN, Powell AE, Rodriguez C, Cochran JR, Kim PS (2021) Neutralizing antibodies targeting the SARS-CoV-2 receptor binding domain isolated from a naïve human antibody library. Protein Sci 30:716–727. https://doi.org/10.1002/pro.4044
46. Sim DS, Shukla M, Mallari CR, Fernández JA, Xu X, Schneider D, Bauzon M, Hermiston TW, Mosnier LO (2023) Selective modulation of activated protein C activities by a nonactive site–targeting nanobody library. Blood Adv 7:3036–3048. https://doi.org/10.1182/bloodadvances.2022008740
47. Chen Y-L, Lin J-J, Ma H, Zhong N, Xie X-X, Yang Y, Zheng P, Zhang L-J, Jin T, Cao M-J (2022) Screening and characterization of shark-derived VNARs against SARS-CoV-2 spike RBD protein. Int J Mol Sci 23:10904. https://doi.org/10.3390/ijms231810904
48. Avsar SY, Kapinos LE, Schoenenberger C-A, Schertler GFX, Mühle J, Meger B, Lim RYH, Ostermaier MK, Lesca E, Palivan CG (2020) Immobilization of arrestin-3 on different biosensor platforms for evaluating GPCR binding. Phys Chem Chem Phys 22:24086–24096. https://doi.org/10.1039/D0CP01464H
49. Podolnikova NP, Hlavackova M, Wu Y, Yakubenko VP, Faust J, Balabiyev A, Wang X, Ugarova TP (2019) Interaction between the integrin Mac-1 and signal regulatory protein α (SIRPα) mediates fusion in heterologous cells. J Biol Chem 294:7833–7849. https://doi.org/10.1074/jbc.RA118.006314
50. Dorgham K, Murail S, Tuffery P, Savier E, Bravo J, Rebollo A (2022) Binding and kinetic analysis of human protein phosphatase PP2A interactions with caspase 9 protein and the interfering peptide C9h. Pharmaceutics 14:2055. https://doi.org/10.3390/pharmaceutics14102055
51. Pinkaew D, Martinez-Hackert E, Jia W, King MD, Miao F, Enger NR, Silakit R, Ramana K, Chen S-Y, Fujise K (2022) Fortilin interacts with TGF-β1 and prevents TGF-β receptor activation. Commun Biol 5:157. https://doi.org/10.1038/s42003-022-03112-6
52. Morla S, Deguchi H, Zilberman-Rudenko J, Gruber A, McCarty OJT, Srivastava P, Gailani D, Griffin JH (2022) Skeletal muscle myosin promotes coagulation by binding factor XI via its A3 domain and enhancing thrombin-induced factor XI activation. J Biol Chem 298:101567. https://doi.org/10.1016/j.jbc.2022.101567
53. Gong SY, Chatterjee D, Richard J, Prévost J, Tauzin A, Gasser R, Bo Y, Vézina D, Goyette G, Gendron-Lepage G, Medjahed H, Roger M, Côté M, Finzi A (2021) Contribution of single mutations to selected SARS-CoV-2 emerging variants spike antigenicity. Virology 563:134–145. https://doi.org/10.1016/j.virol.2021.09.001
54. Vogel M, Augusto G, Chang X, Liu X, Speiser D, Mohsen MO, Bachmann MF (2022) Molecular definition of severe acute respiratory syndrome coronavirus 2 receptor-binding domain mutations: receptor affinity versus neutralization of receptor interaction. Allergy 77:143–149. https://doi.org/10.1111/all.15002
55. Simpson JD, Ray A, Marcon C, Dos Santos NR, Dorrazehi GM, Durlet K, Koehler M, Alsteens D (2023) Single-molecule analysis of SARS-CoV-2 binding to C-type lectin receptors. Nano Lett 23:1496–1504. https://doi.org/10.1021/acs.nanolett.2c04931
56. Casalino L, Gaieb Z, Goldsmith JA, Hjorth CK, Dommer AC, Harbison AM, Fogarty CA, Barros EP, Taylor BC, McLellan JS, Fadda E, Amaro RE (2020) Beyond shielding: the roles of glycans in the SARS-CoV-2 spike protein. ACS Cent Sci 6:1722–1734. https://doi.org/10.1021/acscentsci.0c01056
57. Zhou W, Lin S, Chen R, Liu J, Li Y (2018) Characterization of antibody-C1q interactions by Biolayer Interferometry. Anal Biochem 549:143–148. https://doi.org/10.1016/j.ab.2018.03.022

58. Mostafa M, Elsadek NE, Emam SE, Ando H, Shimizu T, Abdelkader H, Ishima Y, Aly UF, Sarhan HA, Ishida T (2022) Using bio-layer interferometry to evaluate anti-PEG antibody-mediated complement activation. Biol Pharm Bull 45:129–135. https://doi.org/10.1248/bpb.b21-00772
59. Kojima T, Nakane A, Zhu B, Alfi A, Nakano H (2019) A simple, real-time assay of horseradish peroxidase using biolayer interferometry. Biosci Biotechnol Biochem 83:1822–1828. https://doi.org/10.1080/09168451.2019.1621156
60. De Silva ARI, Shrestha S, Page RC (2023) Real-time bio-layer interferometry ubiquitination assays as alternatives to western blotting. Anal Biochem 679:115296. https://doi.org/10.1016/j.ab.2023.115296

7 Studying Macromolecular Interactions of Cellular Machines by the Combined Use of Analytical Ultracentrifugation, Light Scattering, and Fluorescence Spectroscopy Methods

Carlos Alfonso, Marta Sobrinos-Sanguino, Juan Román Luque-Ortega, Silvia Zorrilla, Begoña Monterroso, Oscar M. Nuero, and Germán Rivas

Abstract

Cellular machines formed by the interaction and assembly of macromolecules are essential in many processes of the living cell. These assemblies involve homo- and hetero-associations, including protein-protein, protein-DNA, protein-RNA, and protein-polysaccharide associations, most of which are reversible. This chapter describes the use of analytical ultracentrifugation, light scattering, and fluorescence-based methods, well-established biophysical techniques, to characterize interactions leading to the formation of macromolecular complexes and their modulation in response to specific or unspecific factors. We also illustrate, with several examples taken from studies on bacterial processes, the advantages of the combined use of subsets of these techniques as orthogonal analytical methods to analyze protein oligomerization and polymerization, interactions with ligands, hetero-associations involving membrane proteins, and protein-nucleic acid complexes.

Keywords

Sedimentation velocity · Sedimentation equilibrium · Dynamic light scattering · Multi-angle light scattering · Fluorescence anisotropy · Fluorescence correlation spectroscopy · Protein-protein interactions · Protein-DNA interactions

7.1 Introduction

Most of the biochemical reactions within living cells, both under physiological conditions and in pathological processes, involve the formation of macromolecular complexes. Understanding the processes where these complexes take part as cellular machines requires an in-depth analysis of the interactions leading to their assembly, which can be approached by using different analytical techniques. In this chapter we describe several well stablished biophysical techniques for the analysis of these complexes: analytical ultracentrifugation (AUC), size exclusion chromatography coupled to multi-angle light scattering (SEC-MALS), composition gradient multi-angle light scattering (CG-MALS), dynamic

C. Alfonso (✉) · S. Zorrilla · B. Monterroso · G. Rivas
Structural and Chemical Biology Department, Centro de Investigaciones Biológicas Margarita Salas, Consejo Superior de Investigaciones Científicas, Madrid, Spain
e-mail: carlosa@cib.csic.es

M. Sobrinos-Sanguino · J. R. Luque-Ortega · O. M. Nuero
Molecular Interactions Facility, Centro de Investigaciones Biológicas Margarita Salas, Consejo Superior de Investigaciones Científicas, Madrid, Spain

M. C. Vega, F. J. Fernández (eds.), *Advanced Technologies for Protein Complex Production and Characterization*, Advances in Experimental Medicine and Biology 1453,
https://doi.org/10.1007/978-3-031-52193-5_7

light scattering (DLS), fluorescence anisotropy and fluorescence correlation and cross-correlation spectroscopy (FCS and FCCS), that can be individually used or combined orthogonally. Thus, the AUC technique is a reference method for the quantitative analysis of macromolecules and their interactions in solution. When applied to the study of homo- and hetero-associations leading to the formation of macromolecular complexes, stoichiometry, cooperativity and equilibrium binding affinity can be determined [1–4]. On the other hand, size exclusion chromatography (SEC) and light scattering techniques (MALS and DLS) are high-throughput techniques for the characterization and quality control of macromolecules in biological and biopharmaceuticals studies. These techniques allow the automation of processes and therefore they are preferred over other methods in many industrial procedures [5]. The characterization of the mass and size of macromolecules, and their interaction to form complexes, has been extensively studied using the methods of MALS and DLS [6]. We will also briefly introduce here fluorescence spectroscopy methods, which endowed with an excellent sensitivity, are suitable for the analysis of high affinity interactions in solution and in complex media, including reconstituted systems and live cells [7]. These methods have frequently been employed to develop screening assays with biotechnological or medical purposes, as most fluorescence-based measurements can be conducted simultaneously on hundreds of samples of a few microliters in short times, by using plate readers. The complementary use of various AUC, light scattering and fluorescence techniques allows obtaining a more robust description of the molecular mechanisms involved in a biological process compared with the application of a single method with its associated limitations and uncertainties (Fig. 7.1), as will be shown with several examples presented below. Furthermore, these techniques have been of great utility to differentiate between several possible models of interaction in hetero-associations, strengthening the results obtained by structural techniques such as NMR or EM [8–10].

7.2 Analytical Ultracentrifugation (AUC)

Analytical ultracentrifugation is a powerful technique amenable for the quantitative characterization of macromolecular associations in solution. AUC has been successfully applied to the study of proteins, nucleic acids, polysaccharides and nanoparticles, among others and is a standard method to determine sample purity, association state and molar mass of individual macromolecules. It is based on the application of a centrifugal force and the real-time monitoring of the subsequent spatial macromolecular redistribution, followed by the quantitative analysis of the recorded raw data [1, 4, 11]. It is aimed for the characterization of molecular complexes in terms of molecular size and shape, stoichiometry and thermodynamic binding constants. Experiments are performed without the need for labelling or any other chemical modification of the sample, and there is no interaction with any matrix or surface. AUC can be used within a great range of temperatures (0–40 °C) with a great variety of buffers, including high concentrations and macromolecular crowding conditions [12]. Comprehensive basic principles and common experimental procedures may be found in several previous publications [11, 13–15].

7.2.1 Instrumentation and General Experimental Considerations

Analytical ultracentrifuges are equipped with an optical system that allows the observation of the sample while it is being sedimented. For this purpose, the analytical rotor contains 4–8 holes where the sample holders (cells) are aligned. Each cell has a 3–12 mm pathlength epon-charcoal double-sector centerpiece with two chambers (80–400 μL sample volume), one for the sample and one for the reference buffer, flanked and sealed by quartz or sapphire windows that allow the passage of light across the sectors and hence the monitoring of the migration of macromolecules during the sedimentation process. There are three

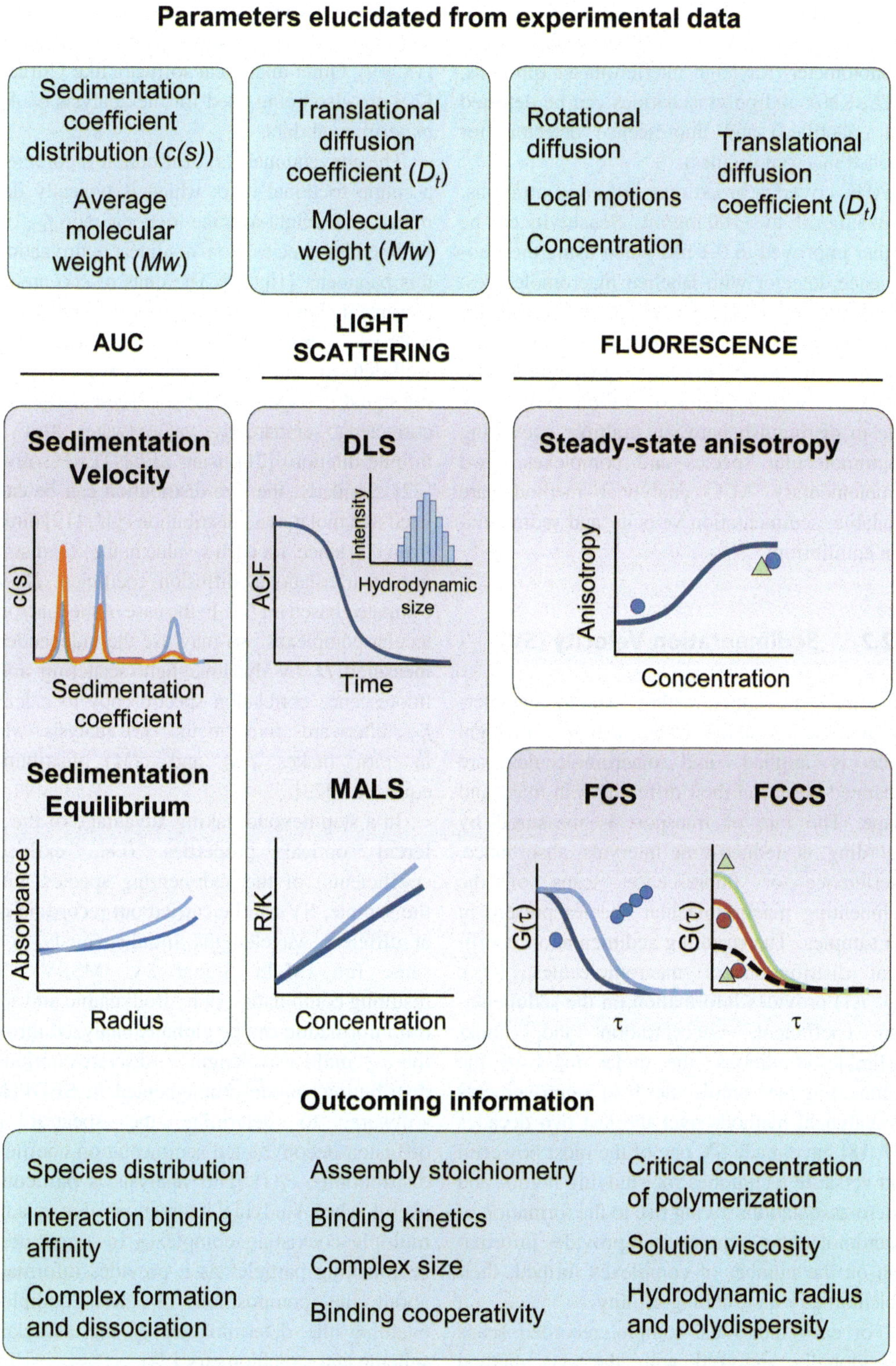

Fig. 7.1 Illustration of the different biophysical methods described in this chapter for the characterization of molecular machines. Parameters and information derived from the individual or orthogonal application of these analytical methods

commercially available optical detectors for analytical ultracentrifuges: UV-VIS absorbance spectrophotometer, Rayleigh interferometer (proteins, DNA, RNA and polysaccharides can be detected and quantified) and fluorescence detector (for labelled macromolecules).

AUC covers a broad range of concentrations: from 5 μg/mL to >100 mg/mL. Sensitivity can be further improved to 0.1 nM when using the fluorescence detector with labelled macromolecules. The range of molecular weights suitable for AUC analysis varies from low molecular weight peptides (1 kDa) to large macromolecular structures (several hundred million Da), being able to distinguish between multiple coexisting macromolecular species and complexes. Two complementary AUC analytical methods are available: sedimentation velocity and sedimentation equilibrium.

7.2.2 Sedimentation Velocity (SV)

SV is a hydrodynamic method ideal to characterize biological systems, where a high centrifugal force is applied, and macromolecules are separated based on their differences in mass and shape. The rate of transport is measured by recording, at defined time intervals, absorbance, interference or fluorescence scans of the sedimenting macromolecular species present in the samples. The resulting sedimentation coefficient distribution of macromolecules, ($c(s)$; Fig. 7.1) provides information on the sedimentation coefficient, concentration and, under favorable conditions, the molar mass of the sedimenting macromolecules [16]. Improvements in analytical methods over the last two decades [17, 18] have made SV one of the most powerful and versatile techniques for studying homo- and hetero-associations giving rise to the formation of macromolecular complexes. It provides information on the number of complexes formed, their stoichiometry and binding affinity.

For each individual sample, recorded scans are globally analyzed with the $c(s)$ method implemented in SEDFIT software [18] that uses finite element solutions of the Lamm equation combined with size distribution analysis techniques by maximum entropy regularization [18, 19]. Other analytical software like UltraScan [20] are also being used for the analysis of AUC experimental data.

The conventional $c(s)$ distribution is confined to a single frictional ratio, which is typically determined as a weight-average frictional ratio $f_{r,w}$ of all sedimenting species, from nonlinear optimization of this parameter [16, 18]. The units of $c(s)$ are such that integration over a peak gives the total signal of material sedimenting within the peak and a well-defined and precise weight-average s-value (or signal average s-value) [3]. These s-values are corrected to standard $s_{20,w}$ values (water, 20 °C, and infinite dilution) [21] using SEDNTERP software [22]. Similarly, the size distribution can be calculated as a molar mass distribution $c(M)$ [19] directly from $c(s)$ since, for each s-value in the $c(s)$ distribution, a translational diffusion coefficient $D_t(s)$ is estimated based on $f_{r,w}$. In the case of the macromolecular complexes, we may use the independently measured D_t by dynamic light scattering and/or fluorescence correlation spectroscopy to calculate $f_{r,w}$, afterwards fixed in the $c(s)$ analysis, which in turn makes $c(s)$ and $c(M)$ distributions equivalent [23].

In a step beyond, taking advantage of the different optical properties (*i.e.*, extinction coefficients) of the sedimenting species within the sample, SV can be carried out recording data at different wavelengths simultaneously in the same run (Multi Signal SV, MSSV). The resulting combination of hydrodynamic and spectral information can be globally analyzed through the "multi-wavelength discrete/continuous distribution" model implemented in SEDPHAT software, to determine the spectral and diffusion-deconvoluted sedimentation coefficient distributions, $c_k(s)$. This analysis is particularly useful when studying interactions that result in multiple coexisting complexes in a mixture of sedimenting particles as it provides information about the composition of these complexes enabling the determination of the association scheme and stoichiometry [24].

Up to eight samples may be run at the same time in a single experiment for 2–7 h, depending on the molar mass of the sedimenting species. Sample volumes loaded into analytical cells are 400 μL for standard cells (1.2 cm pathlength) and 100 μL for narrow cells (0.3 cm pathlength). Most of the buffers commonly used for *in vitro* analysis of macromolecules may be used, though high absorbance additives (as reducing agents) and viscous protectants should be maintained at low concentration for better results (for example, concentrations lower than 0.1 mM for 2-mercaptoethanol and 5% for glycerol are preferred). Samples should be equilibrated with their reference buffer (dialysis), which is especially important when using interference optics for detection.

When using an absorbance detector, sample absorbance should range from 0.1 to 1.5 OD at the selected wavelength.

7.2.3 Sedimentation Equilibrium (SE)

SE methods are specially adapted for the detection and quantitative analysis of interactions (stoichiometry, affinity, reversibility) leading to the formation of macromolecular complexes, including protein-protein, DNA-protein, and receptor-ligand [1, 13, 25]. The SE technique employs low-to-moderate speeds and longer time periods than SV for the runs. When equilibrium is reached the molecular movement in the gravitational field is balanced with diffusion, there is no net transport of molecules and a concentration gradient within the cell length is formed (Fig. 7.1). The analysis of this gradient provides information on the molecular weight (Mw) of the macromolecules present in each sample [4, 11], enabling the absolute measurement of their average molecular weights. Furthermore, sedimentation equilibrium studies can be used to analyze interacting systems. It is one of the best methods for the detection and characterization of reversible macromolecular interactions and the formation of complexes, allowing the determination of the average molecular weight, equilibrium constants and stoichiometry [1]. In this technique, at a given rotor speed, a concentration gradient is formed (10 to 1000-fold) that only depends on the molecular weight of the species present and their concentration. Experimental SE gradients may be analyzed with HETEROANALYSIS software [26].

The molar masses of the macromolecules and their complexes suitable for their characterization by this technique can range from less than 2×10^3 to more than 10^6 Da and, in reversible homo- and hetero-associations, affinities in the range of 10^4–10^8 M^{-1} can be determined. For the analysis of homo- and hetero-associations, data collected at different speeds and different loading concentrations are globally analyzed in terms of different association models using SEDPHAT software [27]. Protein and protein domains, as well as detergent-solubilized membrane proteins may be characterized by this method [28–30].

Up to 21 samples may be analyzed in a single SE experiment, using special six-hole analytical cells that allow three samples plus their references per cell. Two or three running speeds are selected according to the molecular weight of the macromolecules to be tested and of the complexes that could be originated in case we are studying a macromolecular interaction. After collecting the equilibrium scans, a final high-speed centrifugation run (50,000 rpm) is conducted to estimate the corresponding baseline offsets. In SE it is recommended that a previous gel-filtration of the macromolecules be performed to remove low *Mw* contaminants and big aggregates. It is also highly recommended to run previous SV experiment of the samples to check the homogeneity and number of present species, which will determine the probability of success of the SE experiment, as samples with multiple species are difficult to analyze.

Buffer requirements are similar to those described for SV experiments. Sample volumes are 160–180 μL for long column experiments using standard cells (1.2 cm pathlength) or 45 μL using narrow cells (0.3 cm pathlength). SE experiments last longer than SV ones because reaching the concentration gradient at equilibrium, for each running speed tested, typically requires long periods of time, and a typical multi-speed experiment takes several days. Some proteins are not stable over these long periods of time and short columns (100 μL sample), that significantly reduce the equilibration time, have been used satisfactorily, but with the drawback of fewer data points and less accuracy.

7.3 Light Scattering (LS) Techniques

7.3.1 Dynamic Light Scattering (DLS)

Dynamic light scattering is a non-invasive technique for measuring the size of particles or molecules in solution diffusing due to Brownian motion. Brownian motion is the random movement of particles in suspension, and is influenced by particle size, sample viscosity and temperature. A characteristic of this Brownian motion is that large molecules move slower than small ones. DLS is also known as quasi-elastic light scattering (QELS). In a DLS experiment, samples are illuminated with a monochromatic light, and the variation of the intensity of scattered light through time is registered. The timescale dependence of this variation contains information concerning the Brownian motion of the molecules in the sample and is a measure of their translational diffusion coefficient (D_t). D_t is inversely related to the hydrodynamic size of macromolecules (named hydrodynamic radius, R_h). DLS provides information regarding the size and polydispersity index of molecules and particles in solution (Fig. 7.1). DLS is routinely used to determine the size of proteins, nucleic acids, vesicles and macromolecular complexes and is very well suited for the detection of small amounts of aggregates in a sample [31].

From the sedimentation coefficient obtained by SV and the independently measured diffusion coefficient by DLS or FCS the *Mw* of macromolecular species can be determined by using the Svedberg equation [32]:

$$\frac{s}{D_t} = \frac{Mw\,(1 - \nu\rho)}{RT}$$

where ν is the partial specific volume of the macromolecule; ρ is the solvent density; R is the gas constant, and T is the absolute temperature.

DLS technique allows a great variety of sample buffers and a wide range of temperatures. The instrument is composed of a laser light source, a thermostated sample cell, a detector placed at a fixed or variable angle (90° is the standard but other angles can be used), a photomultiplier that amplifies the signal, and a correlator. Samples may be measured after fractionation by size exclusion chromatography (SEC-DLS) or non-fractionated in a cuvette (Batch-DLS). In SEC-DLS the D_t of the different species separated by the column may be precisely determined (e.g., monomers and dimers, that otherwise are indistinguishable in terms of diffusion, can be characterized provided the selected column has the adequate segregation capacity). In the case of Batch-DLS a size distribution may be obtained, but two particles must differ at least twice in their R_h to be differentiated (monomers, dimers, trimers and tetramers appears as a single peak in the size distribution and cannot be distinguished).

DLS buffers must be filtered and degassed, and samples should be filtered or centrifuged for 10–15 min at low speed (11,000 g), to eliminate bubbles and very big aggregates. In Batch-DLS, automated plate readers suitable for the simultaneous analysis of many samples, are available. Only a few microliters per sample (4–15 μL) are needed and samples may be recovered after measurement, if necessary. The minimal concentration of molecules required for a good signal to noise ratio depends on the molecular mass of the macromolecule (e.g., for lysozyme with a *Mw* of 14,700 Da, using a cuvette with 1 cm pathlength, a concentration of at least 0.2 mg/mL is needed). Molecular lower and upper size limits are in the order of 0.5 and 1000 nm, respectively.

7.3.2 Multi-angle Light Scattering (MALS)

When light passes through a sample, some of the incident light is scattered by the particles into new directions. A careful analysis of the scattered light at different angles yields detailed information about the scattering particle. The intensity of scattered light is directly proportional to the mass and concentration of the particles. Multi-angle light scattering (MALS) is one of the most reliable methods for the determination of the mass of the macromolecular complexes [33–38]. It measures the intensity of light scattered by macromolecules in solution detected by different angles (from 2 to 18), allowing the direct determination of their molecular weight. Two main analytical MALS methods are available: size exclusion chromatography coupled to MALS (SEC-MALS) and composition gradient MALS (CG-MALS).

7.3.2.1 Size Exclusion Chromatography Coupled to Multi-angle Light Scattering (SEC-MALS)

SEC-MALS allows the determination of the absolute mass of macromolecules fractionated according to their size, and, for big molecules with scattering angular dependence, also an estimation of the radius of gyration [36, 38]. With the standard configuration, a SEC column is coupled in-line with refractive index (RI), multi-angle light scattering (MALS) and UV-VIS detectors, though they may be also used in parallel to minimize sample dilution. The column must be equilibrated in the running buffer and the signal of the detectors must be stabilized before starting the experiment. Flow rate depends on the column size (e.g., 0.5 mL/min is used for a semi-preparative 10/300 mm column) and the system may be at room temperature, though experiments can be conducted at other temperatures using a thermostated system. A typical experiment usually lasts from 40 to 60 min for a 10/300 column (24 mL bed volume). Depending on the column characteristics, proteins with molecular masses from 200 Da to 10 MDa can be analyzed.

Running buffers must be previously degassed and filtered and should be compatible with the SEC column employed. To avoid unspecific interactions with the matrix of the column, some salt should be added to the running buffer (e.g., 150 mM of NaCl). Samples are equilibrated in the running buffer and sample volume depends on the dimensions of the column. For 10/300 columns, the usual volumes are 20–200 μL (50–500 μg of protein).

SEC-MALS has been successfully used for the characterization, among others, of membrane proteins solubilized with detergents [39], adeno-associated virus [40], and polysaccharide–protein complexes [41]. SEC-MALS is used for protein and polymer characterization in industry [5] for its high reproducibility, short running time and the possibility to automate the whole procedure.

7.3.2.2 Composition Gradient Multi-angle Light Scattering (CG-MALS)

Composition-gradient MALS (CG-MALS) is a technique for quantifying the affinity and stoichiometry of macromolecular interactions. In CG-MALS simultaneous measurements of the intensity of scattered light and of the species concentration are collected (Fig. 7.1), and we can obtain the weight-averaged molar mass of the species in equilibrium, and so the stoichiometry and affinity of the interaction. With that purpose, light scattering and concentration data from the RI detector are continuously acquired from a solution whose composition is being varied with time in a controlled and known fashion. The resulting time-dependent scattering and composition profiles are globally modeled in the context of molecular models for these composition-dependent scattering data [42].

A pump system with three syringes injects, in a sequence and at time intervals required for the programmed concentration gradient, three different components (interacting species A and B, and buffer) into the multi-angle light scattering and concentration detectors, that could be an UV-VIS detector or a RI detector.

CG-MALS set-up is especially useful for the analysis of complex stoichiometries and when the interacting macromolecules exhibit self-association in addition to hetero-association [43]. K_d values from 0.1 μM up to mM may be determined by this technique.

7.4 Fluorescence Spectroscopy Approaches

Fluorescence spectroscopy methods have been widely used to interrogate biomolecular interactions involving proteins, nucleic acids, lipid membranes and small ligands [7, 44, 45]. These methods are generally characterized by their outstanding sensitivity, reaching the detection of single molecules. Because of this notable sensitivity, fluorescence techniques allow assessing the energetic parameters of high affinity interactions under true equilibrium conditions (*i.e.*, at concentrations of the target fluorescent species below the dissociation constant). Complexes requiring high concentrations of their integrating elements to be assembled can also be studied by fluorescence, though. For this purpose, the addition of a small fraction of one of the reactants labeled with a fluorophore, which will act as a tracer representing the whole amount of this species (that is, including the unlabeled fraction) is often a useful strategy (see, for example, [46]). This allows overcoming limitations imposed by the upper detection limit and by the lack of linearity between the fluorescent signal and the fluorophore concentration above a certain concentration threshold.

Fluorescence methods have been applied to study protein dynamics and interactions making efficient use of the intrinsic fluorescence displayed by many proteins harboring tryptophan residues [47]. In other instances, labeling of the protein, nucleic acid, ligand or lipid with an extrinsic dye is either necessary or more convenient, depending on the kind of study intended [45, 48]. The possibility of including labeled species is particularly attractive for measurements in complex media and, in fact, it is one of the reasons of the success of fluorescence approaches for quantitative determinations in crowding solutions, reconstituted cell-like systems, fixed or live cells and even in tissues and multicellular organisms [49–52]. Among the extrinsic dyes, many different small organic molecules are commercially available, modified with suitable reactive groups to be conjugated with proteins, usually at amino groups (N-terminus or lysine residues) or at cysteine residues. Additionally, fluorescently labeled oligonucleotides, lipids and certain small ligands can be purchased from various companies. Proteins labeled by fusion with a fluorescent protein are routinely used for measurements in cells although, occasionally, this type of labeling has also been used for *in vitro* measurements in solution and in reconstituted cell-like systems. Selection of the fluorophore or fluorophores is a key step in the design of a fluorescence-based assay and it should consider the specific requirements of the fluorescence method (number of fluorophores, excitation and emission wavelengths of each fluorophore, level of photostability, quantum yield, etc.) but also the type of interaction under study and the complexity of the system [48, 53]. This is important to minimize the contribution of background species to the fluorescence signal, and to avoid alterations in the target interaction introduced by the dye.

There are many different fluorescence methods that can be applied individually or, more efficiently, combined with each other or with additional biophysical or structural methods to characterize biomolecular complexes. Changes in the fluorescence emission subsequent to the interaction of a molecule bearing a fluorophore with a partner can be of use to monitor this interaction employing a regular spectrofluorometer [45]. When complexes larger than the free species are formed, methods such as fluorescence anisotropy or fluorescence correlation spectroscopy (FCS) can be applied to determine the affinity and the possible cooperativity of the interaction.

Fluorescence anisotropy [54] requires an instrument equipped with polarizers and it is very sensitive to size changes as it depends on the rotational diffusion of the species. Aside from the global tumbling, local motions of the dye would also contribute to the measured steady-state anisotropy and, by performing time-resolved measurements, these kinds of motions can be identified and quantified. Anisotropy methods have been profusely utilized to unravel the molecular mechanisms of processes in which protein-protein, protein-nucleic acid or protein-ligand complexes intervene. A powerful approach to this end entails the parallel application of analytical ultracentrifugation or light scattering methods for the independent determination of stoichiometries [55, 56]. Suitable models to analyze the anisotropy isotherms (Fig. 7.1) based on these stoichiometries could be subsequently built by the use of software such as BIOEQS [57–59].

FCS measurements, on the other hand, rely on the analysis of the fluctuations in the intensity arising from the fluorescent species within a small open volume created by a laser beam focused into the sample through the objective of a microscope [60–62]. From these fluctuations, autocorrelation curves are obtained (Fig. 7.1) that contain information about the concentration and translational diffusion of the fluorescent species under study. FCS is very useful for the analysis of protein aggregation and polymerization events, for which anisotropy is more limited. In contrast, because it depends on the translational diffusion, FCS sensitivity to the formation of complexes moderately larger than the free species is, as with DLS, lower than in the case of anisotropy. Interactions can be also detected and quantified by using the two-color version of FCS, fluorescence cross-correlation spectroscopy (FCCS [63, 64]) with the only requirement that the interacting species, labeled with spectrally different dyes, move together upon complexation (*i.e.*, there is no minimum size change needed for the complexes to be detected). In addition to the autocorrelation curves corresponding to the fluorescence fluctuations in each channel, a cross-correlation curve is obtained in FCCS upon formation of complexes bearing the two fluorophores (Fig. 7.1). In the absence of artifacts, the appearance of this cross-correlation curve proves interaction, and its amplitude, relative to that of the autocorrelation curves, allows affinity evaluation as well as assessment of the stoichiometry of the complexes formed [65–67].

For additional details on the above mentioned fluorescence methodologies, their experimental requirements and their application to study interactions, the readers are referred to a comprehensive and didactic review [7], in which another popular fluorescence method, Förster Resonance Energy Transfer (FRET) is also discussed.

Aside from the usefulness of fluorescence methods to unravel mechanisms of interaction in the context of fundamental research schemes, they also find many applications in biotechnology and biomedicine. Thus, fluorescence approaches are among those of choice for the development of screening assays, due to the low amount of sample required and the rapidity of the measurements, which can often be performed using plate readers [45]. Many examples of the use of anisotropy and FCS to develop systematic screening assays *in vitro* and *in vivo* can be found in the literature [68–71].

7.5 Global Application of these Techniques to the Study of Complex Formation

7.5.1 Quantitative Analysis of the Self-Association and Activation of the Bacterial Chaperone ClpB

In the following example, modulation of the functional oligomerization of the protein ClpB was addressed by using a combination of analytical ultracentrifugation and light scattering techniques. ClpB is a molecular chaperone that belongs to the Hsp100 family of ring-forming heat-shock proteins involved in the protein quality control system in bacteria. This protein, in combination with the DnaK system, disaggregates unfolded and aggregated proteins reactivating protein aggregates. ClpB was previously described to self-associate from monomers to form hexamers (the functional unit) but it was not known how factors such as protein and salt concentration and natural ligand (ADP and ATP) binding modulate this equilibrium. The techniques SE, SV and CG-MALS were applied to get some insight into these aspects [72].

SV was first applied to characterize samples of ClpB at two different salt conditions and to evaluate their polydispersity. Profiles showed that at low ionic strength (50 mM KCl) 10 μM ClpB sedimented as a single peak compatible with the protein hexamer, and at high ionic strength (500 mM KCl) as a single peak compatible with protein monomer. The presence of nucleotide (ADP or ATP) induced a slight decrease of the sedimentation coefficient but had no significant effect on the association state that at low ionic strength remained mainly hexameric. To further analyze the association scheme of the protein and its modulation by ligands, two techniques were employed: SE and CG-MALS. These techniques allowed direct determination of the masses, overcoming the limitations of SV for this purpose, especially in case several species are present, and determination of the molecular masses is not possible. Application of these methods benefited from previous analysis by SV, according to which different species populated the solutions under each condition tested, meaning that the average masses retrieved could be assigned to a defined species or a mixture of them. The dependence of the oligomerization state of ClpB on protein and salt concentration (50–500 mM KCl) as well as on the presence of the natural ligands were subsequently determined by SE and CG-MALS. Decrease of ionic strength induced displacement of the equilibrium isotherms toward the formation of hexamers.

The standard Gibbs free energy change calculated from the ClpB hexamerization constant showed that ATP-bound ClpB was 10 kJ/mol more stable than the ADP state. Nucleotide exchange might promote the conformational change of the protein that drives its functional cycle.

7.5.2 Untangling the Central Bacterial Division Protein FtsZ Polymers

The characterization of the self-association of FtsZ [23], the major component of the bacterial division ring whose reversible GTP-induced polymerization plays a key role in cytokinesis [73, 74] represents a good example of how the application of multi-parametric approaches combining several orthogonal methodologies can aid in the challenging investigation of fibrils of medium-large size. The difficulty in the analysis of FtsZ assemblies is further enhanced by the variability in the degree of polymerization and the final arrangement of these dynamic polymers, very sensitive to the specific conditions under which they are formed.

Initial studies on FtsZ polymers triggered by GTP or GMPCPP (a slowly hydrolysable analogue of GTP), conducted at 500 mM KCl and 5 mM $MgCl_2$, showed single peaks in SV profiles, with faster sedimentation for the GMPCPP polymers (Fig. 7.2a). This was compatible with either a larger mass, or an equal mass but a more compact shape or a more flexible structure for the polymers induced by the analog. The ambiguity was solved through complementary FCS and DLS measurements (Fig. 7.2c), rendering D_t-values that together with the s-values were used to calculate, via Svedberg equation (see above), the molecular mass of the polymers. This estimation indicated that the difference observed in the sedimentation of the polymers was related to their different mass. The joint application of FCS and DLS allowed bypassing the uncertainties associated with each method in the non-straightforward analysis of solutions containing polymers. Thus, while FCS measured the polymer diffusion, the profiles were also sensitive to the presence of unassembled protein, which besides providing additional characterization of the system, increases the number of fitting parameters and hence the uncertainty of the analysis. DLS autocorrelation functions, on the other hand, were mostly contributed by the bigger species (given the substantially smaller size of unassembled species), hence providing accurate measurements of the diffusion of the polymer. However, DLS measurements are highly sensitive to dust and any kind of unwanted aggregates that might be present in the solution. The good agreement between DLS and FCS data allowed validating the models used for analysis in each case and discarding an influence of the dye used to label FtsZ on the FCS determinations.

Multi-angle light scattering measurements, in the concentration-gradient setup, were undertaken to obtain an independent and direct determination of the molar mass of the polymers. In this case, experimental data analysis involved applying a model that considers the dependence of the intensity of the scattered light with the protein concentration and detection angle (Fig. 7.2b). The obtained masses were similar to those from the combination of s- and D_t-values, confirming the higher mass for the GMPCPP polymers. This dependence analysis also allowed

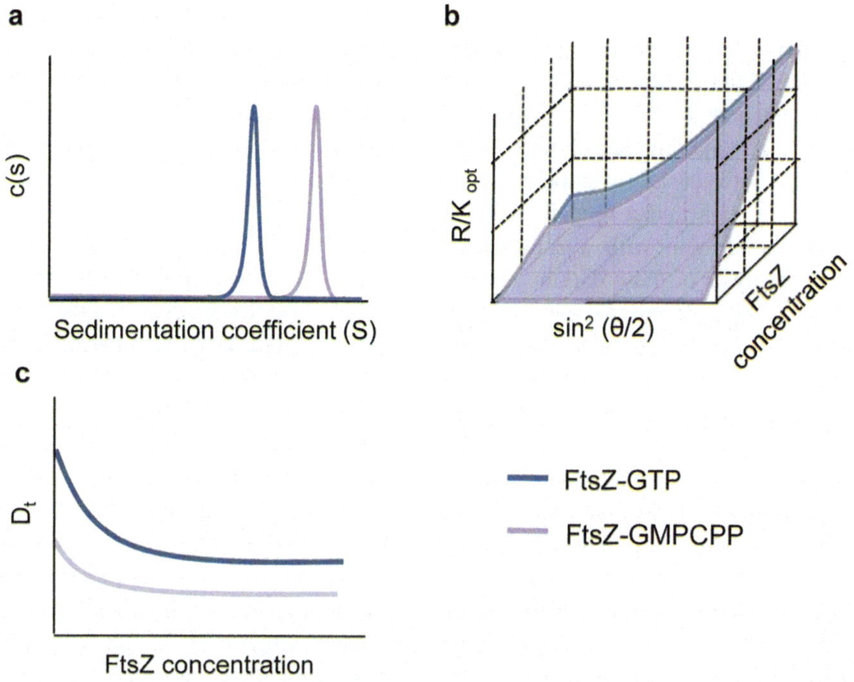

Fig. 7.2 Characterization of FtsZ polymers in the presence of GTP and GMPCPP using several biophysical techniques. Illustration of (**a**) FtsZ sedimentation velocity profiles in the presence of either GTP or GMPCPP, (**b**) angular and concentration dependence of scattering of FtsZ polymers and (**c**) dependence with FtsZ concentration of translational diffusion properties of FtsZ polymers formed by GTP or GMPCPP

getting an estimate of the dimensions of FtsZ polymers by the calculation of the radius of gyration, somewhat larger for GMPCPP-FtsZ polymers. The molecular weight distributions calculated from *c*(*s*) distributions and externally determined frictional ratios via Svedberg [18] were in good agreement with the results derived from the other techniques employed.

The described orthogonal approach was then used to determine the impact of protein, salt and magnesium concentrations on the FtsZ polymers. Within the 0.4–1.5 g/L protein concentration interval, at 5 mM $MgCl_2$ and 500 mM KCl, the polymers triggered by GTP and GMPCPP displayed constant *s*- and D_t-values and hence constant mass, suggesting a concerted formation of preferred FtsZ fibrils and highlighting the qualitatively similar self-association schemes for both kinds of polymers despite of their different size. At lower KCl concentrations the GTP polymers were smaller but still narrowly distributed in size, in excellent agreement with reported electron microscopy (EM) data, while the GMPCPP polymers increased in size and heterogeneity [75]. The GTP polymers exhibited their characteristic narrow size distribution in the 5–0.25 mM $MgCl_2$ interval. However, at lower $MgCl_2$ concentrations, SV showed two main species with *s*-values smaller than those of the polymer and a significant increase in D_t was found by DLS and FCS, indicating a smaller size. The GMPCPP polymers were less sensitive to the $MgCl_2$ concentration, and no polymers were detected in solutions lacking Mg^{2+} regardless which nucleotide was present.

An increase in the amount of unassembled FtsZ was observed by SV and FCS upon lowering Mg^{2+} concentration when polymerization was elicited by GTP. This suggested an influence of the cation on the critical concentration of assembly (C_c), which is a threshold concentration above which the protein assembles into polymers. 90° scattering and fluorescence anisotropy were independently applied to verify this hypothesis (Fig. 7.3). The 90° scattering data showed a concerted transition between a low-slope region at low FtsZ concentration and a second region at higher FtsZ concentration with a substantially greater slope. Analysis of these data in terms of a solubility model considering two main scatterer species (a narrow distribution of low molecular weight scatterers at concentrations below the solubility value, and a high molecular weight scatterer of concentration-independent mean size at concentrations above it) determined a drastic drop in the C_c values for GTP-FtsZ upon increasing $MgCl_2$ concentration (Fig. 7.3a). The impact of Mg^{2+} on the C_c was further explored by a fluorescence anisotropy assay in which fluorescently labeled FtsZ was titrated with increasing concentration of the unlabeled protein [46]. The formation of polymers resulted in an anisotropy increase due, among other factors, to their larger size compared with that of the unassembled protein. Analysis of the anisotropy dependence with protein concentration allowed determination of the C_c, indicating a substantial decrease with higher magnesium concentration (Fig. 7.3b), in good agreement with 90° scattering analysis.

This example shows how the use of hydrodynamic and thermodynamic methods in a complementary fashion allows overcoming the uncertainties associated to each method when individually applied to the system, thus providing very reliable quantitative information on FtsZ polymers. These detailed analyses were crucial for the subsequent characterization of the mechanisms used by different modulators of FtsZ assembly in the context of division, for example, antagonists involved in the positioning of the FtsZ ring at mid-cell [76], like MinC and SlmA [56, 77]. They also represented a starting point to analyze the influence of factors such as the crowded and heterogeneous nature of the bacterial cytoplasm on FtsZ polymerization. Along this line, a study based on fluorescence anisotropy and light scattering evidenced an enhancement of the FtsZ tendency to polymerize in the presence of single crowders or mixtures mimicking the heterogeneity of the intracellular environment. The effect was more pronounced with negatively charged proteins or DNA than with neutral polymers or positively charged proteins, and mostly nonadditive for mixed crowding agents [78].

7.5.3 Unravelling the Interactions of the Bacterial Division Protein FtsZ with the Membrane Anchor ZipA Solubilized in Nanodiscs

Biophysical techniques such as the ones described here can be applied to study the interactions of membrane proteins by using phospholipid bilayer nanodiscs to keep them in solution. Nanodiscs are highly soluble and stable particles formed by a phospholipids bilayer surrounding a target membrane protein, the whole structure being stabilized by a membrane scaffold protein [79] providing a native-like membrane environment.

In this example the membrane protein ZipA, a component of the division machinery that provides membrane tethering to the central cell division FtsZ protein [76], was included in nanodiscs (Nd-ZipA) where the phospholipid bilayer consisted in a mixture matching the lipid composition of the *E. coli* inner membrane. The interactions of Nd-ZipA with FtsZ in the absence of GTP, under conditions in which the protein is forming oligomers (GDP-FtsZ) and with the FtsZ polymers elicited by GTP were analyzed by SV, SE and FCS techniques [80].

Nd-ZipA were purified by SEC and the incorporation of ZipA to nanodiscs was initially verified by electrophoretic analysis. To ascertain the incorporation of a single copy of ZipA into the biomimetic membrane, empty nanodiscs (Nd) and Nd-ZipA were analyzed by analytical ultracentrifugation. SV showed that both sedimented as single species with different sedimentation coefficients (Fig. 7.4a). SE experiments were performed in parallel to determine the stoichiometry of the Nd-ZipA complexes. The comparative analysis of empty

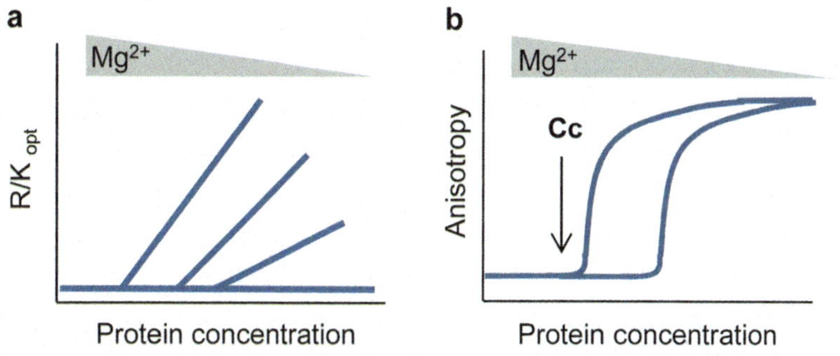

Fig. 7.3 Determination of the critical concentration (C_c) of polymerization of FtsZ. Illustration of (**a**) concentration dependence of GTP-FtsZ static light scattering at 90° and (**b**) steady-state fluorescence anisotropy measurements of FtsZ-GTP as a function of protein concentration

and ZipA nanodiscs demonstrated that the Nd-ZipA complex contained one molecule of ZipA incorporated in each nanodisc. The empty nanodiscs and Nd-ZipA were further characterized by FCS and DLS, which showed a slight decrease in the translational diffusion coefficient upon incorporation of ZipA (Fig. 7.4b). In both cases, the profiles obtained were compatible with single homogeneous species. For FCS measurements, a small fraction of lipids labeled with a fluorophore was incorporated in the lipid mixture used to generate the nanodiscs.

A previous study of the hetero-association of a soluble mutant of ZipA, lacking the transmembrane region (sZipA) to GDP-FtsZ [43], indicated that independently of the oligomer size the binding affinity was moderate, and only one sZipA was bound to any of the different FtsZ species present in the solution (monomer to hexamer). The analysis by SE showed that the addition of increasing concentrations of GDP-FtsZ oligomers to fluorescently labeled Nd-ZipA resulted in complexes of higher molar mass with a similar association model as the one describing the binding of GDP-FtsZ to sZipA, a single Nd-ZipA molecule binding to one FtsZ oligomer independently of the oligomer size.

The interaction of Nd-ZipA with FtsZ polymers was assessed by FCS, using nanodiscs labeled in the lipids and unlabeled FtsZ polymers. Formation of complexes was evidenced by the large decrease in the speed of diffusion of Nd-ZipA, subsequent to the interaction with the FtsZ polymers triggered by GTP (Fig. 7.4b). Titration of FtsZ polymers into solutions containing Nd-ZipA allowed quantification of the apparent affinity of binding, which turned out to be similar to that for GDP-FtsZ oligomers. Physical interaction between ZipA and FtsZ in the complexes was proved by FCCS, using Nd-ZipA and FtsZ labeled with spectrally different fluorophores, namely lissamine rhodamine B and Alexa Fluor 488. FCCS experiments showed significant cross-correlation for Nd-ZipA but not for the empty nanodiscs (Fig. 7.4c), confirming an interaction involving FtsZ/ZipA contacts without interference from the lipids and/or the scaffold protein in the nanodiscs. Sedimentation velocity experiments using the labeled Nd-ZipA also evidenced an interaction with FtsZ polymers, resulting in a notable increase in the sedimentation coefficient of the reconstituted membrane protein (Fig. 7.4a). The usefulness of the assays developed to study the ZipA/FtsZ interactions for the identification of molecules interfering with them was showed by including in the reaction mixture a peptide known to disrupt the complexes through competition with FtsZ for the interaction with ZipA.

This work evidenced the little effect, if any, that the transmembrane region of ZipA has on the formation of the complex with FtsZ. Likewise, it

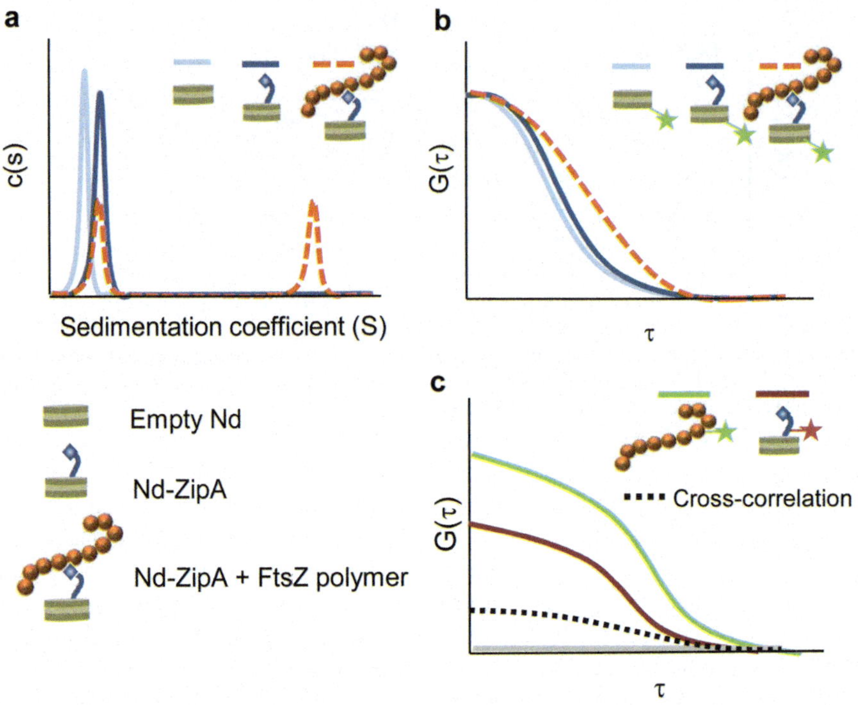

Fig. 7.4 Characterization of the interaction between ZipA embedded in nanodiscs and FtsZ polymers. Illustration of (**a**) sedimentation velocity analysis of empty Nd and of Nd-ZipA, in the absence and presence of GTP-induced FtsZ polymers, (**b**) normalized FCS autocorrelation profiles for Nd and for Nd-ZipA in the absence and presence of GTP-FtsZ polymers and (**c**) FCCS analysis of the interaction of Nd-ZipA with FtsZ polymers labeled with spectrally different dyes. The flat curve indicates the absence of cross-correlation from samples containing empty nanodiscs and FtsZ polymers. Stars indicate fluorophores

shows that FtsZ interaction with ZipA is the same disregarding the nucleotide present, despite the different tendency of GDP-FtsZ and GTP-FtsZ to self-assemble.

7.5.4 Protein-DNA Complexes: The Interaction Between the Repressor Protein Reg576 and Its DNA Operator

An explanatory example of the power of AUC to characterize protein-DNA hetero-associations in solution is provided by the study of the oligomerization state of the *Bacillus pumilus* repressor protein Reg_{576} alone and once bound to its DNA operator sequence [81]. The study was approached combining MSSV, SE and DLS techniques. In a first step SV and DLS were used in parallel to ascertain the oligomerization state of Reg_{576} alone. At different concentrations Reg_{576} behaved in SV assays as a single species with an experimental sedimentation coefficient (*s*) compatible with the theoretical mass of the nearly globular Reg_{576} dimer. DLS analysis of Reg_{576} yielded a translational diffusion coefficient, which once introduced with the obtained *s* coefficient into the Svedberg equation, resulted in an apparent molar mass very close to the molecular mass of the Reg_{576} dimer. This mass was confirmed by SE assays with different concentrations of Reg_{576} that showed a mass matching the molecular weight of the protein dimer.

Once established the oligomerization state of Reg_{576} in solution, the next step was to ascertain the way Reg_{576} dimers bind to their DNA operator. With this aim and taking advantage of the different extinction coefficients of Reg_{576} and the DNA, MSSV experiments using samples containing only the DNA fragment, Reg_{576} alone and the DNA fragment with an excess of Reg_{576} were carried out. Three types of DNA fragments were used: the promoter with its two intact flanking Reg_{576} operators, or derivatives in which only one or both flanking operators were mutated.

As expected, the different DNA fragments in the absence of Reg_{576} showed the same *s*-value, because they had nearly the same molecular mass. The presence of Reg_{576} did not modify the *s*-value of the DNA fragment when both operators were mutated, indicating the absence of interaction. On the contrary, in the presence of Reg_{576}, a major *s*-value increment occurred for the DNA fragment having both operators intact, and a moderate increment for the fragment with a single operator mutated (Fig. 7.5a). The differential increase in the *s*-value suggested that the fragment with intact operators bound more Reg_{576} dimers than the fragment with one mutated operator. To address this issue the simultaneous absorbance data acquisition at 230 and 260 nm was globally analyzed through SEDPHAT to get the diffusion-deconvoluted sedimentation coefficient distributions with spectral deconvolution of the absorbance signals, $c_K(s)$. The MSSV analysis of Reg_{576}-Intact DNA complex indicated that the areas under the peaks corresponded to a

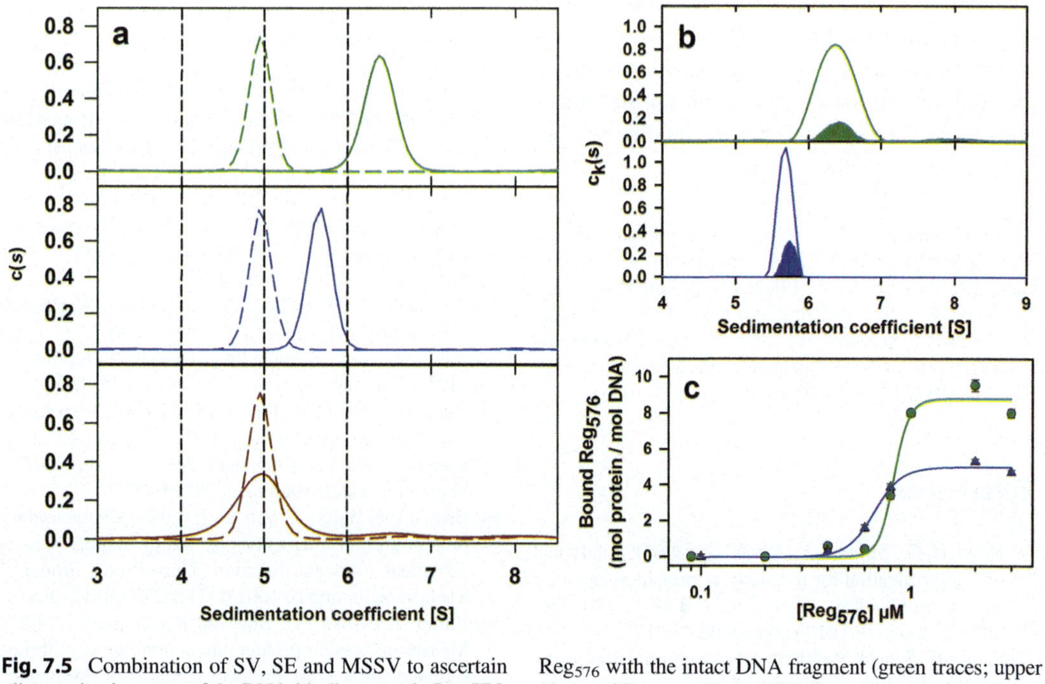

Fig. 7.5 Combination of SV, SE and MSSV to ascertain oligomerization state of the DNA-binding protein Reg576. (**a**) Sedimentation coefficient distributions, *c*(*s*), obtained from SV assays at 260 nm for the DNA fragment with intact operators (green traces; upper plot), the DNA with one mutated operator (blue traces; middle plot) and the DNA fragment with both operators mutated (brown traces; lower plot) showing the shift in the *s*-value of Reg_{576}-DNA complexes (solid trace) relative to the corresponding DNA alone (dashed trace). (**b**) Global multi-wavelength analysis of the complexes resulting from the interaction of Reg_{576} with the intact DNA fragment (green traces; upper plot) and Reg_{576} with the DNA fragment with one mutated operator (blue traces; lower plot) and decomposition into component sedimentation coefficient distributions, $c_K(s)$, for Reg_{576} (empty peaks) and the different DNA fragments (solid peaks). (**c**) Binding isotherms for the interaction of Reg_{576} with the DNA fragment with intact operators (green circles) and the DNA fragment with one mutated operator (blue triangles). The solid curves represent the best fit of the three-parameters Hill equation to the SE experimental data

stoichiometry of 7.9 moles of Reg_{576} bound per mol of DNA. The ratios of Reg_{576} moles with respect to the single mutated DNA fragment was 4.3 (Fig. 7.5b).

To corroborate the calculated stoichiometries determined by MSSV, sedimentation equilibrium experiments were carried out with a fixed DNA concentration titrated with increasing Reg_{576} concentrations. This technique allows determining the exact molecular weight of the complexes formed and, since the molecular weight of the DNA and protein components are known, the obtained masses can be used directly to calculate the number of Reg_{576} molecules bound to a DNA fragment. Figure 7.5c shows the binding isotherms built from the experimental molar mass increments obtained by SE at low speed and 260 nm, through an empirical three parameters Hill plot. These results, together with those obtained by MSSV, clearly confirm that two Reg_{576} dimers bind to one functional operator. In addition, these results confirm that mutation of one operator abolishes the binding of both Reg_{576} dimers to this operator.

Acknowledgments This work was supported by the Spanish Ministerio de Ciencia e Innovacion (grant numbers 2023AEP105 and PID2019-104544GB-I00/AEI/10.13039/501100011033 to C.A., B.M., G.R. and S.Z.). M.S.S. was supported by the Agencia Estatal de Investigación and the European Social Fund (grant number PTA2020-018219-I/AEI/10.13039/501100011033).

References

1. Howlett G, Minton A, Rivas G (2006) Analytical ultracentrifugation for the study of protein association and assembly. Curr Opin Chem Biol 10:430–436. https://doi.org/10.1016/j.cbpa.2006.08.017
2. Scott DJ, Schuck P (2005) A brief introduction to the analytical ultracentrifugation of proteins for beginners. In: Scott DJ, Harding SE, Rowe AJ (eds) Analytical ultracentrifugation: techniques and methods. Royal Society of Chemistry, Cambridge
3. Schuck P (2003) On the analysis of protein self-association by sedimentation velocity analytical ultracentrifugation. Anal Biochem 320:104–124. https://doi.org/10.1016/S0003-2697(03)00289-6
4. Cole JL, Lary JW, Moody T, Laue TM (2008) Analytical ultracentrifugation: sedimentation velocity and sedimentation equilibrium. In: Methods in cell biology. Elsevier, pp 143–179
5. Some D, Razinkov V (2019) High-throughput analytical light scattering for protein quality control and characterization. In: Vincentelli R (ed) High-throughput protein production and purification. Springer, New York, pp 335–359
6. Minton AP (2016) Recent applications of light scattering measurement in the biological and biopharmaceutical sciences. Anal Biochem 501:4–22. https://doi.org/10.1016/j.ab.2016.02.007
7. Royer CA, Scarlata SF (2008) Fluorescence approaches to quantifying biomolecular interactions. In: Methods in enzymology. Elsevier, pp 79–106
8. Rivera-Calzada A, Pal M, Muñoz-Hernández H, Luque-Ortega JR, Gil-Carton D, Degliesposti G, Skehel JM, Prodromou C, Pearl LH, Llorca O (2017) The structure of the R2TP complex defines a platform for recruiting diverse client proteins to the HSP90 molecular chaperone system. Structure 25:1145–1152.e4. https://doi.org/10.1016/j.str.2017.05.016
9. Alvira S, Cuéllar J, Röhl A, Yamamoto S, Itoh H, Alfonso C, Rivas G, Buchner J, Valpuesta JM (2014) Structural characterization of the substrate transfer mechanism in Hsp70/Hsp90 folding machinery mediated by Hop. Nat Commun 5:5484. https://doi.org/10.1038/ncomms6484
10. Bartual SG, Straume D, Stamsås GA, Muñoz IG, Alfonso C, Martínez-Ripoll M, Håvarstein LS, Hermoso JA (2014) Structural basis of PcsB-mediated cell separation in Streptococcus pneumoniae. Nat Commun 5:3842. https://doi.org/10.1038/ncomms4842
11. Lebowitz J, Lewis MS, Schuck P (2002) Modern analytical ultracentrifugation in protein science: a tutorial review. Protein Sci Publ Protein Soc 11:2067–2079. https://doi.org/10.1110/ps.0207702
12. Alfonso C, Del Castillo U, Martín I, Muga A, Rivas G (2015) Sedimentation equilibrium analysis of ClpB self-association in diluted and crowded solutions. In: Methods in enzymology. Elsevier, pp 135–160
13. Rivas G, Stafford W, Minton AP (1999) Characterization of heterologous protein–protein interactions using analytical ultracentrifugation. Methods 19:194–212. https://doi.org/10.1006/meth.1999.0851
14. Brown PH, Balbo A, Schuck P (2008) Characterizing protein-protein interactions by sedimentation velocity analytical ultracentrifugation. Curr Protoc Immunol 81. https://doi.org/10.1002/0471142735.im1815s81
15. Balbo A, Brown PH, Braswell EH, Schuck P (2007) Measuring protein-protein interactions by equilibrium sedimentation. Curr Protoc Immunol 79. https://doi.org/10.1002/0471142735.im1808s79
16. Dam J, Schuck P (2004) Calculating sedimentation coefficient distributions by direct modeling of sedimentation velocity concentration profiles. In: Methods in enzymology. Elsevier, pp 185–212
17. Schuck P, Demeler B (1999) Direct sedimentation analysis of interference optical data in analytical ultracentrifugation. Biophys J 76:2288–2296. https://doi.org/10.1016/S0006-3495(99)77384-4
18. Schuck P (2000) Size-distribution analysis of macromolecules by sedimentation velocity

ultracentrifugation and lamm equation modeling. Biophys J 78:1606–1619. https://doi.org/10.1016/S0006-3495(00)76713-0

19. Schuck P, Perugini MA, Gonzales NR, Howlett GJ, Schubert D (2002) Size-distribution analysis of proteins by analytical ultracentrifugation: strategies and application to model systems. Biophys J 82: 1096–1111. https://doi.org/10.1016/S0006-3495(02)75469-6
20. Demeler B (2005) UltraScan – a comprehensive data analysis software package for analytical ultracentrifugation experiments. In: Scott D, Harding SE, Rowe A (eds) Analytical ultracentrifugation: techniques and methods. The Royal Society of Chemistry, pp 210–230
21. van Holde KE (1985) Sedimentation. In: Physical biochemistry. Prentice-Hall, Englewood Cliffs, NJ, pp 112–114
22. Laue TM, Shah B, Ridgeway TM, Pelletier SL (1992) Computer-aided interpretation of sedimentation data for proteins. In: Harding SE, Horton JC, Rowe AJ (eds) Analytical ultracentrifugation in biochemistry and polymer science. Royal Society of Chemistry, Cambridge, pp 90–125
23. Monterroso B, Alfonso C, Zorrilla S, Rivas G (2013) Combined analytical ultracentrifugation, light scattering and fluorescence spectroscopy studies on the functional associations of the bacterial division FtsZ protein. Methods 59:349–362. https://doi.org/10.1016/j.ymeth.2012.12.014
24. Balbo A, Minor KH, Velikovsky CA, Mariuzza RA, Peterson CB, Schuck P (2005) Studying multiprotein complexes by multisignal sedimentation velocity analytical ultracentrifugation. Proc Natl Acad Sci 102:81–86. https://doi.org/10.1073/pnas.0408399102
25. Alcorlo M, Jiménez M, Ortega A, Hermoso JM, Salas M, Minton AP, Rivas G (2009) Analytical ultracentrifugation studies of phage ϕ29 protein p6 binding to DNA. J Mol Biol 385:1616–1629. https://doi.org/10.1016/j.jmb.2008.11.044
26. Cole JL (2004) Analysis of heterogeneous interactions. In: Methods in enzymology. Elsevier, pp 212–232
27. Vistica J, Dam J, Balbo A, Yikilmaz E, Mariuzza RA, Rouault TA, Schuck P (2004) Sedimentation equilibrium analysis of protein interactions with global implicit mass conservation constraints and systematic noise decomposition. Anal Biochem 326:234–256. https://doi.org/10.1016/j.ab.2003.12.014
28. Schuck P, Schubert D (1991) Band 3-hemoglobin associations the band 3 tetramer is the oxyhemoglobin binding site. FEBS Lett 293:81–84. https://doi.org/10.1016/0014-5793(91)81156-3
29. Fleming KG, Ackerman AL, Engelman DM (1997) The effect of point mutations on the free energy of transmembrane α-helix dimerization. J Mol Biol 272: 266–275. https://doi.org/10.1006/jmbi.1997.1236
30. Le Maire M, Champeil P, Møller JV (2000) Interaction of membrane proteins and lipids with solubilizing detergents. Biochim Biophys Acta BBA Biomembr 1508:86–111. https://doi.org/10.1016/S0304-4157(00)00010-1
31. Stetefeld J, McKenna SA, Patel TR (2016) Dynamic light scattering: a practical guide and applications in biomedical sciences. Biophys Rev 8:409–427. https://doi.org/10.1007/s12551-016-0218-6
32. van Holde KE, Johnson WC, Ho PS (2006) Principles of physical biochemistry, 2nd edn. Pearson/Prentice Hall, Upper Saddle River
33. Yamaguchi T, Adachi K (2002) Hemoglobin equilibrium analysis by the multiangle laser light-scattering method. Biochem Biophys Res Commun 290:1382–1387. https://doi.org/10.1006/bbrc.2002.6362
34. Attri AK, Minton AP (2005) New methods for measuring macromolecular interactions in solution via static light scattering: basic methodology and application to nonassociating and self-associating proteins. Anal Biochem 337:103–110. https://doi.org/10.1016/j.ab.2004.09.045
35. Kameyama K, Minton AP (2006) Rapid quantitative characterization of protein interactions by composition gradient static light scattering. Biophys J 90:2164–2169. https://doi.org/10.1529/biophysj.105.074310
36. Folta-Stogniew E, Williams KR (1999) Determination of molecular masses of proteins in solution: implementation of an HPLC size exclusion chromatography and laser light scattering service in a core laboratory. J Biomol Tech 10:51–63
37. Folta-Stogniew E (2006) Oligomeric states of proteins determined by size-exclusion chromatography coupled with light scattering, absorbance, and refractive index detectors. In: New and emerging proteomic techniques. Humana Press, Totowa, pp 97–112
38. Wyatt PJ (1993) Light scattering and the absolute characterization of macromolecules. Anal Chim Acta 272:1–40. https://doi.org/10.1016/0003-2670(93)80373-S
39. Slotboom DJ, Duurkens RH, Olieman K, Erkens GB (2008) Static light scattering to characterize membrane proteins in detergent solution. Methods 46:73–82. https://doi.org/10.1016/j.ymeth.2008.06.012
40. McIntosh NL, Berguig GY, Karim OA, Cortesio CL, De Angelis R, Khan AA, Gold D, Maga JA, Bhat VS (2021) Comprehensive characterization and quantification of adeno associated vectors by size exclusion chromatography and multi angle light scattering. Sci Rep 11:3012. https://doi.org/10.1038/s41598-021-82599-1
41. Tao Y, Zhang L (2008) Characterization of polysaccharide–protein complexes by size-exclusion chromatography combined with three detectors. Carbohydr Res 343:2251–2257. https://doi.org/10.1016/j.carres.2008.04.030
42. Attri AK, Minton AP (2005) Composition gradient static light scattering: a new technique for rapid detection and quantitative characterization of reversible macromolecular hetero-associations in solution. Anal Biochem 346:132–138. https://doi.org/10.1016/j.ab.2005.08.013

43. Martos A, Alfonso C, López-Navajas P, Ahijado-Guzmán R, Mingorance J, Minton AP, Rivas G (2010) Characterization of self-association and Heteroassociation of bacterial cell division proteins FtsZ and ZipA in solution by composition gradient–static light scattering. Biochemistry 49:10780–10787. https://doi.org/10.1021/bi101495x
44. Deshayes S, Divita G (2013) Fluorescence technologies for monitoring interactions between biological molecules in vitro. In: Progress in molecular biology and translational science. Elsevier, pp 109–143
45. Eccleston JF, Hutchinson JP, Jameson DM (2005) Fluorescence-based assays. In: Progress in medicinal chemistry. Elsevier, Amsterdam, pp 19–48
46. Reija B, Monterroso B, Jiménez M, Vicente M, Rivas G, Zorrilla S (2011) Development of a homogeneous fluorescence anisotropy assay to monitor and measure FtsZ assembly in solution. Anal Biochem 418:89–96. https://doi.org/10.1016/j.ab.2011.07.001
47. Royer CA (2006) Probing protein folding and conformational transitions with fluorescence. Chem Rev 106: 1769–1784. https://doi.org/10.1021/cr0404390
48. Toseland CP (2013) Fluorescent labeling and modification of proteins. J Chem Biol 6:85–95. https://doi.org/10.1007/s12154-013-0094-5
49. Zorrilla S, Lillo M (2009) Quantitative investigation of biomolecular interactions in crowded media by fluorescence spectroscopy, a good choice. Curr Protein Pept Sci 10:376–387. https://doi.org/10.2174/138920309788922225
50. Yamamoto J, Matsui A, Gan F, Oura M, Ando R, Matsuda T, Gong JP, Kinjo M (2021) Quantitative evaluation of macromolecular crowding environment based on translational and rotational diffusion using polarization dependent fluorescence correlation spectroscopy. Sci Rep 11:10594. https://doi.org/10.1038/s41598-021-89987-7
51. Macháň R, Wohland T (2014) Recent applications of fluorescence correlation spectroscopy in live systems. FEBS Lett 588:3571–3584. https://doi.org/10.1016/j.febslet.2014.03.056
52. Royer CA (2019) Characterizing proteins in their cellular environment: examples of recent advances in quantitative fluorescence microscopy. Protein Sci 28: 1210–1221. https://doi.org/10.1002/pro.3630
53. Zhang Z, Yomo D, Gradinaru C (2017) Choosing the right fluorophore for single-molecule fluorescence studies in a lipid environment. Biochim Biophys Acta BBA Biomembr 1859:1242–1253. https://doi.org/10.1016/j.bbamem.2017.04.001
54. Jameson DM, Ross JA (2010) Fluorescence polarization/anisotropy in diagnostics and imaging. Chem Rev 110:2685–2708. https://doi.org/10.1021/cr900267p
55. Zorrilla S, Doan T, Alfonso C, Margeat E, Ortega A, Rivas G, Aymerich S, Royer CA, Declerck N (2007) Inducer-modulated cooperative binding of the tetrameric CggR repressor to operator DNA. Biophys J 92:3215–3227. https://doi.org/10.1529/biophysj.106.095109
56. Cabré EJ, Monterroso B, Alfonso C, Sánchez-Gorostiaga A, Reija B, Jiménez M, Vicente M, Zorrilla S, Rivas G (2015) The nucleoid occlusion SlmA protein accelerates the disassembly of the FtsZ protein polymers without affecting their GTPase activity. PLoS One 10:e0126434. https://doi.org/10.1371/journal.pone.0126434
57. Rosales T, Royer CA (2008) A graphical user interface for BIOEQS: a program for simulating and analyzing complex biomolecular interactions. Anal Biochem 381: 270–272. https://doi.org/10.1016/j.ab.2008.05.056
58. Royer CA (1993) Improvements in the numerical analysis of thermodynamic data from biomolecular complexes. Anal Biochem 210:91–97. https://doi.org/10.1006/abio.1993.1155
59. Royer CA, Smith WR, Beechem JM (1990) Analysis of binding in macromolecular complexes: a generalized numerical approach. Anal Biochem 191: 287–294. https://doi.org/10.1016/0003-2697(90)90221-T
60. Weidemann T, Mücksch J, Schwille P (2014) Fluorescence fluctuation microscopy: a diversified arsenal of methods to investigate molecular dynamics inside cells. Curr Opin Struct Biol 28:69–76. https://doi.org/10.1016/j.sbi.2014.07.008
61. Kitamura A, Kinjo M (2018) State-of-the-art fluorescence fluctuation-based spectroscopic techniques for the study of protein aggregation. Int J Mol Sci 19: 964. https://doi.org/10.3390/ijms19040964
62. Kim SA, Heinze KG, Schwille P (2007) Fluorescence correlation spectroscopy in living cells. Nat Methods 4:963–973. https://doi.org/10.1038/nmeth1104
63. Schwille P, Meyer-Almes FJ, Rigler R (1997) Dual-color fluorescence cross-correlation spectroscopy for multicomponent diffusional analysis in solution. Biophys J 72:1878–1886. https://doi.org/10.1016/S0006-3495(97)78833-7
64. Bacia K, Kim SA, Schwille P (2006) Fluorescence cross-correlation spectroscopy in living cells. Nat Methods 3:83–89. https://doi.org/10.1038/nmeth822
65. Kim SA, Heinze KG, Bacia K, Waxham MN, Schwille P (2005) Two-photon cross-correlation analysis of intracellular reactions with variable stoichiometry. Biophys J 88:4319–4336. https://doi.org/10.1529/biophysj.104.055319
66. Declerck N, Royer CA (2013) Interactions in gene expression networks studied by two-photon fluorescence fluctuation spectroscopy. In: Methods in enzymology. Elsevier, pp 203–230
67. Zorrilla S, Ortega Á, Chaix D, Alfonso C, Rivas G, Aymerich S, Pilar Lillo M, Declerck N, Royer CA (2008) Characterization of the control catabolite protein of gluconeogenic genes repressor by fluorescence cross-correlation spectroscopy and other biophysical approaches. Biophys J 95:4403–4415. https://doi.org/10.1529/biophysj.108.135863

68. Zhang H, Wu Q, Berezin MY (2015) Fluorescence anisotropy (polarization): from drug screening to precision medicine. Expert Opin Drug Discov 10:1145–1161. https://doi.org/10.1517/17460441.2015.1075001
69. Wachsmuth M, Conrad C, Bulkescher J, Koch B, Mahen R, Isokane M, Pepperkok R, Ellenberg J (2015) High-throughput fluorescence correlation spectroscopy enables analysis of proteome dynamics in living cells. Nat Biotechnol 33:384–389. https://doi.org/10.1038/nbt.3146
70. Vinegoni C, Fumene Feruglio P, Brand C, Lee S, Nibbs AE, Stapleton S, Shah S, Gryczynski I, Reiner T, Mazitschek R, Weissleder R (2017) Measurement of drug-target engagement in live cells by two-photon fluorescence anisotropy imaging. Nat Protoc 12:1472–1497. https://doi.org/10.1038/nprot.2017.043
71. Hendrickson OD, Taranova NA, Zherdev AV, Dzantiev BB, Eremin SA (2020) Fluorescence polarization-based bioassays: new horizons. Sensors 20:7132. https://doi.org/10.3390/s20247132
72. Del Castillo U, Alfonso C, Acebrón SP, Martos A, Moro F, Rivas G, Muga A (2011) A quantitative analysis of the effect of nucleotides and the M domain on the association equilibrium of ClpB. Biochemistry 50:1991–2003. https://doi.org/10.1021/bi101670s
73. Du S, Lutkenhaus J (2019) At the heart of bacterial cytokinesis: the Z ring. Trends Microbiol 27:781–791. https://doi.org/10.1016/j.tim.2019.04.011
74. Krupka M, Margolin W (2018) Unite to divide: oligomerization of tubulin and actin homologs regulates initiation of bacterial cell division. F1000Research 7: 235. https://doi.org/10.12688/f1000research.13504.1
75. Ahijado-Guzmán R, Alfonso C, Reija B, Salvarelli E, Mingorance J, Zorrilla S, Monterroso B, Rivas G (2013) Control by potassium of the size distribution of Escherichia coli FtsZ polymers is independent of GTPase activity. J Biol Chem 288:27358–27365. https://doi.org/10.1074/jbc.M113.482943
76. Haeusser DP, Margolin W (2016) Splitsville: structural and functional insights into the dynamic bacterial Z ring. Nat Rev Microbiol 14:305–319. https://doi.org/10.1038/nrmicro.2016.26
77. Hernández-Rocamora VM, García-Montañés C, Reija B, Monterroso B, Margolin W, Alfonso C, Zorrilla S, Rivas G (2013) MinC protein shortens FtsZ protofilaments by preferentially interacting with GDP-bound subunits. J Biol Chem 288:24625–24635. https://doi.org/10.1074/jbc.M113.483222
78. Monterroso B, Reija B, Jiménez M, Zorrilla S, Rivas G (2016) Charged molecules modulate the volume exclusion effects exerted by crowders on FtsZ polymerization. PLoS One 11:e0149060. https://doi.org/10.1371/journal.pone.0149060
79. Bayburt TH, Sligar SG (2010) Membrane protein assembly into nanodiscs. FEBS Lett 584:1721–1727. https://doi.org/10.1016/j.febslet.2009.10.024
80. Hernández-Rocamora VM, Reija B, García C, Natale P, Alfonso C, Minton AP, Zorrilla S, Rivas G, Vicente M (2012) Dynamic interaction of the Escherichia coli cell division ZipA and FtsZ proteins evidenced in nanodiscs. J Biol Chem 287:30097–30104. https://doi.org/10.1074/jbc.M112.388959
81. Val-Calvo J, Luque-Ortega JR, Crespo I, Miguel-Arribas A, Abia D, Sánchez-Hevia DL, Serrano E, Gago-Córdoba C, Ares S, Alfonso C, Rojo F, Wu LJ, Boer DR, Meijer WJJ (2018) Novel regulatory mechanism of establishment genes of conjugative plasmids. Nucleic Acids Res 46:11910–11926. https://doi.org/10.1093/nar/gky996

The Complementarity of Nuclear Magnetic Resonance and Native Mass Spectrometry in Probing Protein-Protein Interactions

8

Elisabetta Boeri Erba and Annalisa Pastore

Abstract

Nuclear magnetic resonance (NMR) and native mass spectrometry (MS) are mature physicochemical techniques with long histories and important applications. NMR spectroscopy provides detailed information about the structure, dynamics, interactions, and chemical environment of biomolecules. MS is an effective approach for determining the mass of biomolecules with high accuracy, sensitivity, and speed. The two techniques offer unique advantages and provide solid tools for structural biology. In the present review, we discuss their individual merits in the context of their applications to structural studies in biology with specific focus on protein interactions and evaluate their limitations. We provide specific examples in which these techniques can complement each other, providing new information on the same scientific case. We discuss how the field may develop and what challenges are expected in the future. Overall, the combination of NMR and MS plays an increasingly important role in integrative structural biology, assisting scientists in deciphering the three-dimensional structure of composite macromolecular assemblies.

Keywords

Biophysics · Nuclear magnetic resonance (NMR) · Native mass spectrometry (MS) · Protein interactions · Structural biology

8.1 Introduction

The three-dimensional shape of a protein is described by its tertiary structure. In some cases, this structure is well defined and rigid, as in globular proteins. In others, it is flexible and covers several quite different conformations, as in intrinsically disordered proteins. Other proteins, probably the majority, include a mixture of the two situations. In all cases, structure determines protein function, including enzymatic, structural, transport, and regulatory functions [1]. However, proteins do not usually act alone but form large macromolecular complexes that coordinate and perform diverse molecular functions within the cell [2, 3]. Understanding the composition, stoichiometry, affinity, and structure of such complexes is a primary goal of structural biology. Two techniques have

E. B. Erba
CEA, CNRS, IBS, Université Grenoble Alpes, Grenoble, France
e-mail: elisabetta.boeri-erba@ibs.fr

A. Pastore (✉)
European Synchrotron Radiation Facility, Grenoble, France
e-mail: annalisa.pastore@esrf.fr

M. C. Vega, F. J. Fernández (eds.), *Advanced Technologies for Protein Complex Production and Characterization*, Advances in Experimental Medicine and Biology 1453,
https://doi.org/10.1007/978-3-031-52193-5_8

increasingly emerged to assist structural biologists in this endeavor: Mass Spectrometry (MS) and Nuclear Magnetic Resonance (NMR) spectroscopy. These techniques have a long history. Development of MS goes all the way back to the end of the nineteenth century and was utilized to ionize samples in a magnetic field and separate the resulting ions according to their mass-to-charge ratio (*m/z*) [4]. The first NMR spectrum was collected in 1949 by physicists in search of a proof of the nuclear spin [5]. Both techniques are now widely exploited by chemists, biologists, biotechnologists, and medical doctors with applications in chemistry, analytics, biotechnology, materials, diagnostics, and biology.

The two techniques provide different information at different levels of resolution depending on the specific application, but they both have become essential tools of structural biology that can go hand-in-hand to understand the assembly of even large protein complexes. In this review, we will focus on the use of these techniques in view of assessing molecular interactions. Since both fields are huge, we will restrict our analysis to liquid phase NMR applied to the detection of molecular interactions and native MS to limit the scope of our review. It is worth mentioning that, although the widely spread use of native MS is relatively recent [6, 7], the technique was conceived and used already in the 1990s to study protein folding [8]. We will discuss the advantages and limitations of each of the two techniques in the detection of molecular interactions, stress their complementarity and provide examples in which they have been used together to provide a deeper understanding of a specific biological problem.

8.2 Native MS Applications in Structural Biology

In the last 30 years, native MS has emerged as an important tool for the investigation of molecular complexes because it preserves non-covalent interactions between biomolecules [9–13]. Using native MS, we assess the mass and stoichiometry of intact non-covalent complexes, identify direct interactions between their components, characterize stable subcomplexes and assign the relative position (core *vs.* periphery) of subunits (Fig. 8.1). We determine the hierarchy of an assembly pathway by mixing subunits in a stepwise manner [14] or by using different pH values to induce a change in oligomeric state [15, 16]. Native MS has been combined with ion mobility (IM) to investigate the shape of macromolecular assemblies [17–23]. Native MS coupled with bioinformatics has also been very powerful for studying the evolutionary history of protein complexes [24–27].

During native MS investigations, macromolecular complexes are subjected to a gentle ionization that preserves non-covalent interactions. Electrospray ionization (ESI) is the main soft ionization method for native MS experiments. ESI-MS experiments may be influenced by the nature of the intermolecular interactions, by the composition, ionic strength, and pH of the sample buffer, and by the voltages and pressures within the mass spectrometer. Consequently, it is important to take these parameters into account when the native MS data are acquired and analyzed. Unlike other types of ESI-MS analysis, neither acidic conditions nor organic solvents are used. Instead, native MS experiments are usually performed using volatile buffers such as ammonium acetate [28, 29], ethylenediammonium diacetate [30] or alkylammonium acetate [31]. Typically, the buffer is exchanged immediately prior to native MS analysis [32].

Native MS provides important structural information when high-resolution data are not available [33]. For example, a bacterial complex involved in chromosome organization and segregation, called MukBEF, was investigated by native MS, size exclusion chromatography, multi-angle light scattering (SEC-MALS), isothermal titration calorimetry (ITC), epifluorescence microscopy and in vivo functional studies [34]. The MukBEF complex contains MukB (an ATPase), MukE, and MukF. Native MS showed that three major complexes were detected [a 2-mer (MukB2), 6-mer (MukE4:MukF2) and 8-mer (MukB2: MukE4:MukF2)] in the presence of ADP. An

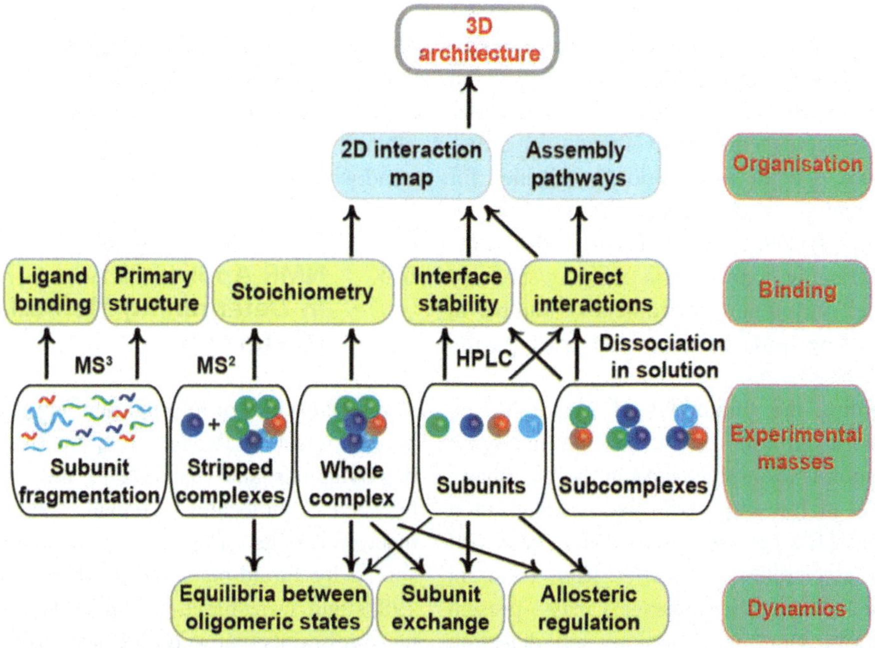

Fig. 8.1 The different levels of information provided by native MS about non-covalent assemblies. When a macromolecular complex is investigated, the various components (the whole complex, stripped complexes, individual components, and subcomplexes) are released and analyzed (white boxes). MS can provide information related to binding (yellow boxes) such as stoichiometry, interface stability, allostery, and equilibria. It can also provide information related to organization (light blue boxes). MS measurements provide insight into dynamics, binding interactions, structural organization, and dynamics of a macromolecular complex (green boxes). Arrows show the flow of information. MS^2 and MS^3 stand for experiments that generate first and second generation product ions

additional 10-mer (MukB4:MukE4:MukF2) was observed in the presence of a non-hydrolysable ATP analogue. Combined with biophysical information, these data offered a first glimpse of the MukBEF organization and its changes after ATP/ADP binding [34].

Native MS provides important information on unexpected complex components. For instance, in the case of the investigation of membrane protein complexes, native MS allows the identification of unknown lipids bound to an assembly. This is particularly important because lipids are normally characterized with great difficulty with other techniques, such as high-resolution cryo-Electron Microscopy (EM), because of their small size [35].

Native MS allows the study of the dynamics of macromolecular assemblies. For instance, an amyloidogenic protein called transthyretin (TTR) has been extensively investigated by native MS [36–40]. Exchange of unlabeled and labeled TTR subunits was monitored over time by native MS [36, 38–40]. These experiments allowed the assessment of the effect of point mutations on subunit exchange. MS-based information combined with neutron crystallography data and modelling studies led to propose a novel mechanism of TTR fibrillation [39].

8.3 Advantages and Limitations of Native MS

Native MS presents several advantages compared to other structural approaches. First, native MS does not require samples to be crosslinked or labeled because the experimental conditions preserve non-covalent interactions. Second, it allows the study of a wide variety of biological samples that differ in mass, symmetry, dynamics, flexibility, and heterogeneity [9]. For instance, different oligomeric states can be characterized simultaneously. Since data are not averaged over different species, specific information is obtained for each individual species. Thus, the dynamics of quaternary structure is studied in real time [10, 11]. Finally, native MS is a highly sensitive technique: only a few microliters of sample at relatively low (μM) concentration are typically required.

Regarding the limitations of native MS, it should be taken into account that the macromolecular complexes are detected in the gas phase. Since MS analyses are performed under vacuum conditions [12, 13], the relative abundances of detected assemblies may differ from those in solution. Distinct complexes may present distinct ionization, transmission, and detection probabilities [14]. For instance, electrostatic interactions are stronger in the gas phase than in solution. Conversely, hydrophobic interactions become weaker. On the other hand, the literature shows that the transition from solution to the gas phase does not drastically alter biomolecules [15, 16]: low-energy electron holography confirmed the stability of folded biomolecules in ultra-high vacuum [17]. It was also proven that the behavior of biomolecules in the gas phase resembles the one in solution: when the solution conditions (e.g., pH and concentrations) were modified, the gas-phase spectra changed correspondingly [14]. As an example, homo-complexes were engineered to be stable above pH 6.5 [18]. When the pH was lowered, buried histidine residues became protonated and these assemblies underwent cooperative, large-scale conformational changes detected by size-exclusion chromatography and native MS [18].

Another limitation of native MS could be that the experiments are performed using volatile buffers (e.g., ammonium acetate). Therefore, the buffer is exchanged prior to native MS analysis [19]. In some cases, the biochemical steps involved in sample preparation require optimization to ensure that the native state of an assembly is preserved during buffer exchange.

Overall, native MS is extremely beneficial to the investigation of macromolecular assemblies thanks to its selectivity, accuracy, sensitivity, and analysis speed. Nonetheless, these inherent limitations discussed above should be born in mind when data are evaluated.

8.4 NMR Applications in Determining Molecular Interactions

Three main lines of information may be obtained by biological NMR: the structure of a protein, the dynamics and mapping of the interactions between molecules (small ligands or macromolecules) (Fig. 8.2). All the first structures solved by liquid-phase NMR date in the early 1980s and were based on the collection of distance restraints that were then used to reconstruct the overall three-dimensional structure [20–24]. Since then, the field has matured so that solving the structure of proteins has become routine provided that the protein size is within NMR reach. This limit has been constantly pushed forward and it is now possible to afford also molecular weights around 40–50 kDa [25]. Dynamics information may be obtained from relaxation measurements, which may provide details on flexible, yet rather rigid regions and residues in conformational or chemical exchange.

Where, however, NMR remains of particular importance is the determination of interactions. In complexes of sufficient stability (nanomolar), it is possible to obtain the structure of the complex by collecting intermolecular distances as obtained experimentally or predicted [26, 27]. However, other much simpler and more general approaches may be used to study protein interactions, which yet provide information at atomic resolution. Here, we will focus on these, given that they might be more directly complementary to native MS.

Detection of interactions by NMR relies on changes in the local electron density that are induced by the spatial proximity of another molecule which will influence the most easily observable NMR parameter, the chemical shift

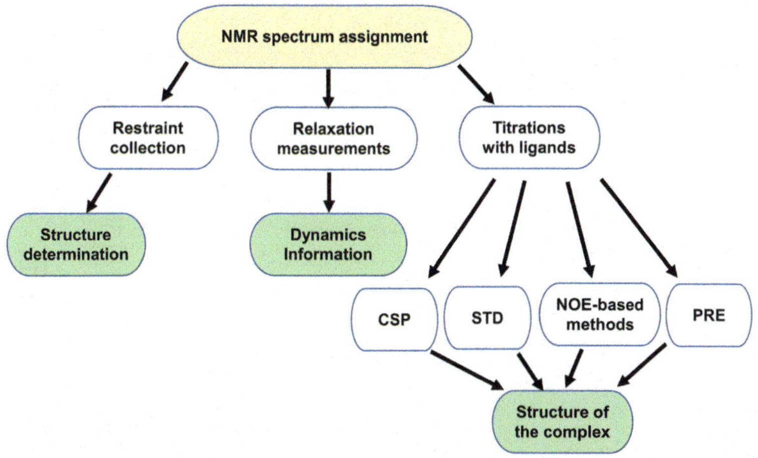

Fig. 8.2 Summary of the different layers of information obtainable by NMR on proteins. All starts with the assignment of the NMR spectrum (yellow box). This step is not indispensable but necessary when residue-specific information is wanted. There are then three main resulting routes (green boxes): determination of the protein structure, dynamics, or interactions. The scheme is meant to provide an overall picture but should by no means be considered exhaustive

[28]. Spatial proximity of groups with magnetic susceptibility anisotropies like aromatic rings can also induce large changes in chemical shifts leading to what is called ring current shift effect. Detection of the variations of chemical shift by an interaction is called chemical shift perturbation (CSP) [29]. Typically, CSP is measured by recording a series of NMR spectra, usually ^{15}N-Heteronuclear Single Quantum Correlation (HSQC) experiments of the ^{15}N labeled protein in the absence and in the presence of varying amounts of the binding partner. It is of course essential that both components are dissolved in exactly the same buffer and measured under the same conditions, since chemical shifts, especially those of amide protons, are very sensitive to differences in pH, temperature, and buffer composition. If there is an interaction, the chemical shifts of the residues involved in the complex are displaced from their original position. There are, however, two limiting cases (Fig. 8.3). In the fast exchange regime, the two signals collapse into one, whose chemical shift represents the population averaged value of the free and complex saturated protein. This means that, depending on the relative amount of protein, interactor, and the value of the dissociation constant (K_d), the resulting signal will have a chemical shift in between those of the free and bound states, following a linear trajectory between the chemical shifts of the two states. In the slow exchange regime, we observe instead both the signals of the bound and free states, and the peak volumes represent their relative amounts. Departures of the HN chemical shifts in the ^{15}N HSQC from a linear trajectory may be seen when the interaction involves multiple binding sites with different affinities. In this case, linearity will be seen only until the primary stronger binding site is saturated. After this point the chemical shift variation might change direction to fill the second weaker binding site. A classic example of this behavior is the interaction between the TAZ2 domain of a transcriptional coactivator titrated with the tumor suppressor p53 domain AD1 domain in which the resonances of the ^{15}N labeled TAZ2 domain completely change directionality [30].

A quantitative estimate of the K_d value can be obtained from plotting the chemical shift variations in the titration. This very simple method is particularly powerful when, for instance, the interaction involves two proteins one of which is above the NMR size limits. In

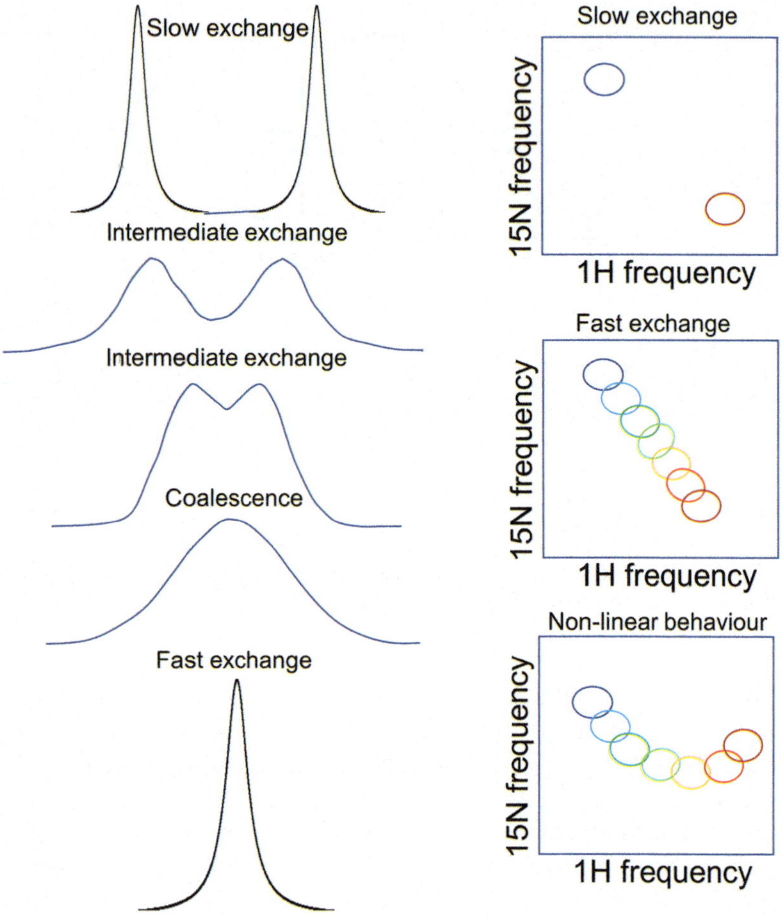

Fig. 8.3 The different exchange regimes in NMR detection. Left panel: During the titration the resonances of the two species (free and bound) can be in slow exchange (top), fast exchange (bottom), or various degrees of intermediate exchange. Right panel: Slow and fast regimes as observed in a two-dimensional HSQC spectrum. In the slow exchange regime, the free and bound peaks have distinct chemical shifts (top). In fast exchange, the peak moves gradually from the chemical shift position of the free to the bound form (middle). Usually, the trajectory is linear. However, in some cases, the trajectory is non-linear (bottom). In these cases, there are more than one binding site: linearity is seen until the primary stronger binding site is saturated. After this point the chemical shift variation might change direction to fill the second weaker binding site

this case and provided that the interaction is not so tight to provoke the immediate disappearance of the spectrum, it is possible to map the surface of interaction on the small component [31].

An alternative to CSP is saturation-transfer difference NMR (STD-NMR) [32, 33]. This method, which relies on the Nuclear Overhauser Effect (NOE), is carried out by irradiating a target protein by a radiofrequency in a region solely populated by it. A second spectrum in which saturation takes place off-resonance (I_0) in an empty region of the spectrum, is then acquired and the difference spectrum ($I_{STD}=I_0 - I_{SAT}$) is plotted that will contain only signals from the saturated ligand that interact with the protein. The method typically works for interactions in the fast exchange regime, with K_d values in the range 10^{-8}–10^{-3} mol L^{-1} [34]. When probing the interaction between two proteins, the experiment typically requires the uniformly isotopic labeling of one of the components with ^2H and ^{15}N, leaving the other component unlabeled, and using 10/90% H_2O/D_2O mixtures. Under these conditions the labeled protein has low proton density thus suppressing most of the spin diffusion [35]. The complex is irradiated at a frequency that affects only the unlabeled protein, for instance, choosing aliphatic proton resonances. In this way, spin diffusion from the high proton density unlabeled protein leads to cross-saturation transfer to the interacting partner. When one of the partners is RNA, irradiation may take place in the region of the spectrum where the exchangeable protons of the nucleic acids are

(10–14 ppm). In this region of the spectrum, only the resonances of RNA are typically present [36].

Other equally important approaches to determine interactions by NMR, which will not be discussed here, are paramagnetic resonance enhancement, chemical exchange saturation transfer (CEST), transferred NOE (for small ligands), and NOE editing/filtering experiments [26, 27]. The information obtained in all these experiments allows to obtain distance restraints which can then be translated into the structure of the complex by classical restraint-driven simulated annealing or docking approaches.

8.5 Advantages and Limitations of NMR in Determining Interactions

One of the main advantages of liquid-phase NMR, as it may easily be deduced from Session 8.4, is that it may provide information on interactions at the single amino acid level whereas many other techniques work averaging the signal over the whole protein or reporting only on specific groups, like with fluorescence. It is also important to understand that, depending on the affinity of the interaction and the exchange regime, NMR can provide distance restraints, which may be used either much in the same way used to derive the three-dimensional structure of an individual protein or to guide an experimentally driven docking procedure. It is however important to remember that shifting of a particular signal in CSP experiments does not always indicate that the corresponding residue is in the interface between the two molecules: shifting can be caused indirectly by conformational changes induced by binding that propagate throughout the protein. However, since this limitation does not hold for STD-NMR because it works only for residues directly involved in the interaction, it is possible to perform both experiments and extract more information from their complementarity.

A second important advantage of NMR is that this technique works within a large range of affinities, which can go up to values in the millimolar range. This means that NMR is a technique able to report on a quite large range of affinities, even though it might not be the optimal method to follow the full titration for very tight complexes. On the other end of the spectrum, NMR allows to observe also transient interactions. This feature places NMR in a unique place, since many techniques, such as isothermal calorimetry or fluorescence spectroscopy, may only be suitable in very specific ranges of affinities.

The third, but not less important, advantage of NMR lies in the possibility of obtaining information on flexible or unstructured proteins (the so-called intrinsically disordered proteins or IDPs [37, 38], which are beyond the reach of other high-resolution structural techniques, such as X-ray crystallography and cryo-EM.

At the same time, while offering unique advantages, NMR has also undeniable limitations.

First of all, NMR provides little information about stoichiometry. It is certainly possible to obtain estimates on the molecular weight of a complex from the correlation time [39] or simply from T_1 relaxation values [40] and obtain from these observables the stoichiometry. However, the method provides only a rough estimate, which, for small proteins, is affected by an error too large to be reliable. The method is also more reliable for globular proteins.

A second area that might be problematic concerns limitations in obtaining information on systems outside the NMR size range. This number is not easily definable because the limits are constantly pushed forward and they strongly depend on the type of expected information, complex affinities, exchange regime, and labeling scheme or, ultimately, on the operator's creativity. We have, for instance, used a hybrid methodology based on a mixture of NMR, small-angle X-ray scattering (SAXS), site-directed mutagenesis, and molecular docking to determine the structure of a weakly interacting 110 kDa complex in which one component was a small protein, while the other component was over 90 kDa [31, 41]. In these studies, we applied the technique to gain insights into the structure of complexes formed among proteins involved in the molecular machine that produces the essential iron-sulfur cluster prosthetic groups.

8.6 The Successful Combination of the Two Techniques

NMR and native MS have been successfully combined to thoroughly investigate non-covalent interactions, characterizing binding sites, and monitoring dynamic events [42–44]. The two methods have for instance been employed to study the specificity of the interaction between a bacterial ABC transporter (McjD) and a ribosomally synthesized and post-translationally modified peptide (RiPP), called microcin J25 (MccJ25) and related sequences [45]. Organisms synthetize RiPPs to damage closely related species. The RiPP peptides are also harmful to the producing organisms, which have specific ABC transporters to prevent self-toxicity. A primary site of interaction between MccJ25 and McjD was identified by CEST NMR experiments [45]. This is a sensitive approach based on the equilibrium between low molecular weight species (observable with solution NMR) and transiently bound states with slow-tumbling high-molecular-weight complexes, which cannot be observed directly in solution NMR as a result of line broadening. Based on the information obtained by CEST, a MccJ25 deletion variant was generated lacking F10 and V11 (called MccJ25-ΔFV). Using native MS and a ligand-induced ATPase assay, the interaction between MccJ25-ΔFV and McjD was investigated. The data indicated that MccJ25-ΔFV has lower affinity for the transporter McjD than wild-type (WT) MccJ25, revealing the mechanism of the recognition between membrane transporters and RiPPs. This study represents a stepping-stone towards the employment of bacterial cell factories to produce novel bioactive compounds.

Another interesting example of the successful combination of NMR and native MS was the study of two peptides of 37-amino acids which form amyloid fibrils in pancreatic islets of type 2 diabetes patients [46]. In this study, the authors analyzed the human islet amyloid polypeptide (hIAPP) and a more amyloidogenic natural form of hIAPP carrying a point mutation on serine 20 (S20G). Native MS was then applied to screen 1500 molecules from a Protein-Protein Interaction (PPI) library. This allowed the identification of compounds binding hIAPPs. Among them, two small molecules were identified for their considerable effects on hIAPP aggregation. Specifically, one of the compounds delayed the aggregation of WT hIAPP, but not of S20G variant. The other molecule enhanced the rate of aggregation of both peptides. 2D SOFAST-HMQC NMR allowed Xu et al. (2022) to characterize the residues involved in the binding of small molecules to ^{15}N-labeled monomeric IAPP [46].

Combination of native MS and NMR is particularly helpful for the characterization of macromolecular complexes that are small or flexible, "invisible" to EM, and generate no crystals. Recently, NMR spectroscopy, native and crosslinking MS combined with isothermal titration calorimetry, small-angle X-ray scattering and other methods enabled scientists to characterize the interaction of an HIV-1 protein (Rev) with a human nuclear transport factor (Importin-β, Impβ) [47] (Fig. 8.4). Rev is crucial for viral replication because it mediates the nuclear export of intron-containing viral RNA transcripts. Rev contains many intrinsically disordered residues. It is known that Rev. is imported into the nucleus by Impβ. However, it is poorly understood how Rev. and Impβ are associated. The structural investigation of Impβ is particularly challenging. Indeed, this nuclear transport factor is conformationally highly flexible, an important feature that permits the efficient recognition of different cargo molecules. Using

Fig. 8.4 (continued) magenta circles correspond to unbound Impβ. Peaks labelled by blue circles with a single dot or by green circles with two dots correspond to Impβ/Rev. complexes with 1:1 or 1:2 stoichiometry, respectively. (B) Spectrum of Impβ in the absence of Rev. (**c**, **d**) Spectra of Impβ incubated with a two- or fivefold molar equivalents of Rev^{OD} respectively. (**e**) Spectra of Impβ in the presence of the truncated construct Rev^{OD} (residues 4-69) lacking the C-terminal domain. (**f**, **g**) Spectra of Impβ incubated with (**f**) a twofold or (**g**) a fivefold molar equivalent of wild-type Rev (Rev^{WT}). (Adapted from [47])

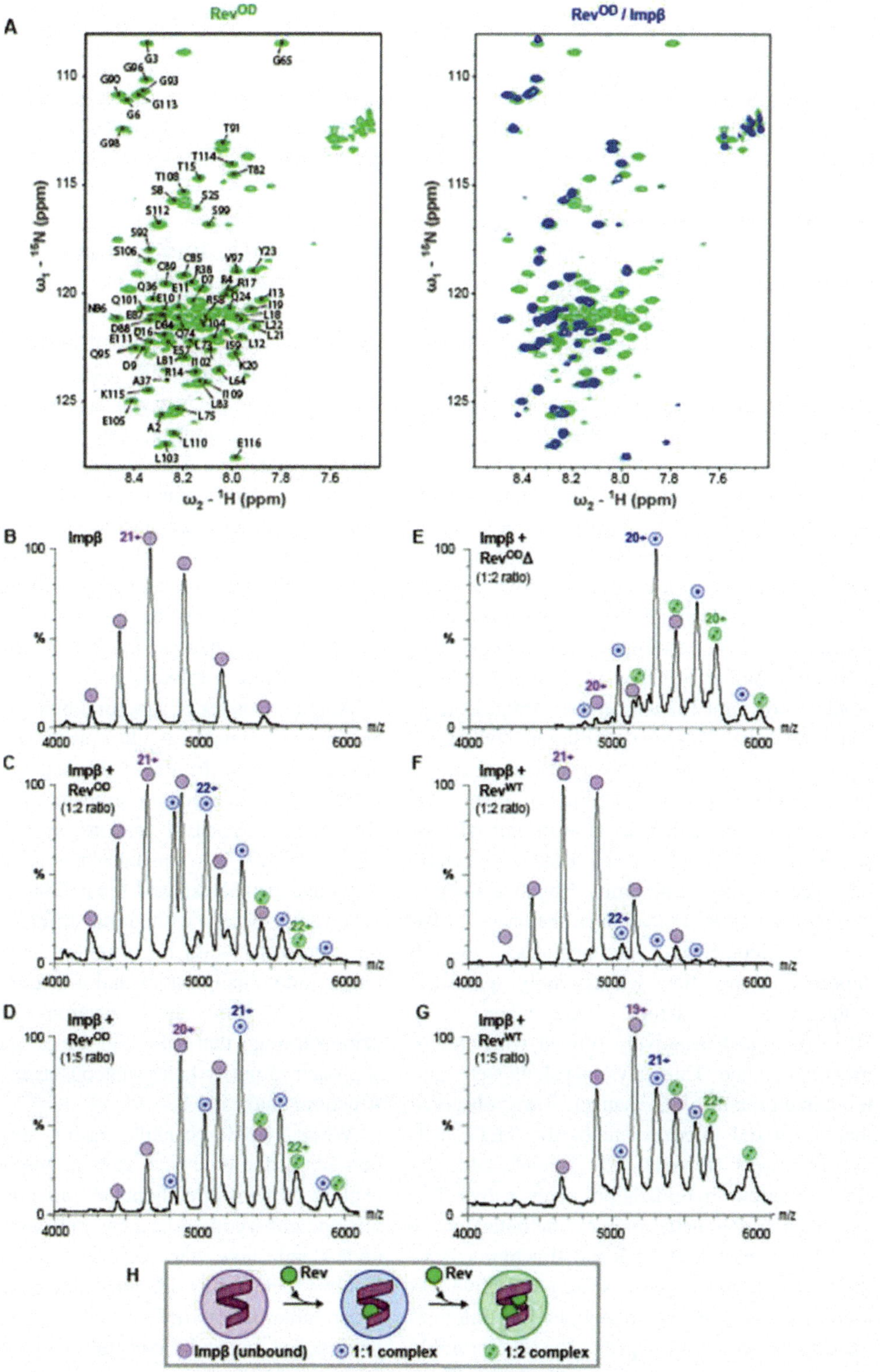

Fig. 8.4 Binding stoichiometryStoichiometries of the HIV-1 Rev/Impβ complex and its interactions. (**a**) NMR analysis of Rev binding by Impβ. (**a**, left panel) ^{1}H,^{15}N-HSQC spectrum of free Rev^{OD} (600 MHz, 283 K). Rev^{OD} is a mutated form of Rev (V16D and I55N), which is oligomerization-deficient and monodisperse. (**a**, right panel) ^{1}H,^{15}N-HSQC spectrum Rev^{OD} (blue) bound to Impβ and superimposed on that of unbound Rev^{OD} (green). (**b**–**g**) Native MSNative Mass Spectrometry (Native MS) spectra. Peaks labelled by

integrative structural biology, this work provided novel insights into the molecular recognition of Rev. by Impβ and an atypical binding behavior of Rev.

The last example that we would like to discuss allowed monitoring of the dynamics of a large macromolecular complex. Through the combined use of native MS, NMR, and EM it was possible to investigate the assembly pathway of tetrahedral aminopeptidase 2 (TET2), a 468 kDa homo-complex [42]. During complex assembly, the subunits undergo appropriate conformational rearrangements to form a supramolecular structure. Tracking such rearrangements represents a challenge because of the low-abundance of intermediate states that change over time. Although the self-assembly is a time-dependent process occurring at the molecular level, understanding how it occurs often originates from static structures, low resolution techniques, and modelling. While NMR is in principle able to monitor structural changes at the atomic level in real time, size and time resolution constraints represent a challenge. In the specific example of TET2, methyl specific labelling of deuterated TET2 combined with relaxation optimized, fast acquisition real-time NMR was used to overcome both size and time scale limits. Negative staining EM allowed characterization of the shape of the large oligomeric intermediates, which appeared along the self-assembly pathway. Aliquots of the sample were taken every 30 seconds and negatively stained with ammonium molybdate. Thus, EM snapshots were captured in a time-resolved manner. It was observed that the initial intermediate of the TET2 self-assembly was of small size, undetectable by EM. "Isotopic hybridization" and native MS were then used to better monitor the initial events of self-assembly. The mass and stoichiometry of the hybrid complex was assessed by native MS thereby allowing the identification of the intermediate at the very early stages of the self-assembly cascade, which consists in the monomer. Thus, native MS complemented well the NMR and EM data used to probe the structural changes of the subunits formed during the assembly of large macromolecular complex.

8.7 Concluding Remarks and Outlook

We have underlined here the individual advantages and limitations of NMR and native MS and discussed applications in which the two techniques were used in a complementary way. We strongly believe that the future of structural biology is in the integration of different techniques, including *in silico* predictions, that can complement and integrate each other [48]. This perspective in turn implies that structural biologists will need to be aware of the complementarities and of the possibility of designing new *ad hoc* modalities to solve specific scientific problems.

Another frontier of structural biology is working in environments as close as possible to the biological ones. In this field, native MS has steadily gained momentum thanks to its wide applicability, speed of analysis, sensitivity, and selectivity [49–51]. This latter feature allows the simultaneous analysis and separation of mixtures of several species with quite different masses. This property was, for instance, exploited to image individual proteins and a macromolecular complex by low-energy electron holography, demonstrating that MS can be a preparative approach to purify heterogeneous assemblies for structural studies [17].

While NMR cannot easily discriminate non-labelled mixtures, it can, in specific cases, explore proteins directly in cell, both of prokaryotic and eukaryotic organisms [52], permitting to understand how the cellular milieu affects features such as the folding/unfolding equilibrium of proteins or interactions between proteins of interest and molecules constituting the intracellular matrix [53].

When working directly with crude cell lysates and culture media (i.e., without the need of purification), the combined use of native MS and NMR can inform on the identity, solubility, oligomeric state, and stability of overexpressed biomolecules [54–56]. This allows the investigation of proteins and macromolecular complexes only minimally purified [57]. Both techniques have been used for the investigation of endogenous macromolecular assemblies in prokaryotic and in eukaryotic hosts. Investigation of endogenous complexes is in fact helping to decipher the function and regulation of post-translational modifications, cofactors and transient complexes assembled in a cellular milieu [57–60]. The added value of the sensitivity of native MS is that it permits the study without the need of recombinant protein over-expression [61, 62]. Once identified the components, it is then possible to study the various components individually and solve the structure by X-ray crystallography, NMR, or cryo-EM. In this endeavor, NMR provides an important contribution especially for IDPs or disordered domains of longer proteins [63].

Finally, native MS and NMR have the potential to become crucial tools for the characterization of biomaterials. A current trend is the investigation of biogenic polymers such as those needed by the industry to replace petroleum-based polymers with materials of biogenic origin to better develop the so-called "circular bioeconomy" [64]. Recent literature has for instance reported the investigation of squid pens (e.g., from *Doryteuthis pealeii* and *Loligo vulgaris*) to better exploit their features such as transparency and flexibility [65, 66]. In a different set of recent studies, it has been possible to characterize and determine the structure of mussel proteins [67, 68]. Some marine organisms are able to resist aqueous tidal environments and adhere tightly on wet surfaces thanks to adherence properties. This behavior has raised increasing attention for potential applications in medicine, biomaterials, and tissue engineering. Among the marine organisms with adhesion properties are mussels that strongly adhere to the rock through the secretion of protein-based stringy structure called byssus. These filaments consist of bundles of fibers made of proteins that are synthesized in the mussel foot, a large organ that in freshwater allows the mussel to pull through the substrate and move. The combination of native MS and NMR will be ideal to determine the structures of the individual proteins and to study the mechanism of their assembly.

While these are exciting new avenues that can be explored in the future, there is still the need to reflect on what are the technical advancements needed in these areas. For native MS, there is the need of further development of novel mass spectrometers, providing even better sensitivity, resolution, and ionization efficiency. Orbitrap instruments already show improvements in sensitivity and resolution as compared to Q-TOF mass spectrometers [69–71]. Novel detectors advance the applicability of MS to study biological nanoparticles. For instance, nanoelectromechanical systems (NEMS) detect large masses with unprecedented sensitivity, requiring only a few hundred single-molecule adsorption events to detect megadalton molecules [72]. Another example of valuable approaches is the charge detection mass spectrometry (CDMS), whereby the m/z and z are simultaneously measured for each ion [73]. CDMS combined with Orbitrap technology [74] was recently utilized to characterize adeno-associated viruses (AAVs), important vectors for gene therapy. The method allows users to quantify different AAV bioprocessing products and to rapidly assess the integrity and amount of genome packed in AAV particles [75].

On the NMR side, it cannot be denied that modern NMR spectroscopy in liquids has reached an unprecedented level of sophistication which covers both the determination of biomolecular structures, the dynamics at atomic resolution, and an incredible range of determination of macromolecular interactions between molecules of very different sizes from proteins/nucleic acids to small molecules (e.g., drug screening) [76]. However, even having gone from the millimolar to the micromolar range of affinities, the sensitivity and specificity of NMR remains low as compared to other techniques, especially MS. This means that some important biological problems remain out of the NMR reach and can beneficially be complemented by other techniques. In the future, higher field magnets and more efficient pulse schemes may help to widen these possibilities.

In conclusion, we hope to have provided an overview, albeit by necessity incomplete, of the complementarities between two techniques of primary importance in structural biology.

Acknowledgements The authors acknowledge support from FRISBI (ANR-10-INSB-05-02) and GRAL (ANR-10-LABX-49-01) within the Grenoble Partnership for Structural Biology (PSB).

References

1. Gu J, Bourne PE (2009) Structural bioinformatics. Wiley-Blackwell, Hoboken
2. Omranian S, Nikoloski Z, Grimm DG (2022) Computational identification of protein complexes from network interactions: present state, challenges, and the way forward. Comput Struct Biotechnol J 20:2699–2712. https://doi.org/10.1016/j.csbj.2022.05.049
3. Spirin V, Mirny LA (2003) Protein complexes and functional modules in molecular networks. Proc Natl Acad Sci U S A 100(21):12123–12128. https://doi.org/10.1073/pnas.2032324100
4. Downard KM (2004) Mass spectrometry: a foundation course. Royal Society of Chemistry, London
5. Becker ED (1993) A brief history of nuclear magnetic resonance. Anal Chem 65(6):295A–302A. https://doi.org/10.1021/ac00054a001
6. Karch KR, Snyder DT, Harvey SR, Wysocki VH (2022) Native mass spectrometry: recent progress and remaining challenges. Annu Rev Biophys 51:157–179. https://doi.org/10.1146/annurev-biophys-092721-085421
7. Wu D, Robinson CV (2022) Understanding glycoprotein structural heterogeneity and interactions: insights from native mass spectrometry. Curr Opin Struct Biol 74:102351. https://doi.org/10.1016/j.sbi.2022.102351
8. Miranker A, Robinson CV, Radford SE, Aplin RT, Dobson CM (1993) Detection of transient protein folding populations by mass spectrometry. Science 262(5135):896–900. https://doi.org/10.1126/science.8235611
9. Kaur U, Johnson DT, Chea EE, Deredge DJ, Espino JA, Jones LM (2019) Evolution of structural biology through the lens of mass spectrometry. Anal Chem 91(1):142–155. https://doi.org/10.1021/acs.analchem.8b05014
10. Yee AW, Aldeghi M, Blakeley MP, Ostermann A, Mas PJ, Moulin M et al (2019) A molecular mechanism for transthyretin amyloidogenesis. Nat Commun 10(1):925. https://doi.org/10.1038/s41467-019-08609-z
11. Yee AW, Moulin M, Breteau N, Haertlein M, Mitchell EP, Cooper JB et al (2016) Impact of deuteration on the assembly kinetics of transthyretin monitored by native mass spectrometry and implications for amyloidoses. Angew Chem Int Ed Engl 55(32):9292–9296. https://doi.org/10.1002/anie.201602747
12. Belov ME, Damoc E, Denisov E, Compton PD, Horning S, Makarov AA et al (2013) From protein complexes to subunit backbone fragments: a multistage approach to native mass spectrometry. Anal Chem 85(23):11163–11173. https://doi.org/10.1021/ac4029328
13. Sobott F, Hernandez H, McCammon MG, Tito MA, Robinson CV (2002) A tandem mass spectrometer for improved transmission and analysis of large macromolecular assemblies. Anal Chem 74(6):1402–1407
14. Boeri Erba E, Barylyuk K, Yang Y, Zenobi R (2011) Quantifying protein-protein interactions within noncovalent complexes using electrospray ionization mass spectrometry. Anal Chem 83(24):9251–9259. https://doi.org/10.1021/ac201576e
15. Meyer T, de la Cruz X, Orozco M (2009) An atomistic view to the gas phase proteome. Structure 17(1):88–95. https://doi.org/10.1016/j.str.2008.11.006
16. Seo J, Hoffmann W, Warnke S, Bowers MT, Pagel K, von Helden G (2016) Retention of native protein structures in the absence of solvent: a coupled ion mobility and spectroscopic study. Angew Chem Int Ed Engl 55(45):14173–14176. https://doi.org/10.1002/anie.201606029
17. Longchamp JN, Rauschenbach S, Abb S, Escher C, Latychevskaia T, Kern K et al (2017) Imaging proteins at the single-molecule level. Proc Natl Acad Sci U S A 114(7):1474–1479. https://doi.org/10.1073/pnas.1614519114
18. Boyken SE, Benhaim MA, Busch F, Jia M, Bick MJ, Choi H et al (2019) De novo design of tunable, pH-driven conformational changes. Science 364(6441):658–664. https://doi.org/10.1126/science.aav7897
19. Boeri Erba E, Signor L, Oliva MF, Hans F, Petosa C (2018) Characterizing intact macromolecular complexes using native mass spectrometry. Methods Mol Biol 1764:133–151. https://doi.org/10.1007/978-1-4939-7759-8_9
20. Havel TF, Wuthrich K (1985) An evaluation of the combined use of nuclear magnetic resonance and distance geometry for the determination of protein conformations in solution. J Mol Biol 182(2):281–294. https://doi.org/10.1016/0022-2836(85)90346-8
21. Mascagni P, Pope M, Gibbons WA, Ciuffetti LM, Knoche HW (1983) The backbone and side chain conformations of the cyclic tetrapeptide HC-toxin.

Biochem Biophys Res Commun 113(1):10–17. https://doi.org/10.1016/0006-291x(83)90424-2
22. Pardi A, Wagner G, Wuthrich K (1983) Protein conformation and proton nuclear-magnetic-resonance chemical shifts. Eur J Biochem 137(3):445–454. https://doi.org/10.1111/j.1432-1033.1983.tb07848.x
23. Wagner G, Braun W, Havel TF, Schaumann T, Go N, Wuthrich K (1987) Protein structures in solution by nuclear magnetic resonance and distance geometry. The polypeptide fold of the basic pancreatic trypsin inhibitor determined using two different algorithms, DISGEO and DISMAN. J Mol Biol 196(3):611–639. https://doi.org/10.1016/0022-2836(87)90037-4
24. Williamson MP, Havel TF, Wuthrich K (1985) Solution conformation of proteinase inhibitor IIA from bull seminal plasma by 1H nuclear magnetic resonance and distance geometry. J Mol Biol 182(2):295–315. https://doi.org/10.1016/0022-2836(85)90347-x
25. Takeuchi K, Baskaran K, Arthanari H (2019) Structure determination using solution NMR: is it worth the effort? J Magn Reson 306:195–201. https://doi.org/10.1016/j.jmr.2019.07.045
26. Gobl C, Madl T, Simon B, Sattler M (2014) NMR approaches for structural analysis of multidomain proteins and complexes in solution. Prog Nucl Magn Reson Spectrosc 80:26–63. https://doi.org/10.1016/j.pnmrs.2014.05.003
27. Yadav DK, Lukavsky PJ (2016) NMR solution structure determination of large RNA-protein complexes. Prog Nucl Magn Reson Spectrosc 97:57–81. https://doi.org/10.1016/j.pnmrs.2016.10.001
28. Becker W, Bhattiprolu KC, Gubensak N, Zangger K (2018) Investigating protein-ligand interactions by solution nuclear magnetic resonance spectroscopy. ChemPhysChem 19(8):895–906. https://doi.org/10.1002/cphc.201701253
29. Williamson MP (2013) Using chemical shift perturbation to characterise ligand binding. Prog Nucl Magn Reson Spectrosc 73:1–16. https://doi.org/10.1016/j.pnmrs.2013.02.001
30. Arai M, Ferreon JC, Wright PE (2012) Quantitative analysis of multisite protein-ligand interactions by NMR: binding of intrinsically disordered p53 transactivation subdomains with the TAZ2 domain of CBP. J Am Chem Soc 134(8):3792–3803. https://doi.org/10.1021/ja209936u
31. Adinolfi S, Iannuzzi C, Prischi F, Pastore C, Iametti S, Martin SR et al (2009) Bacterial frataxin CyaY is the gatekeeper of iron-sulfur cluster formation catalyzed by IscS. Nat Struct Mol Biol 16(4):390–396. https://doi.org/10.1038/nsmb.1579
32. Mayer M, Meyer B (1999) Characterization of ligand binding by saturation transfer difference NMR spectroscopy. Angew Chem Int Ed Engl 38(12):1784–1788. https://doi.org/10.1002/(SICI)1521-3773(19990614)38:12<1784::AID-ANIE1784>3.0.CO;2-Q
33. Takahashi H, Nakanishi T, Kami K, Arata Y, Shimada I (2000) A novel NMR method for determining the interfaces of large protein-protein complexes. Nat Struct Biol 7(3):220–223. https://doi.org/10.1038/73331
34. Meinecke R, Meyer B (2001) Determination of the binding specificity of an integral membrane protein by saturation transfer difference NMR: RGD peptide ligands binding to integrin alphaIIbbeta3. J Med Chem 44(19):3059–3065. https://doi.org/10.1021/jm0109154
35. Edzes HT, Samulski ET (1977) Cross relaxation and spin diffusion in the proton NMR or hydrated collagen. Nature 265(5594):521–523. https://doi.org/10.1038/265521a0
36. Ramos A, Kelly G, Hollingworth D, Pastore A, Frenkiel T (2000) Mapping the interfaces of protein–nucleic acid complexes using cross-saturation. J Am Chem Soc 122:11311–11314. https://doi.org/10.1021/ja002233w
37. Pastore A, Temussi P (2013) When "IUPs" were "BAPs": how to study the nonconformation of intrinsically unfolded polyaminoacid chains. Biopolymers 100(6):592–600. https://doi.org/10.1002/bip.22363
38. Pauwels K, Lebrun P, Tompa P (2017) To be disordered or not to be disordered: is that still a question for proteins in the cell? Cell Mol Life Sci 74(17):3185–3204. https://doi.org/10.1007/s00018-017-2561-6
39. Su XC, Jergic S, Ozawa K, Burns ND, Dixon NE, Otting G (2007) Measurement of dissociation constants of high-molecular weight protein-protein complexes by transferred 15N-relaxation. J Biomol NMR 38(1):65–72. https://doi.org/10.1007/s10858-007-9147-9
40. Varadan R, Assfalg M, Raasi S, Pickart C, Fushman D (2005) Structural determinants for selective recognition of a Lys48-linked polyubiquitin chain by a UBA domain. Mol Cell 18(6):687–698. https://doi.org/10.1016/j.molcel.2005.05.013
41. Yan R, Kelly G, Pastore A (2014) The scaffold protein IscU retains a structured conformation in the Fe-S cluster assembly complex. Chembiochem 15(11):1682–1686. https://doi.org/10.1002/cbic.201402211
42. Macek P, Kerfah R, Boeri Erba E, Crublet E, Moriscot C, Schoehn G et al (2017) Unraveling self-assembly pathways of the 468-kDa proteolytic machine TET2. Sci Adv 3(4):e1601601. https://doi.org/10.1126/sciadv.1601601
43. Rasmussen KK, Frandsen KE, Boeri Erba E, Pedersen M, Varming AK, Hammer K et al (2016) Structural and dynamics studies of a truncated variant of CI repressor from bacteriophage TP901-1. Sci Rep 6:29574. https://doi.org/10.1038/srep29574

44. Rasmussen KK, Palencia A, Varming AK, El-Wali H, Boeri Erba E, Blackledge M et al (2020) Revealing the mechanism of repressor inactivation during switching of a temperate bacteriophage. Proc Natl Acad Sci U S A 117(34):20576–20585. https://doi.org/10.1073/pnas.2005218117
45. Romano M, Fusco G, Choudhury HG, Mehmood S, Robinson CV, Zirah S et al (2018) Structural basis for natural product selection and export by bacterial ABC transporters. ACS Chem Biol 13(6):1598–1609. https://doi.org/10.1021/acschembio.8b00226
46. Xu Y, Maya-Martinez R, Guthertz N, Heath GR, Manfield IW, Breeze AL et al (2022) Tuning the rate of aggregation of hIAPP into amyloid using small-molecule modulators of assembly. Nat Commun 13(1):1040. https://doi.org/10.1038/s41467-022-28660-7
47. Spittler D, Indorato RL, Boeri Erba E, Delaforge E, Signor L, Harris SJ et al (2022) Binding stoichiometry and structural model of the HIV-1 Rev/importin beta complex. Life Sci Alliance 5(10). https://doi.org/10.26508/lsa.202201431
48. Jumper J, Evans R, Pritzel A, Green T, Figurnov M, Ronneberger O et al (2021) Highly accurate protein structure prediction with AlphaFold. Nature 596(7873):583–589. https://doi.org/10.1038/s41586-021-03819-2
49. Faini M, Stengel F, Aebersold R (2016) The evolving contribution of mass spectrometry to integrative structural biology. J Am Soc Mass Spectrom 27(6):966–974. https://doi.org/10.1007/s13361-016-1382-4
50. Mikhailov VA, Mize TH, Benesch JL, Robinson CV (2014) Mass-selective soft-landing of protein assemblies with controlled landing energies. Anal Chem 86(16):8321–8328. https://doi.org/10.1021/ac5018327
51. Robinson CV (2019) Mass spectrometry: from plasma proteins to mitochondrial membranes. Proc Natl Acad Sci U S A 116(8):2814–2820. https://doi.org/10.1073/pnas.1820450116
52. Pastore A, Temussi PA (2017) The Emperor's new clothes: myths and truths of in-cell NMR. Arch Biochem Biophys 628:114–122. https://doi.org/10.1016/j.abb.2017.02.008
53. Luchinat E, Banci L (2018) In-cell NMR in human cells: direct protein expression allows structural studies of protein folding and maturation. Acc Chem Res 51(6):1550–1557. https://doi.org/10.1021/acs.accounts.8b00147
54. Ben-Nissan G, Vimer S, Warszawski S, Katz A, Yona M, Unger T et al (2018) Rapid characterization of secreted recombinant proteins by native mass spectrometry. Commun Biol 1:213. https://doi.org/10.1038/s42003-018-0231-3
55. Cveticanin J, Netzer R, Arkind G, Fleishman SJ, Horovitz A, Sharon M (2018) Estimating interprotein pairwise interaction energies in cell lysates from a single native mass spectrum. Anal Chem 90(17):10090–10094. https://doi.org/10.1021/acs.analchem.8b02349
56. Gan J, Ben-Nissan G, Arkind G, Tarnavsky M, Trudeau D, Noda Garcia L et al (2017) Native mass spectrometry of recombinant proteins from crude cell lysates. Anal Chem 89(8):4398–4404. https://doi.org/10.1021/acs.analchem.7b00398
57. Rogawski R, Sharon M (2022) Characterizing endogenous protein complexes with biological mass spectrometry. Chem Rev 122(8):7386–7414. https://doi.org/10.1021/acs.chemrev.1c00217
58. Chorev DS, Baker LA, Wu D, Beilsten-Edmands V, Rouse SL, Zeev-Ben-Mordehai T et al (2018) Protein assemblies ejected directly from native membranes yield complexes for mass spectrometry. Science 362(6416):829–834. https://doi.org/10.1126/science.aau0976
59. Selenko P (2019) Quo vadis biomolecular NMR spectroscopy? Int J Mol Sci 20(6). https://doi.org/10.3390/ijms20061278
60. van de Waterbeemd M, Tamara S, Fort KL, Damoc E, Franc V, Bieri P et al (2018) Dissecting ribosomal particles throughout the kingdoms of life using advanced hybrid mass spectrometry methods. Nat Commun 9(1):2493. https://doi.org/10.1038/s41467-018-04853-x
61. Domanski M, Molloy K, Jiang H, Chait BT, Rout MP, Jensen TH et al (2012) Improved methodology for the affinity isolation of human protein complexes expressed at near endogenous levels. BioTechniques:1–6. https://doi.org/10.2144/000113864
62. Gavin AC, Maeda K, Kuhner S (2011) Recent advances in charting protein-protein interaction: mass spectrometry-based approaches. Curr Opin Biotechnol 22(1):42–49. https://doi.org/10.1016/j.copbio.2010.09.007
63. Camacho-Zarco AR, Schnapka V, Guseva S, Abyzov A, Adamski W, Milles S et al (2022) NMR provides unique insight into the functional dynamics and interactions of intrinsically disordered proteins. Chem Rev 122(10):9331–9356. https://doi.org/10.1021/acs.chemrev.1c01023
64. Muscat A, de Olde EM, Ripoll-Bosch R, Van Zanten HHE, Metze TAP, Termeer CJAM et al (2021) Principles, drivers and opportunities of a circular bioeconomy. Nat Food 2:561–566. https://doi.org/10.1038/s43016-021-00340-7
65. Messerli MA, Raihan MJ, Kobylkevich BM, Benson AC, Bruening KS, Shribak M et al (2019) Construction and composition of the squid pen from Doryteuthis pealeii. Biol Bull 237(1):1–15. https://doi.org/10.1086/704209
66. Montroni D, Sparla F, Fermani S, Falini G (2021) Influence of proteins on mechanical properties of a natural chitin-protein composite. Acta Biomater 120:81–90. https://doi.org/10.1016/j.actbio.2020.04.039

67. Morando MA, Venturella F, Sollazzo M, Monaca E, Sabbatella R, Vetri V et al (2022) Solution structure of recombinant Pvfp-5beta reveals insights into mussel adhesion. Commun Biol 5(1):739. https://doi.org/10.1038/s42003-022-03699-w
68. Santonocito R, Venturella F, Dal Piaz F, Morando MA, Provenzano A, Rao E et al (2019) Recombinant mussel protein Pvfp-5beta: a potential tissue bioadhesive. J Biol Chem 294(34):12826–12835. https://doi.org/10.1074/jbc.RA119.009531
69. Rosati S, Rose RJ, Thompson NJ, van Duijn E, Damoc E, Denisov E et al (2012) Exploring an orbitrap analyzer for the characterization of intact antibodies by native mass spectrometry. Angew Chem Int Ed Engl 51(52):12992–12996. https://doi.org/10.1002/anie.201206745
70. Rose RJ, Damoc E, Denisov E, Makarov A, Heck AJ (2012) High-sensitivity orbitrap mass analysis of intact macromolecular assemblies. Nat Methods 9(11):1084–1086. https://doi.org/10.1038/nmeth.2208
71. Whitelegge J (2013) Intact protein mass spectrometry and top-down proteomics. Expert Rev Proteomics 10(2):127–129. https://doi.org/10.1586/epr.13.10
72. Dominguez-Medina S, Fostner S, Defoort M, Sansa M, Stark AK, Halim MA et al (2018) Neutral mass spectrometry of virus capsids above 100 megadaltons with nanomechanical resonators. Science 362(6417):918–922. https://doi.org/10.1126/science.aat6457
73. Keifer DZ, Jarrold MF (2017) Single-molecule mass spectrometry. Mass Spectrom Rev 36(6):715–733. https://doi.org/10.1002/mas.21495
74. Worner TP, Aizikov K, Snijder J, Fort KL, Makarov AA, Heck AJR (2022) Frequency chasing of individual megadalton ions in an orbitrap analyser improves precision of analysis in single-molecule mass spectrometry. Nat Chem 14(5):515–522. https://doi.org/10.1038/s41557-022-00897-1
75. Worner TP, Snijder J, Friese O, Powers T, Heck AJR (2022) Assessment of genome packaging in AAVs using orbitrap-based charge-detection mass spectrometry. Mol Ther Methods Clin Dev 24:40–47. https://doi.org/10.1016/j.omtm.2021.11.013
76. Lee JH, Okuno Y, Cavagnero S (2014) Sensitivity enhancement in solution NMR: emerging ideas and new frontiers. J Magn Reson 241:18–31. https://doi.org/10.1016/j.jmr.2014.01.005

9 X-Ray Crystallography for Macromolecular Complexes

Francisco J. Fernández, Javier Querol-García, Sergio Navas-Yuste, Fabrizio Martino, and M. Cristina Vega

Abstract

X-ray crystallography has for most of the last century been the standard technique to determine the high-resolution structure of biological macromolecules, including multi-subunit protein-protein and protein-nucleic acids as large as the ribosome and viruses. As such, the successful application of X-ray crystallography to many biological problems revolutionized biology and biomedicine by solving the structures of small molecules and vitamins, peptides and proteins, DNA and RNA molecules, and many complexes—affording a detailed knowledge of the structures that clarified biological and chemical mechanisms, conformational changes, interactions, catalysis and the biological processes underlying DNA replication, translation, and protein synthesis. Now reaching well into the first quarter of the twenty-first century, X-ray crystallography shares the structural biology stage with cryo-electron microscopy and other innovative structure determination methods, as relevant and central to our understanding of biological function and structure as ever. In this chapter, we provide an overview of modern X-ray crystallography and how it interfaces with other mainstream structural biology techniques, with an emphasis on macromolecular complexes.

F. J. Fernández
Abvance Biotech SL, Madrid, Spain

J. Querol-García · S. Navas-Yuste · M. C. Vega (✉)
Centro de Investigaciones Biológicas Margarita Salas (CIB-CSIC), Madrid, Spain
e-mail: cvega@cib.csic.es

F. Martino
Structural Biology Research Centre, Human Technopole, Milan, Italy
e-mail: fabrizio.martino@fht.org

Keywords

Structural biology · X-ray crystallography · Complement system · Complement receptor CR3 · iC3b complement factor · Synchrotron radiation · Cryo-electron microscopy · X-free electron laser · Electron diffraction

9.1 Introduction

For nearly 60 years, macromolecular X-ray crystallography (MX) has provided a steady stream of high-quality, information-rich structures of biomolecules and of their complexes, to an extent that it is no exaggeration to state that the structural information generated by MX has revolutionized modern biology, biochemistry, and biomedicine. Until quite recently, the only high-resolution structures of macromolecular complexes have come from MX. There is virtually no aspect of

M. C. Vega, F. J. Fernández (eds.), *Advanced Technologies for Protein Complex Production and Characterization*, Advances in Experimental Medicine and Biology 1453,
https://doi.org/10.1007/978-3-031-52193-5_9

modern molecular and cellular biology that has not been transformed at a fundamental level by the possibility to decipher the intricate molecular architecture of macromolecular complexes that participate in those processes. Nowadays, MX is a mature discipline that has spurred the development of methods for data processing, data treatment, structure determination, electron density map interpretation, macromolecular model building and refinement, all of which play crucial roles in the advancement of macromolecular modeling and participate to varying degrees in the recent breakthrough in cryo-electron microscopy. In those various capacities, MX continues to be an active, important, and relevant discipline [1].

Many excellent reviews have been written on the use and practice of MX for the determination of crystal structures of macromolecular complexes [2, 3]. Our goal is to highlight new aspects of MX that facilitate work with macromolecular complexes. From new techniques to produce homogeneous and stable samples of multisubunit complexes to approaches to solve complex, multicomponent structures, this review attempts to offer practical guidance and summarize the main advances made on MX with regards to multisubunit complexes. When introducing a new section, we will make use of particularly interesting examples to make our arguments more memorable while showing real-life applications of the concepts discussed.

9.2 Producing Multisubunit Complexes Discovery

A major hurdle to any structural technique is obtaining sufficient amount of high-quality protein complex samples. This requirement is especially acute for MX since relatively large amounts of the protein sample are necessary to complete the sparse-matrix crystallization screens. This limitation has been alleviated by the arrival of nanocrystallization techniques, which allowed to perform between 3–10 times more crystallization experiments than conventional, manual techniques. By a "sufficient amount" we typically mean several mg of a protein complex at a concentration between 5–15 mg/mL (200-60 μL per mg). By "sufficient quality" we refer to samples that are stable over the course of the crystallization experiment (one day to several weeks to months), ideally lack macro- and microheterogeneity (*i.e.*, proteolytic degradation, chemical modifications), and are devoid of extreme conformational flexibility (which would preclude crystallization).

To reach the quality goal expression systems capable of generating the required amounts of protein complexes are a must. This is no trivial task since multisubunit complexes often have strong requirements with respect to the properties of the expression system. For example, some protein complexes cannot be assembled properly unless certain chaperones are present. Other protein complexes are assembled from individual components which cannot stand or produce diffraction crystals on their own (*e.g.*, because of lacking a defined structure outside their complex) (Fig. 9.1). In other cases, the net expression rates of all individual subunits must be matched to avoid substoichiometric complexes or the coexistence of complexes with varying stoichiometry. Finally, certain subunits have specific requirements that can only be matched by one or a few expression systems (*e.g.*, native system, post-translational modifications by eukaryotic chaperones).

Since the 1990s, the development of ever more powerful expression systems has been tied to the success of structural biology as a discipline. In parallel to the development of other key enabling technologies (*e.g.*, synchrotrons, X-ray detectors), high-throughput expression screening and miniaturization of expression tests has played a transformative effect on how modern structural biology is carried out. In particular, four expression systems stand out as having played important roles in advancing structural biology: (1) *Escherichia coli*; (2) Baculovirus-infected insect cell culture (BVS); (3) Mammalian cell culture; and (4) Cell-free systems (CFS) [5, 6]. In the context of this review, it should suffice to state that animal cell culture has made possible the production of previously inaccessible protein complexes, including transcription factor complexes, cellular receptors,

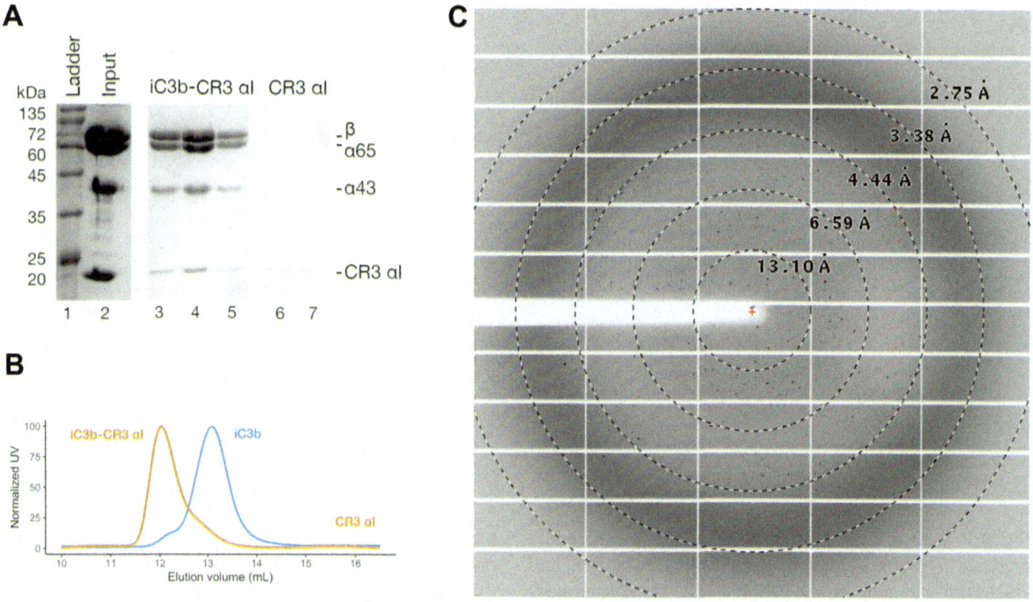

Fig. 9.1 Production and crystallization of the iC3b-CR3 αI heterodimeric complex. (**a**) Heterocomplex assembled *in vitro* from stoichiometric amounts of native source iC3b and recombinant CR3 αI. (**b**) Size-exclusion chromatography allows a complete separation between the heterocomplex and excess free components. (**c**) Diffraction image of iC3b-CR3 αI crystals. (Modified and reproduced with permission from [4])

membrane proteins, and many others, and could therefore be considered as the most disruptive expression technology of the turn of the twenty-first century. In contrast, the *E. coli* expression system has kept on marshaling the production of uncountable proteins and protein complexes for structural biology at a rate and at a cost-efficiency unmatched by any other system. The continuous optimization and fine-tuning of the *E. coli* system has ensured its stability and relevance throughout a period marked by the rise of animal cell culture. Finally, despite the fact that CFS has attracted attention because of its inherent adaptability to high-throughput settings and the promise of simplifying protein expression and processing workflows, its relatively higher costs and less predictable results have imposed severe limitations in its adoption by the community. Before moving on to crystallization, we should say that there are many other useful and productive expression systems that can be even more efficient to produce certain proteins or protein complexes, but their more specialized nature confines them to a narrower range of applications. Amongst these specialized systems are yeasts, protozoans, algal and unconventional bacterial expression systems. The reader is directed to the excellent reviews available on these topics for further information.

9.3 Crystallization and Crystal Handling

Crystallization is a well-known bottleneck for crystallography. Part of the difficulty with crystallization can be explained by the low probability with which large, flexible macromolecular entities settle on a thermodynamically stable 3D arrangement without precipitating or unfolding. Since it was first described, macromolecular crystallization has been considered an art that requires extensive trial-and-error screening to succeed. The transition from an art to a more systematic science, which is clearly not yet complete, has been unfolding slowly but steadily.

The process by which a macromolecule (or a complex) can be crystallized can be divided into nucleation and growth [7–9] (Fig. 9.2).

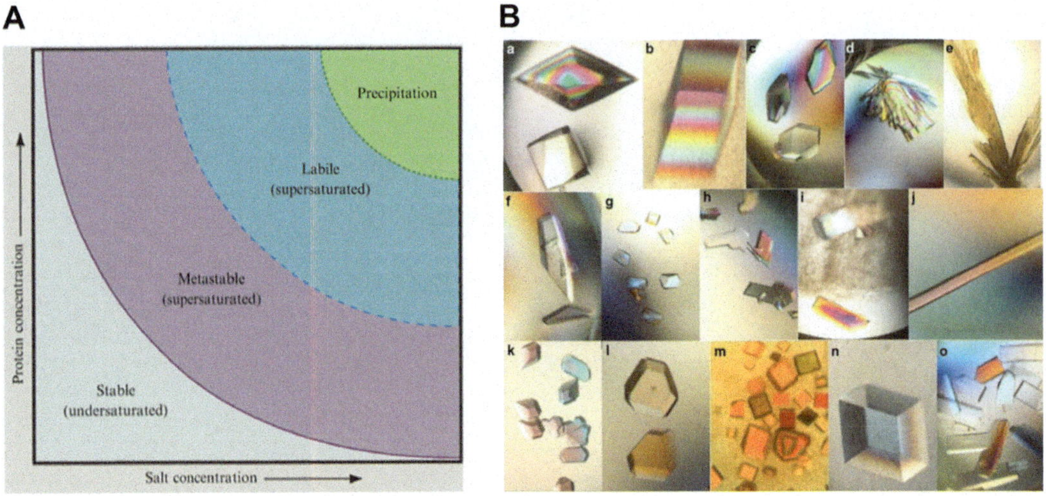

Fig. 9.2 **Macromolecular crystallization.** (**a**) Phase diagram of a protein crystallization. Both nucleation and crystal growth may occur in the labile (supersaturated) zone, while only crystal growth can occur in the metastable (supersaturated) zone. (**b**) Diverse protein crystals displaying the range of habits they may assume. (Reproduced with permission from [9])

Nucleation, the physicochemical process responsible for producing crystal seeds from initially homogeneously dissolved macromolecules, is an intrinsically stochastic process. Although much is now known about nucleation and how to bias the formation of crystal nuclei, finding adequate conditions to start crystal formation remains a trial-and-error process requiring extensive, time- and sample-consuming experimentation. Growth, which can only begin in the presence of crystal nuclei, has distinct requirements from nucleation. Although many times nucleation and growth can be accomplished under a single set of crystallization conditions, decoupling nucleation from growth by using micro- or macroseeding techniques can be advantageous when obtaining large, well-diffracting crystals proves challenging.

Nanocrystallization under regulated humidity conditions can be singled out as one of the transformative advances of crystallography [8, 10]. On the one hand, nanocrystallization allowed reducing the amount of costly protein samples consumed by a factor of 5–10 by simply making technically feasible to miniaturize crystallization experiments from 1–2 μL to 100–400 nL (0.1–0.4 μL). On the other hand, the reduction in sample volume allows to screen a wider range of crystallization conditions with a given amount of sample, hence increasing the chances of finding suitable crystallization conditions for scarce protein samples.

The accumulation of knowledge on the effect of various families of precipitants on the crystallization of macromolecules has permitted the design of more powerful sparse matrix crystallization screens, thus increasing the success of the crystallization experiments and the reduction in the actual chemical space that needs to be explored. Similarly, a greater understanding of the redundancies present in different crystallization screens has led to a more judicious selection of initial crystallization screens and an associated enhanced efficiency and economy on the use of valuable protein complex samples.

Micro- and macroseeding are crystallization techniques that attempt to decouple nucleation from growth in such a way that conditions can be optimized for either process independently [11, 12]. In certain cases, microseeds obtained from low-quality crystals or even microcrystalline precipitate have been used in sparse-matrix crystallization screens with higher success rates than the original macromolecular solution. Micro- and macroseeding have been automated using nanocrystallization robots. A particular application of microseeding using seeds obtained from a different but related macromolecule (heterogenous microseeding) has been successful for certain classes of macromolecules (*e.g.*, antibodies), and offers the promise of aiding the crystallization of otherwise reluctant protein samples.

Issues with cryoprotection of macromolecular crystals have also been tackled as optimization problems [13]. Screens of cryoprotectants have been devised, typically by mixing different families of cryoprotectants in osmotically balanced cocktails.

Although not conventionally regarded as part of the crystallization process, crystal handling and freezing, when properly done, presuppose the application of knowledge gained from the macromolecule's crystallizability properties. Factors such as crystal shape, size, density, and solvent content, in addition to the nature and concentration of the various precipitants and additives present in the crystallization experiment, have to be carefully considered. Lately, the first automated systems for crystal snap freezing have appeared that combine robust cameras and image recognition and processing software with robotic arms capable of looping the crystal and plunging them into liquid nitrogen with surgical precision.

With the advent of X-free electron laser (XFEL) sources (fourth generation synchrotron radiationSynchrotron radiation sources), it has become possible to image the structures of macromolecular complexes which only associate weakly in clusters of a few units, challenging the very naming of crystals for those species [14] (see Chap. 10 by A. Round et al.).

9.4 X-Ray Diffraction Experiment and Data Processing

X-ray diffraction (XRD) requires highly specialized synchrotron beamlines and sophisticated infrastructure, which are themselves adapted to the types of experiments that are performed on each beamline. Planning and executing XRD experiments depend on many considerations and critical decisions must be taken regarding crystal shape and size, alignment, radiation dose tolerance, internal symmetry, presence or absence of certain morphological defects (*e.g.*, twinning, pseudosymmetry), and the purpose of the experiment itself (*e.g.*, single-wavelength native dataset *vs.* multiple anomalous diffraction *vs.* sulfur phasing). The basic XRD experiment at a synchrotron source uses the rotation method of diffraction data collection, which has been used extensively and continues being the preferred method (Fig. 9.3). Excellent reviews have been written that cover some of these aspects and many others, and we will not review them again here. Our focus will be on specific features of the XRD experiments that affect the outcome of the experiments conducted on crystals grown of macromolecular complexes.

To start with, crystals of macromolecular complexes tend to reflect the specific physico-chemical properties of multiprotein complexes when compared to single proteins: they are characteristically more fragile, contain higher solvent content and their crystal unit-cell dimensions are larger. Diffraction patterns therefore become less sharp, reach lower resolution, exhibit a greater variety and intensity of crystal defects, and, overall, they tend to be less tolerant to radiation damage. In addition, the greater size of protein complexes translates into patterns of diffraction spots that are much closer along certain directions and sparser in other directions. The nonrandom distribution of reflection spots, combined with the more blurred point-spread function for protein complex crystals, increases the difficulties associated with collecting quality diffraction patterns.

9.4.1 Data Collection Strategies

Innovative data collection strategies have played fundamental roles in the successful solution of crystals from macromolecular complexes with various crystalline defects or pathologies. For example, mesh screening has allowed to find suitable

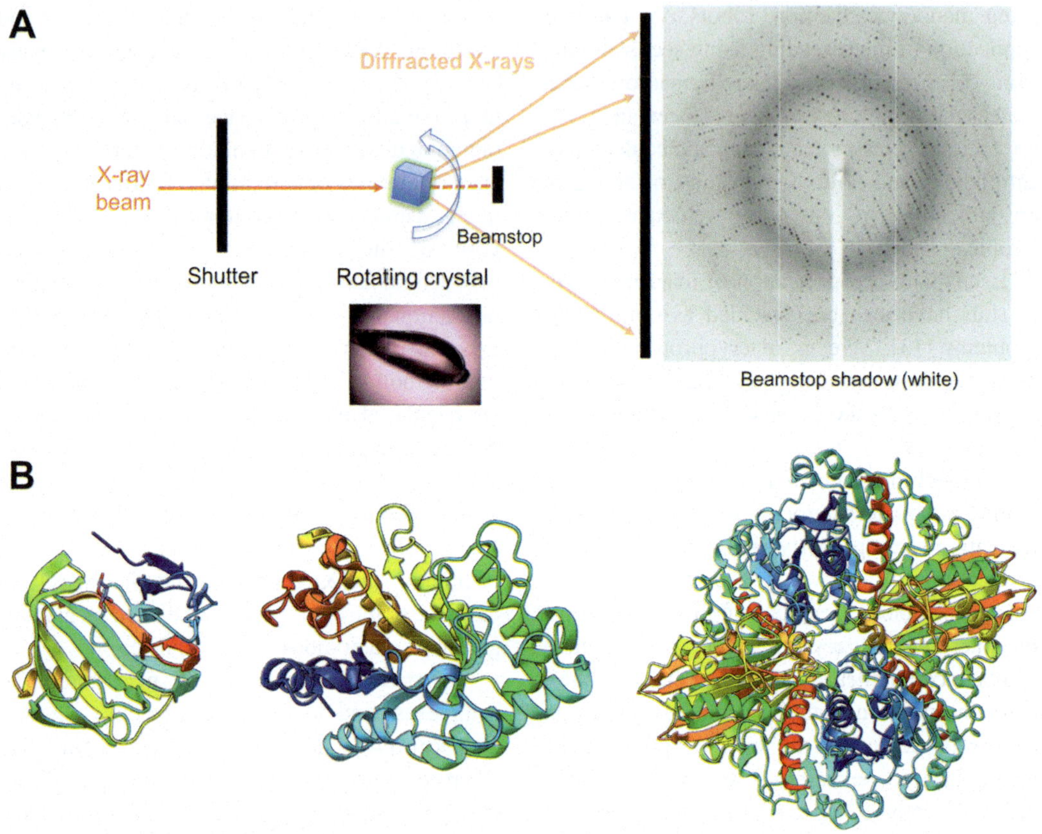

Fig. 9.3 The rotation method of X-ray diffraction data collection. (**a**) Schematic representation of the rotation method in MX. The incoming X-ray beam impinges on the macromolecular crystal as it is rotated around an axis perpendicular to the X-ray beam; after going through the crystal, the beamstop blocks it. The diffracted X-rays continue their journey to the detector, where they are recorded. A cryoprotected crystal inside a small nylon loop and one diffraction image collected from it are shown. (**b**) Most MX structures in the PDB have been solved using complete X-ray datasets acquired by the rotation method. Example MX structures are shown from left to right in cartoon representation and rainbow colors: endo-β-1,4-xylanase from *Fusarium oxysporum* (PDB ID 5JRN) [15], xylanase Xys1Δ from *Streptomyces halstedii* (PDB ID 1NQ6) [16], GAPDH from *Atopobium vaginae* (PDB ID 5LD5) [17]

crystal regions for diffraction in crystals with heterogeneous diffraction properties or where ice or salt precipitates may interfere with protein diffraction. For needle-shaped crystals or crystals with weak diffraction patterns, another non-conventional yet popular strategy consists in overexposing small regions of a larger crystal in narrow overlapping angular wedges and scaling and merging the independent wedges into a single data set. Finally, helical data collection is a refinement of the previous strategy which has facilitated the acquisition of complete data sets for extremely long, often flexible crystals. These approaches have benefitted from the development of more reliable automatic data processing and merging procedures, which have played a central role in successfully processing challenging data sets.

9.4.2 Multi-crystal and Microcrystal Mounts

Although diffracting single crystals is the standard method, there are few theoretical reasons why the crystallographer should mount a single crystal on a loop for diffraction. First and foremost, a single mounted crystal removes ambiguity about the source of the diffraction spots and avoids the need for computer deconvolution of overlapping diffraction patterns. However, for screening of many microcrystals it may be more practical to mount several crystals on one loop and rely on synchrotron beamlines to identify and diffract each crystal independently. Specialized mounts devised for accommodating several crystals in a row are under testing. In its most extreme form, direct diffraction of screening crystallization plates [18] can be done at certain beamlines which have implemented the required hardware and software procedures. Successful in-plate diffraction requires automated light-microscope imaging of the crystallization plate, software tools to mark crystallization drops and specific crystals for XRD, and robotic plate transfer. In the context of macromolecular complexes and membrane protein crystallography, multi-crystal and microcrystal mounts could speed up the process of screening optimal conditions for the diffraction experiment.

9.4.3 Multi-crystal Approaches and Serial X-Ray Crystallography

Alongside conventional single-axis rotation diffraction, other diffraction methods have been adopted to collect XRD data optimally for specific applications. For example, multi-crystal native SAD (MDS) and serial X-ray crystallography (SSX) exploit the fact that low-dose XRD spots can be collected from multiple (isomorphous) crystals, processed, and merged to yield highly complete, highly redundant datasets. These strategies work better than more conventional approaches when crystals are very radiation sensitive or when the anomalous diffraction effects that need to be measured are close to the background intensity signal. Although they still rely on a rotating crystal, the angular wedge that must be sampled per individual crystal can be reduced.

9.4.4 X-Ray Free Electron Laser (XFEL)

Taken to its extreme, XRD data can also be collected from X-ray sources other than synchrotrons and in-house diffractometers. For example, the highly energetic X-ray laser pulses created by X-ray free electron laser sources (XFEL) allow to collect diffraction data from crystallites 100-nm across [19] (see Chap. 10 by A. Round et al. for further information on XFEL). A single shot can be collected from each crystallite, since the large amount of energy absorbed by the diffracted crystallite destroys it a few femtoseconds later—the "diffraction before destruction" method. The intensity and destructive power of the XFEL laser pulse is incompatible with the rotating method. In contrast, XFEL uses many stationary crystals in random orientations to compensate for the few reflections that are in reflecting condition while stationary. To collect a complete dataset, it is therefore necessary to direct the X-ray laser pulses toward a steady stream of crystallites flowing along a direction perpendicular to the direction of the laser pulses. The data collection technique developed for this purpose is called serial femtosecond crystallography (SFX) [19]. Although still in its infancy, XFEL facilities and the SFX collection method show great promise for crystallographic structure determination of large complexes, since the requirement for large crystals, which is generally harder to realize for large multisubunit complexes than for single proteins, is partially relieved. In addition to static structural information, XFEL and related developments can potentially yield information on time-resolved processes.

9.5 Phasing and Structure Determination

As with X-ray data collection, the structure determination methods for macromolecular crystallography have been the subject of rigorous reviews. For this chapter, we would like to emphasize that most protein complex structures require special treatment because of the number of subunits and often the intricacies of protein-protein interfaces. Protein heterocomplex structures are rarely solved exclusively by *ab initio* or *de novo* methods unless the number of subunits is low (<3); homocomplexes are different in that the repetition of identical subunits allows for the application of powerful electron-density averaging methods capable of solving very large structures. Higher-order heterocomplexes typically need additional information in the form of partial molecular replacement searches.

Single-anomalous diffraction (SAD) using the weak anomalous signal extractable from light atoms by diffracting native crystals with long-wavelength X-rays is a powerful technique that does not require heavy-atom derivatization, nor does it introduce artefacts in the structure by virtue of covalent bonding to heavy metals. Long-wavelength SAD is epitomized by sulfur (S)-SAD, although in practice the anomalous scattering from other light atoms contribute to the overall intensity and measurability of the anomalous diffraction effects from native crystals. Besides S, atoms such as P, Cl^-, Br^-, K^+ and Ca^{2+} have measurable anomalous diffraction effects. For years, both the available hardware and analytic techniques have dictated that S-SAD phasing was only feasible for crystals of small proteins diffracting to very high resolution and, hence, for asymmetric units containing small, compact proteins. This consideration alone would suggest that S-SAD is not suited for the study of protein complexes, whose size and less ordered arrangements reduce the diffraction power of their crystals. However, recent breakthroughs in S-SAD phasing from the Hendrickson lab (the multi-crystal native SAD method) [20–22], the conquest of low-diffracting crystals (as low as 4 Å) [23], and the determination of moderately large protein complexes by S-SAD all argue otherwise. The possibility that native SAD is in practice not limited to strongly diffracting crystals and hence applicable to macromolecular complexes has been brilliantly demonstrated by the phasing of the 266-kDa T2R-TTL (αβ-tubulin, stathmin-4 and tubulin-tyrosine ligase) complex [24] and the 132-kDa histidine kinase TorT/TorSS complex [25]. With the recent deployment of long-wavelength SAD dedicated synchrotron beamlines furnished with the most recent technological advancements aimed at boosting the signal-to-noise ratio from the weak anomalous signal from light atoms like S, P, and K^+, we can be confident that many more crystal structures of macromolecular complexes will become solvable by these methods in the coming years [1, 26].

9.6 Macromolecular Refinement

In MX, refinement refers to the process by which the parameters describing the macromolecular structure are iteratively updated to increase the match between the model and the experimental data. Parameters such as the spatial atomic positions (x, y, z) and the atomic displacement parameters (ADP) or B-factors are typically optimized during refinement.

Refining large macromolecular complexes is a challenging task confounded by the sheer size, number of subunits, greater flexibility, and disorder present in those structures. Dedicated refinement algorithms and software implementing them have been developed precisely to circumvent those limitations of conventional macromolecular refinement programs.

Classically, refining multisubunit complexes has relied upon rigid-body refinement (to adjust starting positions) and noncrystallographic symmetry (NCS) averaging across complete models or domains. In both cases, the strategies make use of reduced representations of the highly complex macromolecular models to improve the crystallographic phases at the onset, when the mean phase error of the (incomplete) models is greater. In some cases, these strategies are sufficient to produce electron density maps of sufficiently high quality to allow progressive manual building and refinement cycles until a complete model, with acceptable geometric and stereochemical parameters, finally emerges. In most cases, however, significant portions of the macromolecular models (sometimes entire subunits) cannot be traced to any satisfactory degree. Deficiencies in multisubunit models can also contain imperfections in the tracing of many side chains, secondary structures, and loop regions, in addition to more discrete errors as pointed out before.

To cope with the increased flexibility of complex macromolecular models, several algorithms have been proposed to treat macromolecules (or well-defined parts of them, like helices) as elastic entities that can be continuously bent or twisted to better fit the observed data. Elastic Network (EN) modeling represents an improvement for many-atom models because it finds shortcuts across complicated energy landscapes which would take ages to traverse for conventional algorithms. As rigid-body modeling and NCS averaging, EN modeling has the advantage of reducing the dimensionality of the macromolecular models while allowing a seamless transition between the EN and atomic refinement.

What's the role of *B*-factor modeling for macromolecular complexes? Large complexes often show highly anisotropic distributions of *B*-factors, including significant differences in overall and atomic *B*-factors across subunits. To improve the treatment of these and other limitations of the default isotropic *B*-factor, TLS modeling was invented (TLS, or Translation/Libration/Screw, a mathematical model that predicts the local positional displacement of atoms in a crystal structure). Macromolecular models can exhibit wild variations within and between domains, as well as between side and main-chain atoms. Part of this variation can be accurately modeled by, *e.g.*, assuming a continuous variation of *B* factors along bonded atoms, which can be expressed as *B*-factor restraints.

9.7 Model Building

Macromolecular complexes pose additional challenges to be properly modeled from XRD data. The sheer amount of information that must be captured to duly describe the crystal structure of a multisubunit complex has spurred additions and expansions to the conventional PDB models. For example, the PDB file for the ribosome contains many additional chain labels absent in most other PDB entries. Unconventional techniques for PDB annotation are required to describe the additional complexity present in large complexes. Another format has been developed that removes all restrictions on the size of the structure: the macromolecular Crystallographic Information File (mmCIF). The mmCIF is a well-structured format based on a dictionary, which includes definitions and descriptions about the structure and its experimental determination. Besides the more complicated bookkeeping imposed by the size of macromolecular complex structures, other challenges for modeling of complexes have to do with the increased number of degrees of freedom and the comparatively lower data density of the diffraction patterns.

Using the macromolecular models of complexes by other modeling techniques is also affected by the size and complexity of the models and requires imposing restraints and combining the modeling methods with experimental restraints to maintain reasonable trajectories during modeling.

Rich restraint sets can be obtained from many biochemical, biophysical, and structural methods including nuclear magnetic resonance (NMR), small-angle X-ray scattering (SAXS), and mass spectrometry (MS). SAXS is well-suited as a companion technique to XRD, since it can provide corroboration of the architecture and overall subunit arrangement of the complex as well as lending itself to the quick testing of structural hypotheses (see Chap. 11 by S. Hutin et al.).

9.8 Structure Analysis and Interpretation

The goal of determining the crystal structure of a macromolecular complex is to understand the biological processes the chosen complex participates in. To draw sound inferences about the function of a protein complex, the structure must represent a functional form of the complex, in the correct (physiological) stoichiometry. Otherwise, the inferences drawn might be wrong or be seriously misleading. For complexes, a central concern is how to define the biological assembly starting from the array of experimentally observed crystallographic assemblies (*i.e.*, those defined by crystal contacts regardless of their presence in solution) [27]. Since the precise solution assembly further depends on the protein constructs used for crystallization and the presence of other molecules (*e.g.*, allosteric modulators, binding partners), it is important to recognize the correct biological assembly. Properties that aid the recognition of the correct biological assembly include the stoichiometry, interface sequence and structural conservation, and symmetry.

Often, the process of reconstructing biological function from crystal structures is stepwise and laborious and may need the use of concurrent evidence garnered from structural biology methods or from other disciplines (*e.g.*, cell biology). In all cases, technical accuracy, and a deep understanding of the physicochemical properties of proteins and other macromolecular entities is essential. Wrongly building models that seem to agree with the experimental data but conflict with other pieces of evidence should indicate to the practitioners that something is seriously flawed about the models thus built.

A limitation of many structural techniques is that they can only capture a snapshot of the macromolecules, not the entire biological process as it is supposed to unfold. Crystallography can describe individual lowest-energy conformations one at a time (or, rather, one per crystal). If the crystallographers were lucky, two lowest-energy conformations can be co-crystallized simultaneously, although such cases are statistically improbable and thus are only very rarely found.

Despite that limitation, critical information can be obtained from crystal structures of protein complexes. For example, the organization of protein-protein and protein-nucleic acid interfaces are very informative about the functions performed by the interacting partners. It also frequently suggests ways in which functionally important movements or dynamics could occur by showing the presence of hinges, glide axes or small molecule-binding pockets that may block or impair the interaction. The accuracy of crystallographic models ensures that the conclusions drawn about protein-protein and protein-nucleic acid interfaces have solid grounds.

Especially for novices on MX who try to make sense of complex macromolecular models, it should be noted that the degree of uncertainty in atomic positions (and even of complete side chains and segments of structure) are determined by local and across-the-crystal disorder.

9.9 Power of Two: Combining XRD with SAXS and Cryo-EM

Typically, SAXS and XRD have been used jointly to produce combined interpretations of macromolecular complexes [reviewed in 28]. Typically, SAXS provides pseudo-independent corroboration for unusual or highly novel features observed in XRD models and can be used to prove which one of several possible heteromeric arrangements corresponds to the physiological state (as observed in solution). At a minimum, agreement between the experimental solution scattering and the theoretical 1D scattering calculated from various possible complexes observed in crystalline form can confirm the absence of gross structural artifacts. Beyond this confirmatory role for SAXS, incomplete XRD models due to, *e.g.*, absence of interpretable electron density in loop regions, SAXS data can also be used to reconstruct missing loops or larger parts of the model. SAXS data has also been used in the stepwise reconstruction of ever larger macromolecular complexes, when crystals of the complete complexes cannot be obtained while structures of the individual components or subcomplexes exist. Rigid and flexible fitting of subcomplexes into the SAXS data for complete complexes are powerful methods to discern the interrelationships linking the complex subunits. The one-to-one correspondence between a 3D structure and its theoretical 1D scattering curve ensures that XRD and SAXS can be easily and confidently combined to produce reliable interpretations of the structural complexes.

XRD and cryo-EM are highly compatible techniques [29]. Before the resolution revolution undergone by cryo-EM [30, 31], modeling of supramolecular complexes by cryo-EM had to rely on atomic models reconstructed by XRD. Docking of XRD models into cryo-EM maps, either as rigid bodies or after allowing for limited flexible modifications to the models, represents the most common approach to combine the two techniques. The docking approach can be viewed as the incorporation of high-resolution structural information (on specific components) on a larger complex solved at otherwise moderate resolution. The robustness of the Coulomb density maps built from low-resolution cryo-EM data and the precision of docking algorithms have both been confirmed by the verification of the results when higher-resolution cryo-EM structures became available.

Since the advent of modern electron microscopes, direct detectors and the software advances that have made possible the collection of high-resolution images of macromolecules by cryo-EM, XRD and cryo-EM have assumed an even more complementary role [32] (see Chap. 12 by A. Deniaud et al. and Chap. 13 by J.L. Carrascosa). Software developments made for model building, refinement, and validation for XRD can now be adapted for cryo-EM. In fact, historically crystallography-exclusive software suites like *CCP4* and *PHENIX* have forked out applications to extend their usefulness for cryo-EM models [33–35]. The increase in quality and resolution of cryo-EM maps has facilitated the use of such maps for the phasing and structure determination of XRD structures [36, 37]. This has proven critical in same cases where XRD amplitudes were measured to higher resolution than the cryo-EM images of the sample complex, but the crystal structure could not be phased by XRD alone. The phases measured by cryo-EM, lacking from the crystallographic data, have in some cases allowed the phasing and then the complete structure determination of very intricate complexes. A case in point is the crystal structure of the TLR13-ssRNA complex, which diffracted to 2.3 Å resolution but was phased with a 4.8-Å cryo-EM map of the same complex [38]. Another example is the T7 bacteriophage portal tridecamer (gp8-13mer) and the closed-conformation dodecamer ($gp8_{closed}$) structures. To phase the 3.4-Å gp8-13mer XRD data, a partial model comprising 36% of the structure was first built *ab initio* using a 5.8-Å-resolution cryo-EM map that was successfully used as a molecular replacement model; phase extension from 5.8 to 3.4 Å allowed the tracing of the complete gp8-13mer structure, a monomer of which was then used to solve the 3.6-Å $gp8_{closed}$ crystal structure, again by molecular replacement [37, 39] (Fig. 9.4).

When solving the structure of macromolecular biological assemblies, atomic model building is probably the step where cryo-EM probably has its tightest links to MX. Even though cryo-EM can nowadays produce maps of macromolecular biological complexes at resolution higher than 3 Å thus allowing *de novo* modelling of atomic structures, still the resolution and quality of the maps for many cryo-EM structures is not high enough to support the usage of auto-building software and therefore the building of atomic models relies on the availability of thousands of atomic models deposited in the PDB database. Nevertheless, the quality of the maps generated by cryo-EM is high enough to almost interpret the maps unambiguously. This achievement was supported by hardware development, the most important being the commercialization of direct electron detectors (also called DED) [reviewed in 40]. Direct electron detectors record electrons directly and can assign each electron to a specific pixel. Before DED the detectors recorded photons emitted by the electrons that were hitting a scintillator placed on top of the detectors. DED can also collect movies, allowing the user to correct for beam and stage-induced movements of the samples.

The resolution revolution was possible also thanks to an improvement in the hardware of the electron microscopes. In modern microscopes, the sample is enclosed into the vacuum system of the microscope and is handled by a mechanism called autoloader. Autoloaders allow the screening of up to 12 samples in the same session and importantly increase the stability of the whole microscope. Also, microscope optics have improved in stability. Despite the improvements in the quality of the maps generated by the new

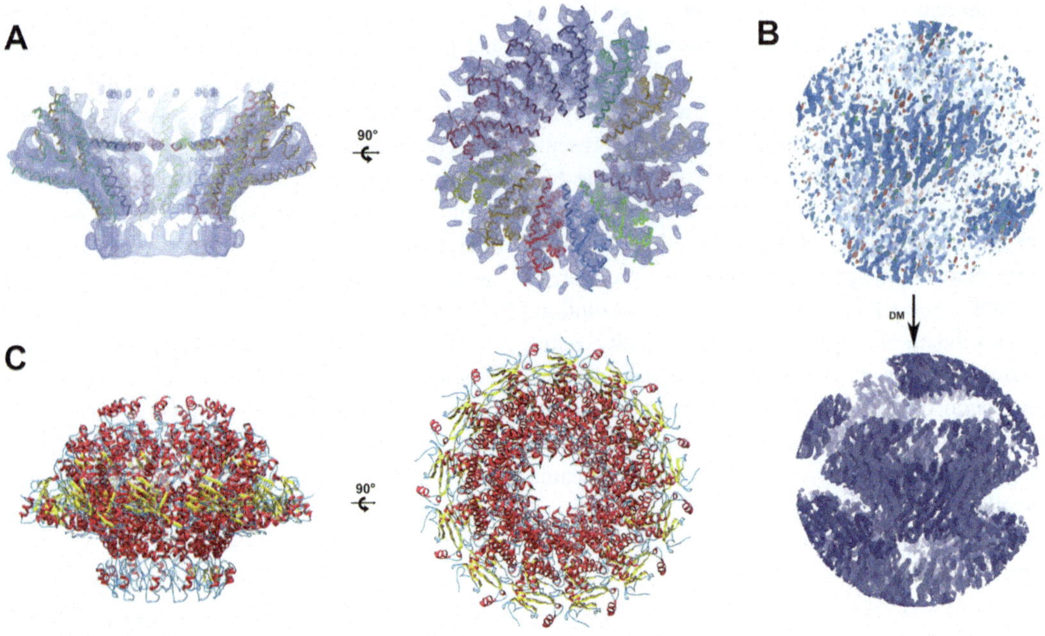

Fig. 9.4 **Combining MX with cryo-EM to solve the crystal structure of T7 bacteriophage gp8-13mer.** (**a**) Poly-Alanine model (36% of the structure) of gp8-13mer built into the 5.8-Å cryo-EM map. This initial model was used as molecular-replacement (MR) search model against a 3.4-Å XRD dataset. (**b**) Density modification dramatically improved the quality of the initial electron density map obtained by MR using the poly-Ala model in (**a**) (top), yielding a readily interpretable map for further model building (bottom). (**c**) Final cartoon model of gp8-13mer (PDB ID 6TJP). (Reproduced with permission from [37])

generation microscopes, for most of the maps generated by cryo-EM *de novo* building has been strongly dependent on homology modelling made possible by software, such as I-TASSER [41] or AlphaFold [42], which rely on the availability of atomic models already deposited in the PDB database that were mainly produced by MX experiments. In the past few years new tools have been developed and old tools have been adapted to the needs of cryo-EM. *Namdinator* is a web tool worth mentioning because it is straightforward to use and very reliable at the same time [43]. *Namdinator* uses molecular dynamics to fit an atomic model into a cryo-EM map. It can be used from initial fitting to regularization of geometries to correct outliers, clashes, and other model problems. Macromolecular refinement and model building programs central to MX like *PHENIX* [35], *REFMAC5* [44], and *Coot* [45] have been adapted to the needs of cryo-EM.

More recently two groups reported cryo-EM maps that have a resolution better than 2.0 Å showing that cryo-EM can now reach real "atomic resolution" and can resolve details at the level of hydrogens, paving the way for automatic *de novo* building also from cryo-EM maps. The Scheres laboratory at the LMB in Cambridge, UK, obtained a 1.7 Å resolution cryo-EM reconstruction for a prototypical human membrane protein, the β3 $GABA_A$ receptor homopentamer, using a cold field emission gun (cold FEG), a new electron source, energy filter, and a new prototype of the Falcon camera called Falcon IV [46]. When they used a protein that is considered a standard for cryo-EM studies (like lysozyme for MX) such as mouse apo-ferritin, they reached 1.2 Å resolution. The Stark laboratory at the Max Planck Institute for Biophysical Chemistry in Göttingen, Germany, reported a 1.25 Å resolution structure of apo-ferritin obtained by cryo-EM with a monochromator and a newly developed spherical aberration corrector [47].

The elevated purchasing and maintenance costs of high-end electron microscopes constitute a bottleneck for the spreading of electron microscopy in all the research centers that would need it. If we compare EM with XRD, what cryo-EM is missing is the equivalent of in-house X-ray sources and detectors to allow the screening and preliminary data collection of samples. Some progress has been made in this direction. For example, Russo and Henderson are working on a microscope that combines low cost and high performance and that would allow the capillary spreading of cryo-EM everywhere [48]. In their proof-of-concept set-up these authors used the widespread 100-keV electron microscope, an instrument that is already used in many institutes for screening cryo-EM grids and for negative staining projects. Henderson and Russo have shown that this low-end microscope can however provide maps at high resolution when equipped with an Eiger detector from Dectris and a FEG as electron source. Such a scope has purchasing and running costs that are at least an order of magnitude lower than those of high-end microscopes. This is an outstanding achievement, because it paves the way for affordable cryo-EM accessible to almost any laboratory. Scientists could use these affordable scopes for preliminary data acquisition and even for solving most of their structures and dedicate the high-end scopes, for example, those present at synchrotrons, for improving resolution or for very challenging cases. This resembles the approach used by MX, where most of the work is done in-house and the synchrotron radiation is used for improvement of the resolution and quality of the data.

It is important to mention that electron diffraction (ED) is gaining more and more visibility in the field of structural biology. The group of Tamir Gonen at UCLA collected ED data from small 3D protein crystals and even extended the technique to "invisible" crystals, as small as 300–800 nm across [49]. Recently, the same group collected ED data from canonical lysozyme crystals. They thinned the crystals with a Focused Ion Beam milling apparatus. Then, they collected high-resolution diffraction data by using a low-intensity beam coupled with a new generation Falcon 4 detector in the modality that ensures counting of single diffraction events. The combination of all the techniques generated such high-quality data that the authors could solve the phases with *ab initio* methods starting from a short peptide with as few as three alanine residues for lysozyme and a 14-residue α-helical fragment in the case of proteinase K crystals [50].

Excitingly, new techniques are being developed to acquire ED data. Several groups were able to set up protocols that allow the continuous (or serial) acquisition of data. Serial/continuous MicroED has rapidly developed into a highly promising field [51], with distinct advantages of MicroED over conventional XRD like less sample consumption, smaller crystals, and the reutilization of XRD software (sometimes with some modifications) to perform data processing and structure determination. By filling the gaps between XRD and cryo-EM, MicroED has the potential to become a game changer in the field of structural biology.

The software for cryo-EM data processing has also improved over the years. It is not the aim of this chapter to list all the new tools and software available, but it is worth mentioning that Dimitry Tegunov, together with Patrik Cramer's and Julia Mahamid's groups, has developed a new software called M, which allowed them to solve *in situ* (*i.e.*, directly in the cell) the structure of ribosomes at 3.7 Å resolution. This is a milestone in the field because it proves that cryo-electron tomography can reach resolutions very close to the ones reached by single particle analysis. It also suggests that the available technology is suited for high-resolution structure investigation of complexes even in vivo, where the bottlenecks are sample preparation and data processing [52].

Many other experimental as well as theoretical approaches have been successfully combined with XRD, including NMR, MS, and others. Furthermore, theoretical approaches like molecular dynamics and quantum mechanical calculations require highly detailed and accurate crystal structures as starting points to ensure trajectory stability and a reasonable probability of finding the correct local minima during calculation.

9.10 Conclusions

- MX is a powerful and relevant technique for structure determination of large multi-subunit protein complexes.
- MX can be combined with other techniques, including cryo-EM.
- Derive static as well as dynamical information, and even time-resolved information.

Acknowledgements This work was supported by Spanish *Ministerio de Ciencia, Innovación y Universidades-FEDER* grant RTI2018-102242-B-I00 (MCV), the Spanish Ministerio de Ciencia e Innovación-Recovery, Transformation and Resilience Plan (PRTR) grant PDC2022-133713-I00 (MCV), grant S2022/BMD-7278 of the Regional Government of Madrid (MCV), the European Commission – NextGenerationEU through CSIC's Global Health Platform ("PTI Salud Global") (SGL2103020) (MCV), and the CSIC Special Intramural Grant PIE201620E064 (MCV). It was additionally supported by the Research Network on Complement in Health and Disease (RED2022-134750-T).

References

1. Grimes JM, Hall DR, Ashton AW, Evans G, Owen RL, Wagner A, McAuley KE, Von Delft F, Orville AM, Sorensen T, Walsh MA, Ginn HM, Stuart DI (2018) Where is crystallography going? Acta Crystallogr Sect Struct Biol 74:152–166. https://doi.org/10.1107/S2059798317016709
2. Ilari A, Savino C (2008) Protein structure determination by X-ray crystallography. Methods Mol Biol 452: 63–87. https://doi.org/10.1007/978-1-60327-159-2_3
3. Ilari A, Savino C (2017) A practical approach to protein crystallography. Methods Mol Biol 1525:27896717. https://doi.org/10.1007/978-1-4939-6622-6_3
4. Fernández FJ, Santos-López J, Martínez-Barricarte R, Querol-García J, Martín-Merinero H, Navas-Yuste S, Savko M, Shepard WE, Rodríguez de Córdoba S, Vega MC (2022) The crystal structure of iC3b-CR3 alphaI reveals a modular recognition of the main opsonin iC3b by the CR3 integrin receptor. Nat Commun 13:1955. https://doi.org/10.1038/s41467-022-29580-2
5. Fernández FJ, Vega MC (2013) Technologies to keep an eye on: alternative hosts for protein production in structural biology. Curr Opin Struct Biol 23:365–373. https://doi.org/10.1016/j.sbi.2013.02.002
6. Fernández FJ, Vega MC (2016) Choose a suitable expression host: a survey of available protein production platforms. In: Vega MC (ed) Advanced technologies for protein complex production and characterization. Springer, Cham, pp 15–24
7. Deivanayagam C, Cook WJ, Walter MR (2007) Protein crystallization. In: Fisher PB (ed) Cancer genomics and proteomics. Humana Press, Totowa, pp 337–349
8. Chayen NE (2009) High-throughput protein crystallization. In: Advances in protein chemistry and structural biology. Elsevier, pp 1–22
9. McPherson A (2017) Protein crystallization. In: Wlodawer A, Dauter Z, Jaskolski M (eds) Protein crystallography. Springer, New York, pp 17–50
10. Lynch ML, Snell ME, Potter SA, Snell EH, Bowman SEJ (2023) 20 years of crystal hits: progress and promise in ultrahigh-throughput crystallization screening. Acta Crystallogr Sect Struct Biol 79:198–205. https://doi.org/10.1107/S2059798323001274
11. D'Arcy A, Villard F, Marsh M (2007) An automated microseed matrix-screening method for protein crystallization. Acta Crystallogr D Biol Crystallogr 63:550–554. https://doi.org/10.1107/S0907444907007652
12. D'Arcy A, Bergfors T, Cowan-Jacob SW, Marsh M (2014) Microseed matrix screening for optimization in protein crystallization: what have we learned? Acta Crystallogr Sect F Struct Biol Commun 70:1117–1126. https://doi.org/10.1107/S2053230X14015507
13. Pflugrath JW (2015) Practical macromolecular cryocrystallography. Acta Crystallogr Sect F Struct Biol Commun 71:622–642. https://doi.org/10.1107/S2053230X15008304
14. Helliwell JR (2017) New developments in crystallography: exploring its technology, methods and scope in the molecular biosciences. Biosci Rep 37. https://doi.org/10.1042/BSR20170204
15. Gómez S, Payne AM, Savko M, Fox GC, Shepard WE, Fernandez FJ, Cristina Vega M (2016) Structural and functional characterization of a highly stable endo-beta-1,4-xylanase from Fusarium oxysporum and its development as an efficient immobilized biocatalyst. Biotechnol Biofuels 9:191. https://doi.org/10.1186/s13068-016-0605-z
16. Canals A, Vega MC, Gomis-Rüth FX, Díaz M, Santamaría RI, Coll M (2003) Structure of xylanase Xys1Δ from *Streptomyces halstedii*. Acta Crystallogr D Biol Crystallogr 59:1447–1453. https://doi.org/10.1107/S0907444903012629
17. Querol-García J, Fernández FJ, Marin AV, Gómez S, Fullà D, Melchor-Tafur C, Franco-Hidalgo V, Albertí S, Juanhuix J, Rodríguez de Córdoba S, Regueiro JR, Vega MC (2017) Crystal structure of Glyceraldehyde-3-phosphate dehydrogenase from the gram-positive bacterial pathogen A. vaginae, an Immunoevasive factor that interacts with the human C5a anaphylatoxin. Front Microbiol 8:541. https://doi.org/10.3389/fmicb.2017.00541
18. le Maire A, Gelin M, Pochet S, Hoh F, Pirocchi M, Guichou JF, Ferrer JL, Labesse G (2011) In-plate protein crystallization, in situ ligand soaking and X-ray diffraction. Acta Crystallogr Sect Biol Crystallogr 67:747–755. https://doi.org/10.1107/S0907444911023249

19. Chapman HN, Fromme P, Barty A, White TA, Kirian RA, Aquila A, Hunter MS (2011) Femtosecond X-ray protein nanocrystallography. Nature 470:73–77. https://doi.org/10.1038/nature09750
20. Liu Q, Dahmane T, Zhang Z, Assur Z, Brasch J, Shapiro L, Mancia F, Hendrickson WA (2012) Structures from anomalous diffraction of native biological macromolecules. Science 336:1033–1037. https://doi.org/10.1126/science.1218753
21. Liu Q, Liu Q, Hendrickson WA (2013) Robust structural analysis of native biological macromolecules from multi-crystal anomalous diffraction data. Acta Crystallogr D Biol Crystallogr 69:1314–1332. https://doi.org/10.1107/S0907444913001479
22. Liu Q, Hendrickson WA (2015) Crystallographic phasing from weak anomalous signals. Curr Opin Struct Biol 34:99–107. https://doi.org/10.1016/j.sbi.2015.08.003
23. El Omari K, Iourin O, Kadlec J, Fearn R, Hall DR, Harlos K, Grimes JM, Stuart DI (2014) Pushing the limits of sulfur SAD phasing: *de novo* structure solution of the N-terminal domain of the ectodomain of HCV E1. Acta Crystallogr D Biol Crystallogr 70:2197–2203. https://doi.org/10.1107/S139900471401339X
24. Weinert T, Olieric V, Waltersperger S, Panepucci E, Chen L, Zhang H, Zhou D, Rose J, Ebihara A, Kuramitsu S, Li D, Howe N, Schnapp G, Pautsch A, Bargsten K, Prota AE, Surana P, Kottur J, Nair DT, Basilico F, Cecatiello V, Pasqualato S, Boland A, Weichenrieder O, Wang B-C, Steinmetz MO, Caffrey M, Wang M (2015) Fast native-SAD phasing for routine macromolecular structure determination. Nat Methods 12:131–133. https://doi.org/10.1038/nmeth.3211
25. Moore JO, Hendrickson WA (2012) An asymmetry-to-symmetry switch in signal transmission by the histidine kinase receptor for TMAO. Structure 20:729–741. https://doi.org/10.1016/j.str.2012.02.021
26. Sanchez-Weatherby J, Sandy J, Mikolajek H, Lobley CMC, Mazzorana M, Kelly J, Preece G, Littlewood R, Sørensen TL-M (2019) VMXi: a fully automated, fully remote, high-flux *in situ* macromolecular crystallography beamline. J Synchrotron Radiat 26:291–301. https://doi.org/10.1107/S1600577518015114
27. Xu Q, Dunbrack RL (2019) Principles and characteristics of biological assemblies in experimentally determined protein structures. Curr Opin Struct Biol 55:34–49. https://doi.org/10.1016/j.sbi.2019.03.006
28. Putnam CD, Hammel M, Hura GL, Tainer JA (2007) X-ray solution scattering (SAXS) combined with crystallography and computation: defining accurate macromolecular structures, conformations and assemblies in solution. Q Rev Biophys 40:191–285. https://doi.org/10.1017/S0033583507004635
29. Wang H-W, Wang J-W (2017) How cryo-electron microscopy and X-ray crystallography complement each other. Protein Sci 26:32–39. https://doi.org/10.1002/pro.3022
30. Bai X, McMullan G, Scheres SHW (2015) How cryo-EM is revolutionizing structural biology. Trends Biochem Sci 40:49–57. https://doi.org/10.1016/j.tibs.2014.10.005
31. Mitra AK (2019) Visualization of biological macromolecules at near-atomic resolution: cryo-electron microscopy comes of age. Acta Crystallogr Sect F Struct Biol Commun 75:3–11. https://doi.org/10.1107/S2053230X18015133
32. Muench SP, Antonyuk SV, Hasnain SS (2019) The expanding toolkit for structural biology: synchrotrons, X-ray lasers and cryoEM. IUCrJ 6:167–177. https://doi.org/10.1107/S2052252519002422
33. Wood C, Burnley T, Patwardhan A, Scheres S, Topf M, Roseman A, Winn M (2015) Collaborative computational project for electron cryo-microscopy. Acta Crystallogr D Biol Crystallogr 71:123–126. https://doi.org/10.1107/S1399004714018070
34. Afonine PV, Klaholz BP, Moriarty NW, Poon BK, Sobolev OV, Terwilliger TC, Adams PD, Urzhumtsev A (2018) New tools for the analysis and validation of cryo-EM maps and atomic models. Acta Crystallogr Sect Struct Biol 74:814–840. https://doi.org/10.1107/S2059798318009324
35. Afonine PV, Poon BK, Read RJ, Sobolev OV, Terwilliger TC, Urzhumtsev A, Adams PD (2018) Real-space refinement in *PHENIX* for cryo-EM and crystallography. Acta Crystallogr Sect Struct Biol 74:531–544. https://doi.org/10.1107/S2059798318006551
36. Zeng L, Ding W, Hao Q (2018) Using cryo-electron microscopy maps for X-ray structure determination. IUCrJ 5:382–389. https://doi.org/10.1107/S2052252518005857
37. Fàbrega-Ferrer M, Cuervo A, Fernández FJ, Machón C, Pérez-Luque R, Pous J, Vega MC, Carrascosa JL, Coll M (2021) Using a partial atomic model from medium-resolution cryo-EM to solve a large crystal structure. Acta Crystallogr Sect Struct Biol 77:11–18. https://doi.org/10.1107/S2059798320015156
38. Song W, Wang J, Han Z, Zhang Y, Zhang H, Wang W, Chang J, Xia B, Fan S, Zhang D, Wang J, Wang H-W, Chai J (2015) Structural basis for specific recognition of single-stranded RNA by toll-like receptor 13. Nat Struct Mol Biol 22:782–787. https://doi.org/10.1038/nsmb.3080
39. Cuervo A, Fàbrega-Ferrer M, Machón C, Conesa JJ, Fernández FJ, Pérez-Luque R, Pérez-Ruiz M, Pous J, Vega MC, Carrascosa JL, Coll M (2019) Structures of T7 bacteriophage portal and tail suggest a viral DNA retention and ejection mechanism. Nat Commun 10:3746. https://doi.org/10.1038/s41467-019-11705-9
40. McMullan G, Faruqi AR, Henderson R (2016) Direct electron detectors. Methods Enzymol 579:1–17. https://doi.org/10.1016/bs.mie.2016.05.056
41. Yang J, Yan R, Roy A, Xu D, Poisson J, Zhang Y (2015) The I-TASSER suite: protein structure and function prediction. Nat Methods 12:7–8. https://doi.org/10.1038/nmeth.3213

42. Jumper J, Evans R, Pritzel A, Green T, Figurnov M, Ronneberger O (2021) Highly accurate protein structure prediction with AlphaFold. Nature 596:583–589. https://doi.org/10.1038/s41586-021-03819-2
43. Kidmose RT, Juhl J, Nissen P, Boesen T, Karlsen JL, Pedersen BP (2019) *Namdinator* – automatic molecular dynamics flexible fitting of structural models into cryo-EM and crystallography experimental maps. IUCrJ 6:526–531. https://doi.org/10.1107/S2052252519007619
44. Kovalevskiy O, Nicholls RA, Long F, Carlon A, Murshudov GN (2018) Overview of refinement procedures within *REFMAC* 5: utilizing data from different sources. Acta Crystallogr Sect Struct Biol 74:215–227. https://doi.org/10.1107/S2059798318000979
45. Brown A, Long F, Nicholls RA, Toots J, Emsley P, Murshudov G (2015) Tools for macromolecular model building and refinement into electron cryo-microscopy reconstructions. Acta Crystallogr D Biol Crystallogr 71:136–153. https://doi.org/10.1107/S1399004714021683
46. Nakane T, Kotecha A, Sente A, McMullan G, Masiulis S, Brown PMGE, Grigoras IT, Malinauskaite L, Malinauskas T, Miehling J, Uchański T, Yu L, Karia D, Pechnikova EV, De Jong E, Keizer J, Bischoff M, McCormack J, Tiemeijer P, Hardwick SW, Chirgadze DY, Murshudov G, Aricescu AR, Scheres SHW (2020) Single-particle cryo-EM at atomic resolution. Nature 587:152–156. https://doi.org/10.1038/s41586-020-2829-0
47. Yip KM, Fischer N, Paknia E, Chari A, Stark H (2020) Atomic-resolution protein structure determination by cryo-EM. Nature 587:157–161. https://doi.org/10.1038/s41586-020-2833-4
48. Naydenova K, McMullan G, Peet MJ, Lee Y, Edwards PC, Chen S, Leahy E, Scotcher S, Henderson R, Russo CJ (2019) CryoEM at 100 keV: a demonstration and prospects. IUCrJ 6:1086–1098. https://doi.org/10.1107/S2052252519012612
49. Nannenga BL, Gonen T (2019) The cryo-EM method microcrystal electron diffraction (MicroED). Nat Methods 16:369–379. https://doi.org/10.1038/s41592-019-0395-x
50. Martynowycz MW, Clabbers MTB, Hattne J, Gonen T (2022) Ab initio phasing macromolecular structures using electron-counted MicroED data. Nat Methods 19:724–729. https://doi.org/10.1038/s41592-022-01485-4
51. Bücker R, Hogan-Lamarre P, Mehrabi P, Schulz EC, Bultema LA, Gevorkov Y, Brehm W, Yefanov O, Oberthür D, Kassier GH, Dwayne Miller RJ (2020) Serial protein crystallography in an electron microscope. Nat Commun 11:996. https://doi.org/10.1038/s41467-020-14793-0
52. Tegunov D, Xue L, Dienemann C, Cramer P, Mahamid J (2021) Multi-particle cryo-EM refinement with M visualizes ribosome-antibiotic complex at 3.5 Å in cells. Nat Methods 18:186–193. https://doi.org/10.1038/s41592-020-01054-7

10 Characterization of Biological Samples Using Ultra-Short and Ultra-Bright XFEL Pulses

Adam Round, E. Jungcheng, Carsten Fortmann-Grote, Klaus Giewekemeyer, Rita Graceffa, Chan Kim, Henry Kirkwood, Grant Mills, Ekaterina Round, Tokushi Sato, Sakura Pascarelli, and Adrian Mancuso

Abstract

The advent of X-ray Free Electron Lasers (XFELs) has ushered in a transformative era in the field of structural biology, materials science, and ultrafast physics. These state-of-the-art facilities generate ultra-bright, femtosecond-long X-ray pulses, allowing researchers to delve into the structure and dynamics of molecular systems with unprecedented temporal and spatial resolutions. The unique properties of XFEL pulses have opened new avenues for scientific exploration that were previously considered unattainable. One of the most notable applications of XFELs is in structural biology. Traditional X-ray crystallography, while instrumental in determining the structures of countless biomolecules, often requires large, high-quality crystals and may not capture highly transient states of proteins. XFELs, with their ability to produce diffraction patterns from nanocrystals or even single particles, have provided solutions to these challenges. XFEL has expanded the toolbox of structural biologists by enabling structural determination approaches such as Single Particle Imaging (SPI) and Serial X-ray Crystallography (SFX). Despite their remarkable capabilities, the journey of XFELs is still in its nascent stages, with ongoing advancements aimed at improving their coherence, pulse duration, and wavelength tunability.

Keywords

Structural biology · X-free electron laser (XFEL) · Crystallography · Single Particle Imaging (SPI) · Serial Femtosecond Crystallography (SFX)

10.1 Introduction

The advances of X-ray sources and their increasing brilliance (rotating anodes through generations of synchrotrons to X-ray free-electron lasers) has brought new insight to biology and is yet to surpass the needs of the life science community. Each advance in X-ray source, instrumentation, automation, data analysis, etc. leads to an increased understanding of the way the natural world operates at the molecular level, to more structures solved and yielding higher resolution in different states or even the intermediate steps through a complex reaction. The ultimate aim is to visualize and document molecular processes, observing individual atoms at a short enough time scale to follow the transfer of charges from the initiation of a reaction to its

A. Round (✉) · E. Jungcheng · C. Fortmann-Grote · K. Giewekemeyer · R. Graceffa · C. Kim · H. Kirkwood · G. Mills · E. Round · T. Sato · S. Pascarelli · A. Mancuso
European XFEL, Schenefeld, Germany
e-mail: adam.round@xfel.eu

M. C. Vega, F. J. Fernández (eds.), *Advanced Technologies for Protein Complex Production and Characterization*, Advances in Experimental Medicine and Biology 1453,
https://doi.org/10.1007/978-3-031-52193-5_10

Fig. 10.1 An XFEL experimental station at the European X-Ray Free-Electron Laser Facility. The experiment SPB/SFX station investigates crystalline and non-crystalline matter, with a particular emphasis on the determination of 3D structures of biological objects. Examples include crystals of macromolecules and macromolecular complexes as well as viruses, organelles, and cells. (Reproduced with permission. © European XFEL/ Jan Hosan)

completion. The desire to make "molecular movies" demonstrating the processes vital to life and abundant in nature offers the possibility for us to harness these tools and put them to use for a wide variety of applications, not only for the understanding and treatment of disease [1], but also for other disciplines such as harnessing photosynthesis for safe, cheap and renewable energy production [2, 3].

The advent of X-ray free-electron Laser (XFEL) sources continues the important role of ever brighter X-ray sources in these fields (Fig. 10.1). The major benefits of XFELs to structural biology fall into two broad experimental classes—serial crystallography and single particle imaging.

10.2 Properties and Benefits of XFEL Sources

The main distinguishing features of XFEL sources over synchrotrons is the pulsed nature of XFELs giving extremely brilliant ultrashort (femtosecond, 10^{-15} s) pulses of X-rays. The intensity of each pulse from an XFEL provides enough scattering to record a useful diffraction pattern from single, sub-micrometer sized protein crystals.

10.2.1 Time Resolution

This pulsed nature of an XFEL source gives the opportunity to see femtosecond snapshots of active processes with variable time delay from the initiation of the reaction, which can be triggered in a variety of ways (mixing, pump-probe, temperature, or pressure jumps) (Fig. 10.2). Measuring full data sets at each time delay with femtosecond resolution gives the opportunity to visualize "molecular" movies at atomic resolution, meaning that the process can be observed through each stage of multi-step reactions (Fig. 10.2). This provides significantly more understanding of the processes than can be achieved by capturing only the longer-lived transition states and interpreting the changes between

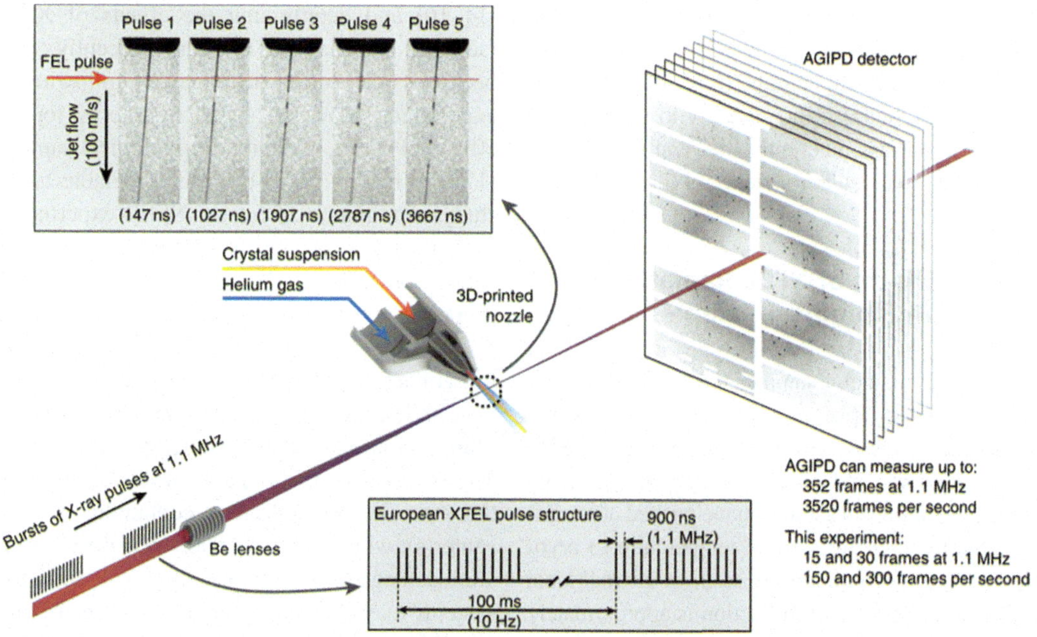

Fig. 10.2 **XFEL time resolution enables megahertz data collection.** Pulses from the European XFEL can be focused on protein crystals in crystallization solution as they are introduced into the focused XFEL beam using a liquid jet moving at speeds between 50 and 100 m/s. Diffraction patterns recorded from the sample are measured using an AGIPD detector, which can measure up to 3520 pulses per second at megahertz frame rates. (Figure reproduced with permission from Ref. [4])

them, giving the possibility to observe and understand the fastest reactions without dependence on trapping intermediates. This provides researchers the ability to monitor and gain new understanding of processes vital for life and the fundamental nature of the biological systems.

10.2.2 Signal From Smaller Crystals

As XFELs produce somewhere on the order of 10^{12} hard X-ray photons per pulse, crystals can be used that are much smaller than required for traditional macromolecular crystallography (MX) at synchrotrons. This relieves the burden of producing "large" crystals for structure determination—a bottleneck in many cases. Serial femtosecond crystallography (SFX) at an XFEL source has been performed for crystals as small as 500 nm^3. The challenge, of course, is then to produce and deliver these small crystals in high enough numbers for SFX structure determination, which requires a shift in the sample optimization strategies, addressed later in this chapter.

10.2.3 Possibility to Follow Irreversible Time-Resolved Processes

Synchrotron crystallography can efficiently follow *reversible* processes by integrating the signal at a given time delay of a process and repeating the excitation and measurement process on the same sample. In contrast, irreversible processes require fresh sample for each measurement, dramatically increasing sample consumption as a function of the signal and measurement time required. Given that single frames of diffraction collected from individual crystals are often interpretable in XFEL serial crystallography, the possibility to follow time-dependent processes that are irreversible becomes not only possible but more sample efficient with XFEL studies than with synchrotrons. This is relevant to so-called pump-probe processes, which are processes started by an optical "pump" (typically an optical laser pulse), but also increasingly to mixing processes which are inherently irreversible and also require small crystals for rapid mixing times [5].

10.3 Effects of XFEL Radiation on Biological Samples

Radiation damage of samples has been a long and deeply studied topic since the beginning of X-ray experiments themselves. Strategies to minimize the adverse effects and observe unaltered data from sensitive biological samples have pushed crystallographic research from room temperature to cryo-conditions. The radiation dose limit of traditional room temperature data collection is approximately 0.2 MGy, whereas under cryo-cooled conditions samples may withstand two orders of magnitude greater dose (30 MGy) [6–8]. However, it has been reported that this dose limit for room temperature experiments is dose rate dependent [9] and can be dramatically increased (up to 150 MGy) when using the highly intense femtosecond pulses at XFEL facilities [10–13].

As shown by the recent resurgence of interest in room temperature data collection, especially in biological research, the active (uncooled) forms are most interesting–especially for health-related research in particular drug design. Such room temperature structural studies are therefore ideally suited for XFEL based experiments, which naturally do not use cryo-cooling. The ability to access unaltered states for sensitive structures is key to understanding their function. This strategy is enabled by XFELs. The ultrashort pulses give rise to diffraction *before* secondary changes to the structure can take place, with the data collected "outrunning" damage to the sample, even though it may be destroyed by the probing pulse (Fig. 10.3). The effects of XFEL radiation (dose rate and exposure time) on biological samples is an ongoing and highly active area of research [14–16] and determining the effects of XFEL radiation on samples has, until recently, only been possible during data analysis following an experiment. However, prediction tools for the effects of XFEL radiation on biological samples [17] are beginning to be made available to aid study of radiation effects in XFEL experiments with the aim to support feasibility and the justification for XFEL experiments in an experiment proposal.

The very short duration of XFEL pulses allows the collection of diffraction data from samples that, if exposed to a longer duration of X-ray illumination, may change structure in the beam. This is particularly valuable when determining the structure of molecules containing a metal center(s), which are particularly sensitive to radiation damage from X-ray beams [18]. Photoactive systems may also show alterations to their structure at very low (0.06 MGy) dose [19] which is not necessarily "damage" but effects on the active site by the X-rays used to probe them.

10.4 Use of XFELs for Life Science Research

10.4.1 Current State of the Art and "Standard" Experimental Options

XFEL experiments are continuing to develop. The capabilities available along with improved ease of use aim to facilitate access to a generation of users new to the field. Standardization and automation are central to this, and relatively "standard" XFEL experiments in structural biology are nowadays performed with high reliability. Users of XFEL facilities are, therefore, strongly recommended to familiarize themselves with the standard options offered (available on the facility webpages) as well as how the instrumentation available has been used successfully by the community (lists of publications are generally available on the instrument webpages). A brief

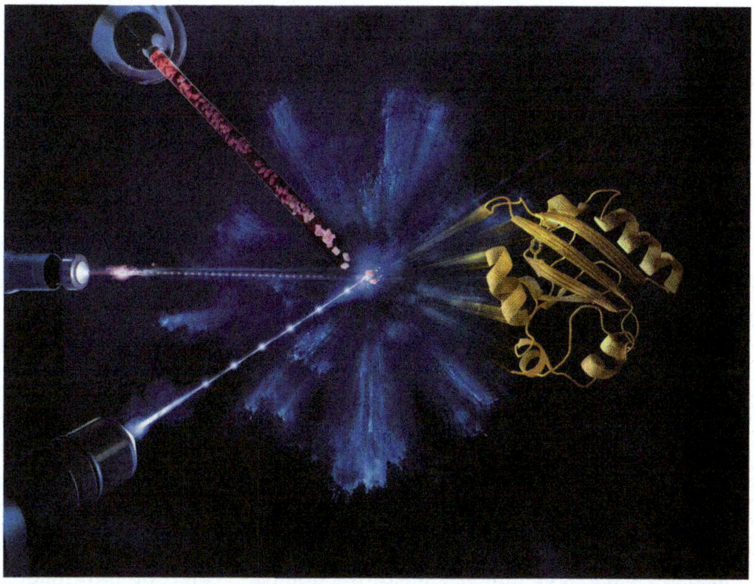

Fig. 10.3 Radiation damage in XFEL sources. Artistic visualization of a serial crystallography experiment showing a stream of crystalline proteins being hit by an optical laser before being struck by the X-ray laser. The microcrystals are destroyed by radiation damage, but the information about arrangement of the atoms in the protein is recorded to reconstruct a model of the structure of the protein. (Reproduced with permission. © European XFEL/Blue Clay Studios)

overview of the possibilities offered at XFELs by technique is provided below.

10.4.2 Crystallography

The acquisition of atomic resolution information of fundamental biological processes has been a significant motivation for developments for life science experiments at XFELs. Structure determination using X-ray FEL sources with small micro- and nanometer crystals of biomolecules [10, 11, 20–23] has broadened of the scope of crystallography for biological structure determination [24]. Building on previous experience of the crystallographic community, data collection strategies to combine partial data sets from a few crystals are now scaled up to many thousands in the case of XFELs. This is a necessity at XFELs, as due to the power of each pulse the sample cannot survive and therefore needs to be replaced by fresh sample for the subsequent pulse (serial crystallography) requiring fast injection speeds for MHz repetition rate experiments [4, 25].

In comparison to synchrotrons, SFX at XFELs address samples that (1) do not form large enough crystals to provide an adequate signal-to-noise ratio [21], (2) contain metal atoms that may be easily altered chemically by longer exposure time (and hence not reflect the native structure of the sample) [26], or (3) time-resolved systems, where either femtosecond time resolution is needed or, for example, irreversible reactions such as mixing are to be studied, requiring small crystals to minimize the mixing time and define a clear t_0 as the start of the reaction [5, 27].

Time-resolved serial femtosecond crystallography, where possible especially at MHz data collection rates and using the initiation of reactions via a laser or mixing, provides access to observe reactions and record "molecular movies" with the ability to observe and understand biochemical reactions involving unstable and short-lived intermediates which cannot be seen via other means.

10.4.3 Single Particle (Coherent Diffraction) Imaging

One of the fundamental limitations of crystallography has always been the need for crystals. Unfortunately, crystallization is not trivial for all samples, and many are either not yet crystalized or simply not amenable to crystallization. The need to obtain information on biological processes where the components cannot be (or at least have not yet) crystallized has driven developments at XFELs [28] as it has with other methods (EM, NMR, SAXS).

The potential for Single Particle Imaging (SPI) of biological macromolecules using the extremely intense and ultrashort pulses from an X-ray Free Electron Laser was introduced to the broad scientific community in a landmark study [29], about a decade before the first hard-X-ray FEL came into operation. The prospects for FEL-based SPI (near-atomic 3D resolution, without the need for crystallization, using single biomolecules in aqueous solution) highlight the motivation to build FELs and the associated infrastructure such as dedicated beamlines [30]. The ultrabright, ultrashort XFEL pulses potentially allow enough diffracted signal to be collected from individual particles. As with SFX many frames can be combined to yield a three-dimensional diffraction pattern that can be interpreted (See Data reduction and analysis: SPI) to yield the three-dimensional electron density of the molecule in question. No other class of X-ray source can attempt such a measurement approaching resolutions relevant for biomolecules, and even at XFELs this method is still a work in progress with ever improving resolutions and smaller and smaller samples probed in each measurement [31–33].

Usually, the particles are injected as a focused stream of aerosolized (See Choice of sample injection: Aerosol), randomly oriented, reproducible or nearly reproducible entities. If then an FEL pulse coincides in space and time with a particle in the X-ray focal plane, scattered radiation in the optical far field can be collected on a two-dimensional detector downstream of the interaction region.

A community-wide Single Particle Imaging Initiative is striving towards the ultimate goal of a 3-Å-resolution structure [34] with a systematic approach to advance FEL-based SPI on smaller particles. A limiting factor towards this goal is the relatively small number of elastically scattered photons resulting from the interaction of a biological macromolecule with even a highly intensive FEL pulse [35, 36]: It becomes an increasingly challenging task to distinguish the resulting very photon-sparse diffraction patterns [37], the so-called "hits", from the majority of diffraction patterns where there was no particle (See Hit finding - Identification of frames with usable diffraction data). With a so-called "hit rate" between 0.1% and 1% [37] the advent of new sources, such as the European XFEL and LCLS-II, with MHz repetition rates, now brings the acquisition of complete SPI datasets within easier reach [38, 39].

10.4.4 Scattering Techniques

Measuring more than a single particle can be used to increase the scattering statistics and is usually performed in liquid jets rather than aerosol (as for SPI). Fluctuation X-ray scattering (FXS) [40, 41] provides size and shape data which takes advantage of information from variation around azimuth (which in traditional solution scattering is uniform due to spherical averaging) and therefore aims to provide higher resolution reconstructions. While solution scattering experiments have been undertaken at FEL sources these have untill recently been limited to difference experiments, comparing ground and activated states [42–44]. However, recent efforts have provided proof of principle of form factor recovery from traditional solution scattering data collected using a FEL source [45] despite the requirements for stable background subtraction with the complications of shot-to-shot variation from SASE sources which has previously complicated these efforts.

10.4.5 Complementarity with Synchrotron Sources

XFEL sources are not alone in the field of life science research and synchrotron experiments are highly complementary to the understanding of biological processes. Research time at any large-scale facility is limited and faces fierce competition during the application process. This is especially true for XFEL sources due to there being fewer XFEL sources than synchrotrons as well as XFEL sources having fewer instruments which, due to the linear nature of these facilities, typically cannot all operate at the same time. Although efficient use of XFEL sources using multiplexing is a high priority for XFEL facilities to maximize the time available at their instruments, they still cannot offer as much experimental time as synchrotrons which are both more numerous and each having many more instruments than any XFEL.

Given the limitation for XFEL experiments it is required (in order to be successful) in the application for beamtime to show the feasibility of the experiment. This feasibility not only covers the concept for the data collection plus the samples to be measured. It must also (ideally) be demonstrated that samples are expected to diffract well enough to achieve the necessary resolution to interpret the subtle differences in the structure. This necessitates that a proposal should include preliminary data collected at a synchrotron source. This should not be seen as the synchrotron facilities being a filter and a route to the XFEL. Instead, synchrotron sources are truly complementary and should be used to learn as much about the sample as possible. For example, synchrotrons are suitable sources for longer time scale serial crystallography (with sufficiently sized crystals) and as the understanding of the systems under study progresses, the XFEL can be used to complement the existing data - for example observing shorter timescale behavior or smaller crystals to acceptable resolution. Optimization of an experiment by combining data collection using a synchrotron, especially with the advent of the ongoing upgrades to increase the brilliance of synchrotron sources worldwide, and an XFEL where appropriate is advised to maximize the chance of success for experiments.

10.5 How to Prepare an XFEL Experiment

XFEL instruments are typically, by design, more flexible owing to the wider experimental scope necessitated by their limited number. As such experimental setups at XFELs often require more preparation compared to, for example, optimized experimental stations for MX at synchrotrons. The experimental setup required is dependent on the sample (delivery method) and data to be collected (instrument configuration, camera length, X-ray energy). Although there is scope for reducing setup time by scheduling similar experiments sequentially, in many cases optimization is still needed for the three essential requirements, beam delivery, detection and sample delivery. In the case of time resolved experiments the activation/triggering method (laser/mixing) must also be set up. To be successful in the highly competitive XFEL application systems one must not only submit a proposal with a highly motivating science case but demonstrate the feasibility of all three key points as well as any activation method (mixing or pump-probe) if required. This means in practice one should ideally show preliminary data and demonstrate that the samples can be prepared in the quality and quantity required (see below) as well as explain why an XFEL is needed instead or as well as synchrotron (or other) experiments. Modeling and simulation packages are also available to aid the experimental design and optimization facilitating the extension of the preliminary data to show what should be practically observable at an XFEL. These simulations and modeling tools become even more important for the more cutting-edge proposals which are extending the capabilities of the instruments.

Choices for the beam delivery are highly dependent on the facility capabilities, X-ray energy, focus size, power (energy and pulse duration) per pulse and sometimes even repetition rate of delivered pulses. Detection is a choice of the kind of detector and most facilities have a very limited choice of detectors. A key parameter for a user is the sample to detector distance, usually as short as possible for SFX and as long as necessary to sample non-crystalline diffraction appropriately for coherent diffractive imaging (CDI) [46]. Additional detection capabilities for complementary measurements can be highly valuable to give additional insights into the sample, for example, X-ray emission spectrometry for detection of the oxidation states in time-resolved studies [47]. The capabilities and scope of each facility and instrument are documented on their respective websites and publications of similar research at these facilities can be highly valuable to the design and optimization of your experiments. Further to this, early inclusion of the instrument scientists in the design and proposal stage of a beamtime application is highly encouraged to facilitate the application and the experiment. XFELs are relatively novel facilities, and the rate of improvement in instrumentation and capability is very rapid in most cases. Discussion with the local experts will enable one to make the best use of each facility's capabilities in any given user experiment.

10.6 XFEL Experiment Simulation

Simulations of X-ray scattering have become an important part of experimental X-ray research. Simulations can be used to explore novel parameter regimes and identify promising areas for experimental research where new phenomena may occur. A more practical aspect of simulations is that they often play a pivotal role during the design, execution, and analysis of an experiment. Recent advances in the development of theoretical methods, and simulation codes, as well as in high performance computing (both software and hardware), have enabled simulations of complete experiments, describing the photon source, X-ray optics, interaction with and scattering from the sample, as well as X-ray registration in an X-ray detector. Such start-to-end simulations allow one to simulate realistic experimental data that take into account various imperfections, *e.g.*, in the source's pointing stability, temporal and spatial structure, optical elements like mirror height and slope errors, artefacts introduced by lenses, radiation damage of the sample during irradiation with the probing X-ray beam, as well as detector noise [1, 36, 48]. All these effects and their impact on the measured data can be studied in isolation or in combination with each other. The simulation suite SIMEX [36] provides a platform for start-to-end simulations for a broad range of X-ray based experiments, including single-particle imaging, SAXS/WAXS, absorption spectroscopy, and inelastic X-ray scattering. It bundles simulation codes for the propagation of X-rays from the source to the interaction point, photon-matter interaction, and detection of scattered or transmitted radiation. For detectors at the European XFEL, these simulations also include the actual detector calibration constants. The SIMEX simulation environment allows X-ray facility users to predict observable data under realistic, experimental conditions, evaluate how the data and data characteristics (*e.g.*, signal-to-noise ratio, hit rates) scale with machine parameters such as the pulse duration, bandwidth, or wavefront profile. This information could then be used to optimize the experimental configuration for maximized data quality. The simulations can be run using simplified models for quick qualitative estimates, *e.g.*, online during an experiment or using more sophisticated *ab initio* methods for accurate predictions and data analysis. Another potential future application of SIMEX is its use during evaluation of beamtime proposals, *i.e.*, reviewers can use the simulations to assess feasibility and potential impact of proposed experiments. In view of the very limited number of XFEL sources and the high costs for operating these facilities and providing beamtime to users, use of simulation and modelling tools become crucial to make the most efficient use of the precious and scarce beamtime. Demonstrating the likeliness of an experiment's success through start-to-end simulations would therefore be highly beneficial to beamtime proposals especially where preliminary data cannot be measured experimentally without an XFEL.

10.7 Sample Characteristics

10.7.1 Crystal Size and Estimation of Amounts Needed

The tendency for XFEL serial crystallography experiments is towards smaller crystals in the order of micron or even nano sized crystals, as the smaller volume is desirable not only for diffusion of ligands and substrates but also for photo excitation so that the entire crystal is excited rather than only the surface (due to limited absorption length). In addition, a small crystal size is better suited for injection to minimize the risk of clogging and diffraction quality may be better in smaller single crystals, as there may be fewer dislocations and defects disrupting the crystal lattice. Crystal concentration is desirable to be as high as possible (40% by volume of crystal to solution) to achieve high data hit rates. However, the higher the concentration the greater the possibility of problems with settling and clogging, which interrupts sample delivery. This can be highly detrimental for reliable high throughput data collection. The sample volumes required will depend on the flow rate needed for stability, jet replenishment in MHz experiments or to achieve the desired time points for a mixing experiment. It is therefore not straightforward to give a single estimate for material required which would be applicable for all cases, though an overly simplified estimate would be that 200–300 μl of concentrated crystal solution can provide a single data set. For these reasons, the sample delivery method (Choice of Sample injection options) should be tested in advance in collaboration with the facility staff to determine the optimal sample conditions and amounts required for any given experiment.

10.7.2 Single Particle Sizes (Easy to Challenging) and Estimation of Amount Needed

Complete 3D reconstructions of single particles have been obtained for viruses in a size range between several hundred nanometer down to 70 nm diameter [49], in the latter case to a full-period resolution of around 10 nm [50]. Single hits from smaller viruses (diameter: 40 nm) have been obtained using hard X-rays (5.5 keV) [50], with signal above background down to about 4 nm resolution. At higher photon energy (7 keV), signal above background was obtained from 70-nm virus particles down to 5.9 Å resolution, indicating the potential for higher-resolution structures than obtained so far, using FEL-based SPI.

Particles have been injected at concentrations between 10^{11} and 10^{12} particles per ml with a gas dynamic virtual nozzle (GDVN) [51–53] using a flow rate between 1 and 3 μl/min. Given that a total dataset can require tens of hours of data collection time, a few milliliters of sample solution are sufficient for a beamtime of several days. Using a GDVN in electroflow focusing mode, a higher concentration (10^{13} particles per ml) and lower flow rate (0.3–1 μl/min) has been used [37]. The sample's size distribution and homogeneity should be checked independently prior to the experiment [52, 53].

10.7.3 On-Site Sample Preparation and Biophysical Characterization

Current trends in life science research require working with biological samples that are increasingly more complex. This is even more critical for samples intended for study using XFEL sources as they are by virtue of needing such a source expected to be functional with strong measurable activity to enable time resolved experiments. Furthermore, as the activity is often light or oxygen induced, they can be highly sensitive and therefore unstable in terms of time scales and temperatures expected during production, shipping or final preparation. Large scale research infrastructures (synchrotrons and XFELs) with applications in structural biology often include sample preparation laboratories "on-site", which are essential to provide the highest quality active and or functional samples and hence give the best possible chance to observe the sensitive biochemistry. In addition, these facilities must also provide adequate complementary sample characterization to enable quality control and sample prioritization for any given measurement with beam. This additional information not only aids making informed decisions during the data collection but also assists and complements data analysis as well as supporting the resulting conclusions.

All large-scale X-ray facilities worldwide with an interest in life sciences provide some form of access to biological laboratories as part of access to the facility for experiments. The scope of the provided facilities is related to their individual research portfolio which is continually being improved. Each facility maintains up to date details on their websites regarding the current techniques and equipment available. Careful preparation during the application process is advised to understand what additional facilities offered will be of benefit to the proposed experiment and for that to be included in the beamtime application. It is therefore suggested to contact facility staff for sample preparation and characterization during the early stages of developing ideas for experiments.

10.7.4 Choice of Sample Injection Options

The choice of sample delivery is complex as it is highly dependent on the sample to be studied. General guidelines and examples are given here, however, we again recommend you speak with the instrument scientists and staff responsible for sample delivery at the facility where you intend to undertake your experiment to find the best solution for your particular sample(s). For most XFELs, the prevalent methods are serial based as the FEL power is not conducive to multiple measurements at the sample position (See above).

FEL experiments employ three main classes of sample delivery: (1) liquid jets for delivering (primarily) small crystals to the X-ray FEL beam for serial crystallography (SX, SFX), (2) focused aerosol beams for delivering (primarily) non-crystalline particles to the X-ray FEL beam for single particle imaging (SPI), and (3) samples arranged on fixed targets, which may be crystalline or non-crystalline.

10.7.5 Liquid Jets

For sample delivery into vacuum, jet producing nozzles are mounted at the end of a hollow rod containing the liquid and gas lines for sample delivery and jetting control. This assembly is inserted via a load lock to position the nozzle above the X-ray focus position allowing exchange without the need to vent the entire chamber. The nozzle rod mates to a catcher centered around the interaction region. The catcher restricts the majority of sample residue to a small easy-to-clean volume and provides a degree of differential pumping of the liquid and gas loads to ensure the pressure of the main interaction chamber (and detector) stays below the required threshold for safe operation [54].

The most frequently used methods to establish jets are electrospinning [55] and gas dynamic virtual nozzles (GDVNs) [56]. Electrospinning uses an electric field to accelerate the liquid whereas a GDVNs use gas (typically He) to apply pressure to a liquid sample jet emerging from a typically 50–100 μm capillary which constricts the flow to a few μm in diameter accelerating the liquid. Both produce a very thin and fast jet capable of delivering crystals in suspension to the interaction region at an optimized rate for data collection. GDVN nozzles can provide fast jets with velocities over 80 m/s [57], recently shown to successfully deliver sample jets compatible with the current operational pulse train structure of the FEL beam at Single Particles, Clusters, and Biomolecules (SPB)/SFX [4, 57].

High viscosity extrusion jet sample delivery [58] works in a similar way to GDVNs but the viscosity limits the restriction and acceleration which can be achieved. This results in lower speed and lower possible frame rates that can be achieved compared to their liquid counterparts. However, the lower speed increases the probability of hitting the crystals and generally results in higher hit rates. As membrane proteins can be directly crystallized in lipidic cubic phase and injected directly these nozzles enable measurement of non-soluble and also very important proteins which could not be investigated otherwise.

10.7.6 Mixing Jets

Mixing experiments provide important insights into the structure-function relationship of proteins *in operando*, by taking sequential snapshots at different time points after mixing [5, 27]. Up to date XFEL mixing experiments on crystal suspensions, allow access to the time scale of seconds [59, 60]. However, the macromolecular conformational rearrangements that underlie signaling, transport, catalysis, and assembly happen on the μs-ms time scales. There is a great interest, therefore, to develop rapid mixing injectors to access such processes. Mixing devices that incorporate the GDVN geometry to reduce sample consumption have been developed and consist of three capillaries arranged in a coaxial configuration [61, 62]. These devices allow two reagents to co-flow and mix in the two inner capillaries before being focused by the gas flowing through the outer capillary. It is possible to probe short reaction times using small crystals (minimizing diffusion times). For example, a crystal with dimensions of $0.5 \times 0.5 \times 0.5\ \mu m^3$ results in a modelled diffusion time of 17 μs, while a $3 \times 4 \times 5\ \mu m^3$ crystal is estimated to take 1 ms, and a large $300 \times 400 \times 500\ \mu m^3$ crystal would take 9.5 s [5]. By adapting the distance and flow rates between the mixing point and the probed position a range of timescales can be probed. The GDVN-principle and the jet-in-jet geometry can be integrated as well into microfluidic devices fabricated by soft-lithography [63]. The advantage of such devices, made of PDMS (polydimethylsiloxane), lies in their microstructure reproducibility. The concept has been extended also to other materials, like glass and silicon, which show a stronger resistance to aggressive chemicals and to high pressure. Another approach is combining 3D printed nozzles [64] and 3D printed mixing regions with capillaries and/or microfluidics.

10.7.7 Aerosol

The small scattering cross section and non-crystalline nature of single particles require that the scattering background is reduced to a minimum. In aerosol-based sample delivery, the aim is to isolate the sample from any surrounding liquid, removing scattering from the delivery medium that would otherwise overwhelm the weak sample scattering signal.

An aerosol of (sub-)micron-sized droplets, each on average containing one sample particle, is generated by GVDN or for smaller particles and therefore droplets, by electrospray [65]. The droplets evaporate, leaving behind isolated sample particles that are funneled into an aerodynamic lens [66, 67]. Within the aerodynamic lens, the particle flow is focused into a narrow beam that intersects the X-ray beam at the exit of the lens stack enabling efficient delivery of particles ranging from 30 to 3000 nm in diameter.

The nozzle-to-interaction particle transmission varies with both particle size and gas flow, and can reach 70% for particles of a few hundred nanometers in diameter. The exit velocity also depends on particle diameter and gas flow, with sub-100 nm particles reaching 200 m/s, while micron sized particles travel slower at approximately 20 m/s [66].

Potential complications of limited liquid jet speeds in the high repetition rate FEL beam, such as sample replacement and jet disruption, can be alleviated with aerosol sample delivery, where no surrounding liquid is present. The lack of surrounding liquid also enables the use of ion time-of-flight spectroscopy as a potential means to provide confirmation that an X-ray pulse intersected with a sample [68]. These features make aerosol sample delivery an intriguing possibility for delivering crystalline samples into the X-ray beam.

10.7.8 Droplet Injection

Performing experiments at atmospheric pressure (usually in a He atmosphere to reduce background scattering) opens up the possibility to use other delivery-methods, like drop-on-demand or acoustic droplet ejection. The droplets can be probed after deposition on a moving belt [69] with the addition of optical laser activation with multiple pulses for complex reactions or passed through an oxygen rich environment. Drops can also be merged with the second drop initiating the reaction. Alternatively, droplets can be probed in flight [69, 70] and the reactions can be triggered by mixing two ballistic droplets [71]. Droplet injection, although it provides access to probe more complicated reactions, is currently limited by the repetition rate of the ejection apparatus and in the case of drops on a belt the speed of the belt. As such these experiments are currently limited in the kHz range and are the limiting factor at MHz repetition rate facilities.

10.7.9 Fixed Targets

Injection methods are inherently inefficient with the sample as most of the sample volume, which must flow constantly, is never exposed due to the pulsed nature of the XFEL source. In contrast, mounting samples on a fixed target is highly sample efficient as using support grids to help align the crystals nearly all crystals or simply scanning through a loop all crystals can be exposed to X-rays and data collected. However, moving the target to a fresh spot takes time and to date only 10–120 Hz repetition rates have been achievable at FEL sources. Using highly optimized systems such as the roadrunner [72] data collection rates in the kHz rate have been reported [73]. Combining fixed target data acquisition with a humidity and temperature-controlled environment also allows additional control of the local sample environment which provides an advantage compared to "in vacuum" data collection.

10.8 Serial Femtosecond Crystallography (SFX)

A simplistic overview of SFX data processing does not necessarily need to deviate significantly from traditional X-ray diffraction solutions. Once a useful data collection strategy is identified, peak finding (identification of Bragg reflections), background subtraction (using the integration of the area around the Bragg peak), indexing, and merging is performed before a structural model can be proposed [74]. Just like traditional methods, most steps can be iteratively improved as more information is revealed. However, the nature of SFX experiments imposes novel hurdles at each of these steps. This short section aims to give the reader an insight into the problems and solutions available with a goal of providing novice crystallographers with an awareness of some of the pitfalls in SFX data analysis. For further details we encourage the reader to follow the references for any method mentioned above.

10.8.1 Comparison with Single Crystal Crystallography

Goniometer-based MX experiments have the benefit of data collection efficiency, as each frame will contain data and each subsequent frame will be related to the former images by some rotation angle, making data analysis relatively simple. Since SFX experiments do not typically use single crystals and often don't control crystal orientation; large quantities of data – and large quantities of crystals – are required to ensure a dataset is complete. In some cases, this can be in the order of millions of images and therefore millions of crystals [75]. Since SFX data collection is stochastic in nature, the final data can contain images of "empty" frames, single crystal reflections, and multi-crystal reflections. For this chapter, we will focus on datasets containing "empty" and single crystal images. Images containing multi-crystal reflections will not be considered here in any detail. Separating the multi datasets is relatively trivial in low numbers and not specific to SFX datasets.

The most significant difference between MX and SFX data sets stems from the "femtosecond" nature of the data collection (Fig. 10.4). It refers to the X-ray pulse length and therefore the length of time each crystal is exposed [12, 29, 75]. As each crystal is only exposed once, the effective rotation of the crystal becomes zero degrees. To the uninitiated this may seem trivial, however, it is in fact one of the biggest considerations in SFX data analysis. The common aim of goniometer-based crystallography experiments is to measure the full intensity of each possible reflection and relate it to the number of the electrons in a given reflection plane [74]. While SFX experiments have the same final goal, limitations in the measurement of full reflections by only sampling

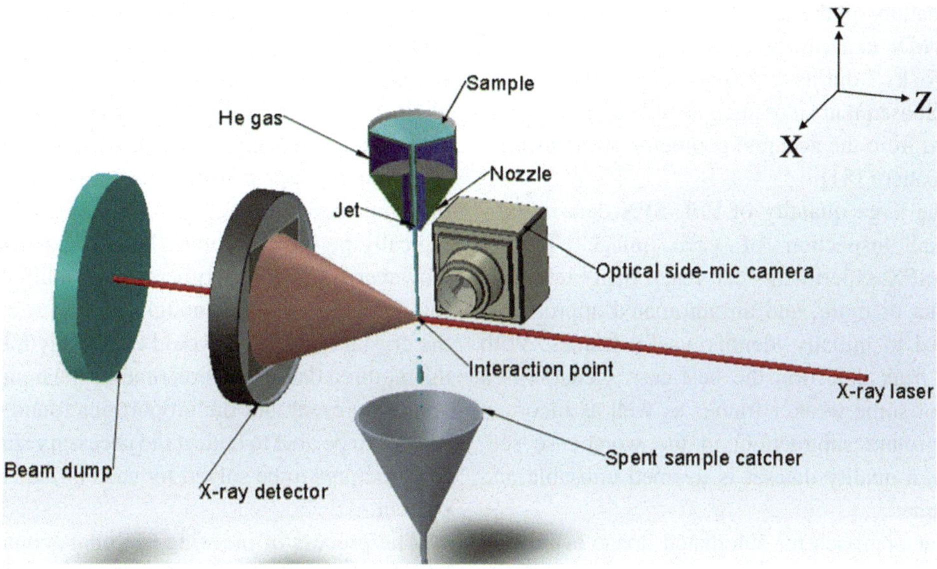

Fig. 10.4 Experimental SFX setup. A schematic diagram of the experimental setup in the vicinity of the sample interaction region on the SPB/SFX beamline. In contrast to the conventional rotation method implemented in X-ray crystallography setups, in the SFX setup the goniometer head does not rotate and the sample is delivered as a continuous jet stream. The optical side microscope camera (Oxford Instruments Andor Zyla 5.5sCMOS) records images at a rate of 10 Hz to match the frequency of pulse trains arriving from the XFEL source. (Figure reproduced with permission from Ref. [76])

limited rotations means that a different analytical approach is needed. The following is an overview of the SFX data reduction and analysis process.

10.8.2 Calibration of Raw Data

Area detectors typically used at FEL facilities have central holes to allow the intense, direct X-ray pulses to pass through the detector. Each sensor module is mobile since the detector geometry should be optimized for each experiment. The position of each pixel in a frame is therefore required to be determined for meaningful data interpretation. For SFX data, well-defined peaks allow for a crosscheck as any errors in the determination of the present detector geometry can be identified and revised during data analysis. However, a verified initial calibration reference is valuable to aid optimization, common methods for this are the use of powder diffraction or small-angle scattering signals of a known calibration sample is measured and analyzed before the main experiment. Lithium titanate or silver behenate are common choices as calibration samples, with the preference depending on the sample-to-detector distance of the experiment. Since XFEL pulse durations are extremely short, ~10 fs, pulse resolved photon counting detectors are not useful for XFEL pulses and only charge integrating detectors, *e.g.*, the Adaptive Gain Integrating Pixel Detector (AGIPD) [77] at the European XFEL, are usable (Fig. 10.2). The measured diffraction intensities by these charge integrating detectors need post correction processes, *e.g.*, baseline correction (detector offset), background subtraction, and analog-to-digital unit (ADU) conversion, which result in photon counts for each pixel in an image [78].

10.8.3 Hit Finding – Identification of Frames with Usable Diffraction Data

Detection of frames containing diffraction peaks or "hits" is arguably the most important part of the data analysis pipeline. Peak detection is the foundation of the crystallographic data, and as with MX experiments it is a priority for rapid feedback during experiments [35, 79, 80]. Subsequent steps such as indexing are determined from the accurate geometry of the diffraction pattern [81].

The large quantity of FEL-SFX data inhibits manual inspection of each image. Typical FEL-SFX experiments can reach many terabytes of data or more, and an automated approach is needed to initially identify useful frames. With poor peak detection the best-case scenario is a loss of some weaker frames as well as incorrect background subtraction; in the worst-case scenario, a quality dataset is deemed unusable and discarded.

One approach for automated image evaluation is radial integration, where a threshold is determined from the radial average and standard deviation of the diffracted intensity. Pixels found above the threshold are excluded and the process iterated several times. Once a reasonable threshold has been determined the outlying pixels above this threshold are used for peak analysis. The outliers (or potential reflections) are checked for a satisfactory number of adjacent outlier pixels. If the outlier is sufficiently connected, then this region of the detector has likely measured a reflection and will be used for further analysis. Background subtraction can be performed by a "3 concentric ring" method where the inner ring contains the expected peak, the middle ring contains an ignored spacer region, while the outer ring is then used for the background measurements.

Once a catalogue of data-containing images and relative peak locations has been determined, the next step is often indexing, the process of relating the identified reflections to a particle space-group.

10.8.4 Indexing

Indexing software commonly used in MX experiments are XDS [82] and MOSFLM [83], which are also utilized in some SFX data analysis software [79, 80, 84]. Each data-containing image can be indexed and will have a set of dimensions related to the unit cell. Plotting a histogram of each dimension determined from each image should result in a distribution around the most likely parameter. Any heterogenous populations of crystals will become evident if multi-modal distributions appear in the histogram. This type of simultaneous collection of "multi-data" is not typically performed with MX style experiments. However, an SFX experiment with a multi-crystal sample would require enough frames for each of the crystal types. This would effectively multiply the required data collection time by the number of "types" present and the proportions found in the sample to be able to collect the necessary statistics for structures to be solved for each of the "types" present.

The process of indexing each data-containing image is subtly different for SFX than MX datasets. It is possible for some crystallographic data to be indexed as its twin if the symmetry of the Bravais lattice is higher than the space-group symmetry. In the case of MX data, this is avoided, as each frame relates to each other geometrically, allowing the user to test all possible indexes and find the best fit. However, SFX does not have this luxury as each image is independent so if one was to continue through the subsequent steps of merging, one would find a perfectly twinned dataset even if the crystal sample is not twinned. This is known as the "indexing ambiguity" [85].

The indexing ambiguity has had some solutions [85, 86] that have remained robust under most circumstances, though it must be noted some modern indexing methods using minimal data still suffer from indexing ambiguities [87]. The final choice of indexing algorithm should be robust against the indexing ambiguity problem otherwise only be used if prior knowledge of the space-group is known.

Once indexing is successful, the detector geometry should be checked for self-consistency. A common feature of FEL detectors is the modular construction and often movable independent active parts. It is therefore possible the detector module positions have not remained at the given positions defined in the analysis software (geometry file) and optimization would be beneficial. Improving detector geometry may improve subsequent iterations of indexing, particularly those with few or low-intensity peaks. Once these iterations are complete, it is possible to merge the indexed frames.

10.8.5 Merging

CrystFEL allows for both Monte-Carlo methods and more modern scaling and post-refinement methods. Once the lattice/lattices of the sample have been found, the task of assigning intensities to each reflection begins. Almost all reflections measured in SFX experiments are partial reflections, so an estimate of the full intensity is required. The increase in the number of partial reflections is due to the femtosecond exposure and consequently the zero-oscillation resulting in few-to-no reciprocal lattice points crossing the Ewald sphere.

In traditional MX diffraction experiments, the oscillation range can be optimized to maximize the number of full reflections in each image without causing overlap. In SFX, the combinations of these hurdles require novel approaches to achieve similar data pipelines as with the traditional approach.

SFX experiments often address projects with small crystals, which often do not crystallize or diffract well. A small crystal (in the order of <1 μm) can significantly change the diffraction patterns in a qualitative way, as smaller crystals provide relatively larger contributions from their shape transform. In fact, the number of features ("fringes") aligned and spaced between neighboring Bragg peaks is proportional to the number of unit cells in the illuminated crystal. As the number of unit cells becomes greater, so does the number of features, though the intensities of these features become negligible, and we return to the regime of traditional Bragg diffraction. As the crystal becomes smaller and smaller, the shape transform becomes more prominent until the regime of single unit cells where data collection moves into the realm of single particle imaging (see SPI). Peak finding algorithms may need to correct for these features before accurate Bragg peak detection is possible.

The features of the shape transform found in small crystal diffraction are not necessarily problematic and in fact can provide information about the size and projected shape of the crystal by using phase retrieval techniques [20]. Isolated integration of single peaks will likely underestimate the structure factor due to the shape transform. Kirian et al. (2010) demonstrates how the Monte-Carlo integration of the features near and around Bragg peaks can result in the square of the structure factor magnitude [88]. To simplify the simulations, Kirian et al. (2010) used a measured lattice and symmetry from a PS1 dataset but reduced the large number of atoms (only changing the Bragg intensities) and then used appropriate scaling so that the modelled structure factors agreed with the measured structure factors for the photosystem 1 (PS1). The Monte-Carlo approach is an excellent way to remove or minimize the variations in X-ray intensity and crystal size from shot-to-shot. However, the Monte Carlo approach alone, without further intervention, may not allow for correctly merged datasets as there may exist significant indexing problems inherent in SFX data, which extremely large amounts of data would be needed to overcome.

Issues arise when the symmetry of the Bravais lattice is greater than the space group, where there exist two or four equally likely indexing options. Direct merging of these data can result in the appearance of perfectly twinned data, despite no actual crystal twinning being present. This can be traditionally resolved with the measurement of full reflections but as mentioned earlier SFX data consist primarily of partial reflections [86]. Of the 65 space groups there are 27 space groups with this indexing ambiguity. Liu et al. (2014) was able to overcome this indexing ambiguity by using an expectation maximization algorithm. They briefly describe the algorithm's operation by comparing the measured diffraction pattern to a three-dimensional model of the full reflection intensities on the reciprocal lattice. Correlation coefficients are then computed between the pattern and all possible indexing possibilities. The indexing model that provides the highest correlation is then used to merge the images into the model. This model is then used in the next iteration and the algorithm continues until convergence [86].

Algorithms published by Brehm and Diederichs (2014) have also solved the indexing ambiguity [85]. The authors focused on a dataset from PS1, incorrectly indexed as $P6_322$, and were able to correctly index it as $P6_3$. Their approach was to consider each shot as a vector (x) in k-dimensional space and by minimizing the difference between ($1 - r_{ij}$) and ($x_i - x_j$). Doing so results in the identification of clusters of "clouds" when inspected graphically. These clouds represent the indexing modes and can be used to merge the data correctly.

10.8.6 Overview of Analysis Procedures

As SFX experiments are in their infancy by comparison to traditional protein crystal diffraction experiments, so too are the programs that handle SFX data. This is not to say that the methods used are primitive, far from it, rather they are not as universally used as compared to such traditional programs for example: XDS [82], Aimless [89], and PHENIX [90]. It should be noted that this chapter does not aim to cover all the available programs and those mentioned herein are not necessarily the definitive standard, rather a way to describe the common processes involved in data processing. Furthermore, there are several research groups who use in-house algorithms optimized for their needs and the reader is encouraged to branch out to whatever algorithms they feel most useful and applicable to their data.

CrystFEL [84], for example, is a useful program suite which covers a vast number of tools for SFX data, including peak finding, beam cantering, indexing, integrating, scaling, merging, and even simple simulations.

10.9 Single Particle Imaging (SPI)

10.9.1 Comparison with SFX

SFX and SPI experiments are strongly related. Both aim to determine 3D structural information from a set of 2D images each with random orientation. SFX and SPI data analysis challenges are therefore also related, at least conceptually. The key difference is that SPI explores structures without the need of crystals, which enables the possibility to observe a wider range of systems than is currently possible for crystallography. However, the lack of the strong diffraction signal observed from crystals presents new challenges.

10.9.2 Overview of Analysis Procedures

The data analysis workflow for FEL-based SPI is similar to SFX and consists also of three major steps: (1) identification of usable frames ("hits") from a very large data set, in this case frames containing diffraction from a single particle, (2) orientation determination, and (3) 3D image reconstruction by phasing. All these steps are accompanied by validation procedures, and they will be discussed step by step in the following paragraphs.

10.9.3 Calibration of Raw Data

The weak signals observed in SPI experiments increase the importance of calibration data to obtain and confirm the geometry of each sensor module prior to the start of data collection. Data pre-processing for SPI is therefore a crucial and sensitive first step for a successful image reconstruction. Use of optimized geometry and calibration information from other techniques (such as SFX) can be used to aid optimization. However, accurate conversion to photon counts is especially important for SPI as the average signal level per pixel is often well below a single photon [36, 91].

10.9.4 Hit Finding – Identification of Single Particles Only

For a successful 3D image reconstruction in an SPI experiment, a massive data set measured from many identical particles is required as the individual 2D diffraction patterns are randomly oriented and the signal from a single particle is relatively weak. The goal of this step is to identify single particle hits out of most non-hits and multiple-hits and (ideally) to also classify the particle-size distribution. Selection is usually based on a per-frame analysis of the calibrated 2D detector data. A basic version of such a procedure sums up the total number of photons in a region of interest on the detector and identifies a frame as a hit if the sum is above a pre-selected threshold [37, 51] – the so-called "lit pixel" method. Any thus selected "initial-hit" then must be classified more specifically, *i.e.*, single particle hits, multiple particle hits, and hits of particles which are not sample (*e.g.*, aggregates, solvent droplets). Currently available classification algorithms are mostly based on a spectral clustering method know as a kernel-based Principal Component Analysis (PCA), which utilizes the nonlinear correlations over a wide range of length scales [37, 51, 92, 93]. Once the single particle hits are identified, the size distribution of particles can be clarified, although the particles are typically already pre-selected in size to some extent, *e.g.*, by using a centrifugal separator. Beyond analyzing the recorded 2D detector data, hit identification may be aided by independent information such as time-of-flight spectroscopy of debris particles generated by a hit [68].

10.9.5 Orientation Determination

Well-defined Bragg peaks exhibiting symmetry as observed in SFX experiments facilitate the determination of the proper orientation of each image. In contrast, the more diffuse and lower intensity diffraction patterns observed in SPI present a greater though not insurmountable challenge. The most direct way to classify the same orientations is via cross correlation analysis, although the costs of direct cross correlation analysis on entire 2D diffraction intensity patterns in computing power is significant. Furthermore, the weak diffraction intensities typically applicable to bio-SPI with photon counting noise make these calculations inaccurate. The so-called EMC (expand-maximize-compress) algorithm, which is based on the expectation maximization (EM) algorithm [94], offers improved outcomes compared to traditional common-line or correlation methods and has become a key orientation determination tool for SPI experiments [91].

The EMC algorithm iteratively estimates a 3D diffraction intensity pattern by maximum likelihood estimation between the estimated and measured diffraction intensity patterns. As it is recognizable from its name, the algorithm consists of three steps: expansion (E), expectation maximization (M), and compression (C) steps. During the expansion step, at first a 3D model diffraction intensity map and an updated 3D diffraction intensity map afterward is expanded to 2D projected diffraction intensity patterns, like tomography data sets, at all possible orientations. Then, the estimated 2D projections are compared with the measured diffraction intensity patterns to update their orientation information during the expectation maximization step. All the intensity patterns with the same orientation are summed up and an updated 3D diffraction intensity pattern is reassembled from them in the compression step. After a few iterations, a final 3D diffraction intensity pattern is ready for the next step.

10.9.6 Phasing

The last step of the data analysis workflow for FEL-based SPI is the phase retrieval process, or phasing since the phase information in the reciprocal space is lost during the measurement. The practical phase retrieval algorithms, *i.e.*, the error reduction (ER) [95] and hybrid input output (HIO) [96], have been expanded over the past decade or so to include some novel variations, *e.g.*, the guided HIO (GHIO) [97], shrinkwrap [98], and the ptychographic iterative engine (PIE) [99]. In SPI data analysis, both the ER and the HIO algorithms, each leveraging two major constraints known as modulus and support constraints, have been widely used. Since small biomolecules can be considered as weak phase objects, a positivity constraint, which only allows real positive numbers in the real space image, may also improve the phase retrieval processes.

The aforementioned algorithms and constraints are applied to the final 3D diffraction intensity pattern obtained by the EMC algorithm and a quantitative 3D image (an electron density) in real space can be achievable.

10.10 Notes on the Need for and Use of Automation

Both SFX and SPI, as discussed above, require many usable frames not only due to the stochastic nature of injection but also enough for Monte-Carlo integration to be useful. Current and foreseeable injection techniques cannot predict or determine sample rotation prior to analysis [58]; hence the data collection process resides in a parameter space much larger than traditional predetermined rotation schemes. Traditional crystallography will often result in a structure from less than 720 frames allowing a crystallographer to check each frame manually if they so desire. In comparison SFX and SPI data often require >10,000 usable frames. As successful SFX and SPI experiments may achieve "hit rates" (the coincidental act of the diffracting object of interest intersecting the X-ray pulse) of ~10% or lower, a usable data set can easily exceed 100,000 frames.

Given these large data sets, it would be unreasonable for anyone to manually check each frame and therefore automation of data analysis is a necessity. Furthermore, as feedback from analysis of collected data to the ongoing data collection (in as close to real time as possible) is of great benefit for optimization and success of the experiment, automation of data analysis to provide this feedback in a structured way is a high priority. Each facility employs their own approach, which is related to their main scientific focus. These data analysis pipelines are constantly being improved with integration of more complex algorithms and possibilities for optimization for a given experiment - often with strong collaboration with the user community. It is therefore strongly recommended to discuss what data analysis and feedback is possible with the staff of the facility where you intend to collect data. These discussions during the planning stages enable not only the possibility of a customized approach if needed but are highly valuable to demonstrating feasibility of a proposal.

References

1. Yoon CH, Yurkov MV, Schneidmiller EA, Samoylova L, Buzmakov A, Jurek Z, Ziaja B (2016) A comprehensive simulation framework for imaging single particles and biomolecules at the European X-ray free-electron laser. Sci Rep. https://doi.org/10.1038/srep24791
2. Keable SM, Kölsch A, Simon PS, Dasgupta M, Chatterjee R, Subramanian SK, Hussein R (2021) Room temperature XFEL crystallography reveals asymmetry in the vicinity of the two phylloquinones in photosystem I. Sci Rep 11:21787. https://doi.org/10.1038/s41598-021-00236-3
3. Cox N, Pantazis DA, Lubitz W (2020) Current understanding of the mechanism of water oxidation in photosystem II and its relation to XFEL data. Annu Rev Biochem 89:795–820. https://doi.org/10.1146/annurev-biochem-011520-104801
4. Wiedorn MO, Oberthür D, Bean R, Schubert R, Werner N, Abbey B, Aepfelbacher M (2018) Megahertz serial crystallography. Nat Commun 9:4025. https://doi.org/10.1038/s41467-018-06156-7
5. Schmidt M (2013) Mix and inject: reaction initiation by diffusion for time-resolved macromolecular crystallography. Adv Condens Matter Phys
6. Owen RL, Rudino-Pinera E, Garman EF (2006) Experimental determination of the radiation dose limit for cryocooled protein crystals. Proc Natl Acad Sci 103:4912
7. Owen RL, Axford D, Nettleship JE, Owens RJ, Robinson JI, Morgan AW, Doré AS (2012) Outrunning free radicals in room-temperature macromolecular crystallography. Acta Crystallogr D Biol Crystallogr. https://doi.org/10.1107/S0907444912012553
8. de la Mora E, Mora E, Coquelle N, Bury CS, Rosenthal M, Holton JM, Carmichael I (2020) Radiation damage and dose limits in serial synchrotron crystallography at cryo- and room temperatures. Proc Natl Acad Sci 117:4142
9. Kern J, Alonso-Mori R, Tran R, Hattne J, Gildea RJ, Echols N, Glockner C (2013) Simultaneous femtosecond X-ray spectroscopy and diffraction of photosystem II at room temperature. Science. https://doi.org/10.1126/science.1234273
10. Boutet S, Lomb L, Williams GJ, Barends TRM, Aquila A, Doak RB, Weierstall U (2012) High-resolution protein structure determination by serial femtosecond crystallography. Science. https://doi.org/10.1126/science.1217737
11. Liu W, Wacker D, Gati C, Han GW, James D, Wang D, Nelson G (2013) Serial femtosecond

crystallography of G protein-coupled receptors. Science. https://doi.org/10.1126/science.1244142
12. Lomb L, Barends TRM, Kassemeyer S, Aquila A, Epp SW, Erk B, Foucar L (2011) Radiation damage in protein serial femtosecond crystallography using an X-ray free-electron laser. Phys Rev B Condens Matter Mater Phys 84:214111. https://doi.org/10.1103/PhysRevB.84.214111
13. Nass K (2019) Radiation damage in protein crystallography at X-ray free-electron lasers. Acta Crystallogr Sect Struct Biol. https://doi.org/10.1107/s2059798319000317
14. Abbey B, Dilanian RA, Darmanin C, Ryan RA, Putkunz CT, Martin AV, Wood D (2016) X-ray laser-induced electron dynamics observed by femtosecond diffraction from nanocrystals of buckminsterfullerene. Sci Adv 2:1601186. https://doi.org/10.1126/sciadv.1601186
15. Amin M, Badawi A, Obayya SS (2016) Radiation damage in XFEL: case study from the oxygen-evolving complex of photosystem II. Sci Rep 6: 36492. https://doi.org/10.1038/srep36492
16. Standfuss J (2019) Membrane protein dynamics studied by X-ray lasers – or why only time will tell. Curr Opin Struct Biol 57:63–71. https://doi.org/10.1016/j.sbi.2019.02.001
17. Dickerson JL, McCubbin PTN, Garman EF (2020) RADDOSE-XFEL: femtosecond time-resolved dose estimates for macromolecular X-ray free-electron laser experiments. J Appl Crystallogr. https://doi.org/10.1107/s1600576720000643
18. Suga M, Shimada A, Akita F, Shen J-R, Tosha T, Sugimoto H (2020) Time-resolved studies of metalloproteins using X-ray free electron laser radiation at SACLA. Biochim Biophys Acta Gen Subj 1864 2:129466. https://doi.org/10.1016/j.bbagen.2019.129466
19. Borshchevskiy V, Round E, Erofeev I, Weik M, Ishchenko A, Gushchin I, Mishin A, Willbold D, Büldt G, Gordeliy V (2014) Low-dose X-ray radiation induces structural alterations in proteins. Acta Crystallogr D Biol Crystallogr 70:2675–2685. https://doi.org/10.1107/S1399004714017295
20. Chapman HN, Fromme P, Barty A, White TA, Kirian RA, Aquila A, Hunter MS (2011) Femtosecond X-ray protein nanocrystallography. Nature 470:73–77. https://doi.org/10.1038/nature09750
21. Redecke L, Nass K, DePonte DP, White TA, Rehders D, Barty A, Stellato F (2013) Natively inhibited trypanosoma brucei cathepsin B structure determined by using an X-ray laser. Science 339: 227–230. https://doi.org/10.1126/science.1229663
22. Barends TRM, Foucar L, Sabine Botha RBD, Shoeman RL, Nass K, Koglin JE (2014) De novo protein crystal structure determination from X-ray free-electron laser data. Nature 505:244–247. https://doi.org/10.1038/nature12773
23. Aquila A, Hunter MS, Doak RB, Kirian RA, Fromme P, White TA, Andreasson J (2012) Time-resolved protein nanocrystallography using an X-ray free-electron laser. Opt Express 20:2706–2716. https://doi.org/10.1364/OE.20.002706
24. Standfuss J, Spence J (2017) Serial crystallography at synchrotrons and X-ray lasers. IUCrJ. https://doi.org/10.1107/S2052252517001877
25. Grünbein ML, Bielecki J, Gorel A, Stricker M, Bean R, Cammarata M, Dörner K, Fröhlich L, Hartmann E, Hauf S, Hilpert M, Kim Y, Kloos M, Letrun R, Messerschmidt M, Mills G, Nass Kovacs G, Ramilli M, Roome CM, Sato T, Scholz M, Sliwa M, Sztuk-Dambietz J, Weik M, Weinhausen B, Al-Qudami N, Boukhelef D, Brockhauser S, Ehsan W, Emons M, Esenov S, Fangohr H, Kaukher A, Kluyver T, Lederer M, Maia L, Manetti M, Michelat T, Münnich A, Pallas F, Palmer G, Previtali G, Raab N, Silenzi A, Szuba J, Venkatesan S, Wrona K, Zhu J, Doak RB, Shoeman RL, Foucar L, Colletier J-P, Mancuso AP, Barends TRM, Stan CA, Schlichting I (2018) Megahertz data collection from protein microcrystals at an X-ray free-electron laser. Nat Commun 9:3487. https://doi.org/10.1038/s41467-018-05953-4
26. Suga M, Akita F, Hirata K, Ueno G, Murakami H, Nakajima Y, Shimizu T (2015) Native structure of photosystem II at 1.95 Å resolution viewed by femtosecond X-ray pulses. Nature 517:99–103. https://doi.org/10.1038/nature13991
27. Pandey S, Bean R, Sato T, Poudyal I, Bielecki J, Villarreal JC, Yefanov O (2020) Time-resolved serial femtosecond crystallography at the European XFEL. Nat Methods 17:73–78. https://doi.org/10.1038/s41592-019-0628-z
28. Miao J, Ishikawa T, Robinson IK, Murnane MM (2015) Beyond crystallography: diffractive imaging using coherent X-ray light sources. Science 348:530–535. https://doi.org/10.1126/science.aaa1394
29. Neutze R, Wouts R, Spoel D, Weckert E, Hajdu J (2000) Potential for biomolecular imaging with femtosecond X-ray pulses. Nature 406:752–757. https://doi.org/10.1038/35021099
30. Altarelli M (2006) XFEL: the European X-ray free-electron laser. Technical design report
31. Sun Z, Fan J, Li H, Jiang H (2018) Current status of single particle imaging with X-ray lasers. Appl Sci. https://doi.org/10.3390/app8010132
32. Giewekemeyer K, Aquila A, Loh N-TD, Chushkin Y, Shanks KS, Weiss JT, Tate MW (2019) Experimental 3D coherent diffractive imaging from photon-sparse random projections. IUCrJ 6:357–365. https://doi.org/10.1107/S2052252519002781
33. Poudyal I, Marius S, Peter S (2020) Single-particle imaging by x-ray free-electron lasers – how many snapshots are needed? Struct Dyn 7:024102. https://doi.org/10.1063/1.5144516
34. Aquila A, Barty A, Bostedt C, Boutet S, Carini G, dePonte D, Drell P (2015) The linac coherent light source single particle imaging road map. Struct Dyn 2:041701. https://doi.org/10.1063/1.4918726

35. Mariani V, Morgan A, Yoon CH, Lane TJ, White TA, O'Grady C, Kuhn M (2016) OnDa: online data analysis and feedback for serial X-ray imaging. J Appl Crystallogr 49:1073–1080. https://doi.org/10.1107/S1600576716007469
36. Fortmann-Grote C, Buzmakov A, Jurek Z, Loh N-TD, Samoylova L, Santra R, Schneidmiller EA (2017) Start-to-end simulation of single-particle imaging using ultra-short pulses at the European X-ray free-electron laser. IUCrJ. https://doi.org/10.1107/s2052252517009496
37. Daurer BJ, Okamoto K, Bielecki J, Maia FRNC, Kerstin Mühlig MMS, Hantke MF (2017) Experimental strategies for imaging bioparticles with femtosecond hard X-ray pulses. IUCrJ 4:251–262. https://doi.org/10.1107/S2052252517003591
38. Bielecki J, Maia FRNC, Mancuso AP (2020) Perspectives on single particle imaging with X rays at the advent of high repetition rate X-ray free electron laser sources. Struct Dyn 7:040901. https://doi.org/10.1063/4.0000024
39. Sobolev E, Zolotarev S, Giewekemeyer K, Bielecki J, Okamoto K, Reddy HKN, Andreasson J (2020) Megahertz single-particle imaging at the European XFEL. Commun Phys 3:1–11. https://doi.org/10.1038/s42005-020-0362-y
40. Pande K, Donatelli JJ, Malmerberg E, Foucar L, Poon BK, Sutter M, Botha S (2018) Free-electron laser data for multiple-particle fluctuation scattering analysis. Sci Data 5. https://doi.org/10.1038/sdata.2018.201
41. Malmerberg E, Kerfeld CA, Zwart PH (2015) Operational properties of fluctuation X-ray scattering data. IUCrJ 2:309–316. https://doi.org/10.1107/S2052252515002535
42. Levantino M, Schirò G, Lemke HT, Cottone G, Glownia JM, Zhu D, Chollet M, Ihee H, Cupane A, Cammarata M (2015) Ultrafast myoglobin structural dynamics observed with an X-ray free-electron laser. Nat Commun 6. https://doi.org/10.1038/ncomms7772
43. Arnlund D, Johansson LC, Wickstrand C, Barty A, Williams GJ, Malmerberg E, Davidsson J (2014) Visualizing a protein quake with time-resolved X-ray scattering at a free-electron laser. Nat Methods. https://doi.org/10.1038/nmeth.3067
44. Lee Y, Kim JG, Lee SJ, Muniyappan S, Kim TW, Ki H, Kim H (2021) Ultrafast coherent motion and helix rearrangement of homodimeric hemoglobin visualized with femtosecond X-ray solution scattering. Nat Commun 12:3677. https://doi.org/10.1038/s41467-021-23947-7
45. Blanchet CE, Round A, Mertens HDT, Ayyer K, Graewert M, Awel S, Franke D, Dörner K, Bajt S, Bean R, Custódio TF, de Wijn R, Juncheng E, Henkel A, Gruzinov A, Jeffries CM, Kim Y, Kirkwood H, Kloos M, Knoška J, Koliyadu J, Letrun R, Löw C, Makroczyova J, Mall A, Meijers R, Murillo GEP, Oberthür D, Round E, Seuring C, Sikorski M, Vagovic P, Valerio J, Wollweber T, Zhuang Y, Schulz J, Haas H, Chapman HN, Mancuso AP, Svergun D (2023) Form factor determination of biological molecules with X-ray free electron laser small-angle scattering (XFEL-SAS) abstract. Commun Biol 6:1057. https://doi.org/10.1038/s42003-023-05416-7
46. Spence JCH, Weierstall U, Howells M (2004) Coherence and sampling requirements for diffractive imaging. Ultramicroscopy 101:149–152. https://doi.org/10.1016/j.ultramic.2004.05.005
47. Fuller FD, Gul S, Ruchira Chatterjee ESB, Young ID, Lebrette H, Srinivas V (2017) Drop-on-demand sample delivery for studying biocatalysts in action at X-ray free-electron lasers. Nat Methods 14:443–449. https://doi.org/10.1038/nmeth.4195
48. Juncheng E, Stransky M, Jurek Z, Fortmann-Grote C, Juha L, Santra R, Ziaja B, Mancuso AP (2021) Effects of radiation damage and inelastic scattering on single-particle imaging of hydrated proteins with an X-ray free-electron laser. Sci Rep 11:1–11. https://doi.org/10.1038/s41598-021-97142-5
49. Ekeberg T, Svenda M, Abergel C, Maia FRNC, Seltzer V, Claverie J-M, Hantke M (2015) Three-dimensional reconstruction of the giant mimivirus particle with an X-ray free-electron laser. Phys Rev Lett 114:098102. https://doi.org/10.1103/PhysRevLett.114.098102
50. Kurta RP, Donatelli JJ, Yoon CH, Berntsen P, Bielecki J, Daurer BJ, DeMirci H (2017) Correlations in scattered X-ray laser pulses reveal nanoscale structural features of viruses. Phys Rev Lett 119:158102. https://doi.org/10.1103/PhysRevLett.119.158102
51. Hantke MF, Hasse D, Ekeberg T, John K, Martin Svenda NDL (2014) High-throughput imaging of heterogeneous cell organelles with an X-ray laser. Nat Photonics 8. https://doi.org/10.1038/nphoton.2014.270
52. Reddy HKN, Yoon CH, Aquila A, Awel S, Ayyer K, Barty A, Berntsen P (2017) Coherent soft X-ray diffraction imaging of coliphage PR772 at the linac coherent light source. Sci Data 4. https://doi.org/10.1038/sdata.2017.79
53. Munke A, Andreasson J, Aquila A, Awel S, Ayyer K, Barty A, Bean RJ (2016) Coherent diffraction of single rice dwarf virus particles using hard X-rays at the linac coherent light source. Sci Data 3. https://doi.org/10.1038/sdata.2016.64
54. Mancuso AP, Aquila A, Batchelor L, Bean RJ, Bielecki J, Borchers G, Doerner K (2019) The single particles, clusters and biomolecules and serial femtosecond crystallography instrument of the European XFEL: initial installation. J Synchrotron Radiat 26:660–676. https://doi.org/10.1107/S1600577519003308
55. Yamashita M, Fenn JB (1984) Electrospray ion source. Another variation on the free-jet theme. J Phys Chem. https://doi.org/10.1021/j150664a002
56. DePonte DP, Weierstall U, Schmidt K, Warner J, Starodub D, Spence JCH, Doak RB (2008) Gas dynamic virtual nozzle for generation of microscopic

droplet streams. J Phys D Appl Phys. https://doi.org/10.1088/0022-3727/41/19/195505
57. Grünbein ML, Shoeman RL, Doak RB (2018) Velocimetry of fast microscopic liquid jets by nanosecond dual-pulse laser illumination for megahertz X-ray free-electron lasers. Opt Express 26:7190–7203. https://doi.org/10.1364/OE.26.007190
58. Weierstall U, James D, Wang C, White TA, Wang D, Liu W, Spence JCH (2014) Lipidic cubic phase injector facilitates membrane protein serial femtosecond crystallography. Nat Commun 5:3309
59. Stagno JR, Liu Y, Bhandari YR, Conrad CE, Panja S, Swain M, Fan L (2017) Structures of riboswitch RNA reaction states by mix-and-inject XFEL serial crystallography. Nature 541:242–246. https://doi.org/10.1038/nature20599
60. Kupitz C, Olmos JL Jr, Holl M, Tremblay L, Pande K, Pandey S, Oberthür D (2017) Structural enzymology using X-ray free electron lasers. Struct Dyn 4:044003. https://doi.org/10.1063/1.4972069
61. Calvey GD, Katz AM, Schaffer CB, Pollack L (2016) Mixing injector enables time-resolved crystallography with high hit rate at X-ray free electron lasers. Struct Dyn 3:054301. https://doi.org/10.1063/1.4961971
62. Wang D, Weierstall U, Pollack L, Spence J (2014) Double-focusing mixing jet for XFEL study of chemical kinetics. J Synchrotron Radiat 21:1364–1366. https://doi.org/10.1107/S160057751401858X
63. Trebbin M, Krüger K, DePonte D, Roth SV, Chapman HN, Förster S (2014) Microfluidic liquid jet system with compatibility for atmospheric and high-vacuum conditions. Lab Chip 14:1733–1745. https://doi.org/10.1039/C3LC51363G
64. Nelson G, Kirian RA, Weierstall U, Zatsepin NA, Faragó T, Baumbach T, Wilde F (2016) Three-dimensional-printed gas dynamic virtual nozzles for X-ray laser sample delivery. Opt Express 24:11515–11530. https://doi.org/10.1364/OE.24.011515
65. Bielecki J, Hantke MF, Daurer BJ, Reddy HKN, Hasse D, Larsson DSD, Gunn LH (2019) Electrospray sample injection for single-particle imaging with X-ray lasers. Sci Adv 5:8801. https://doi.org/10.1126/sciadv.aav8801
66. Hantke MF, Bielecki J, Kulyk O, Westphal D, Larsson DSD, Svenda M, Reddy HKN (2018) Rayleigh-scattering microscopy for tracking and sizing nanoparticles in focused aerosol beams. IUCrJ 5:673–680. https://doi.org/10.1107/S2052252518010837
67. Roth N, Awel S, Horke DA, Küpper J (2018) Optimizing aerodynamic lenses for single-particle imaging. J Aerosol Sci. https://doi.org/10.1016/j.jaerosci.2018.06.010
68. Andreasson J, Martin AV, Liang M, Timneanu N, Aquila A, Wang F, Iwan B (2014) Automated identification and classification of single particle serial femtosecond X-ray diffraction data. Opt Express 22:2497–2510. https://doi.org/10.1364/OE.22.002497
69. Soares AS, Engel MA, Stearns R, Datwani S, Olechno J, Ellson R, Skinner JM, Allaire M, Orville AM (2011) Acoustically mounted microcrystals yield high-resolution X-ray structures. Biochemistry 50: 4399–4401. https://doi.org/10.1021/bi200549x
70. Roessler CG, Agarwal R, Allaire M, Alonso-Mori R, Andi B, Bachega JFR, Bommer M (2016) Acoustic injectors for drop-on-demand serial femtosecond crystallography. Structure 24:631–640. https://doi.org/10.1016/j.str.2016.02.007
71. Graceffa R, Burghammer M, Davies RJ, Riekel C (2012) Probing ballistic microdrop coalescence by stroboscopic small-angle X-ray scattering. Appl Phys Lett. https://doi.org/10.1063/1.4772631
72. Roedig P, Ginn HM, Pakendorf T, Sutton G, Harlos K, Walter TS, Meyer J (2017) High-speed fixed-target serial virus crystallography. Nat Methods 14:805–810. https://doi.org/10.1038/nmeth.4335
73. Tolstikova A, Levantino M, Yefanov O, Hennicke V, Fischer P, Meyer J, Mozzanica A (2019) 1 kHz fixed-target serial crystallography using a multilayer monochromator and an integrating pixel detector. IUCrJ. https://doi.org/10.1107/S205225251900914X
74. Rupp B (2009) Biomolecular crystallography. Garland Science
75. Yefanov O, Oberthür D, Bean R, Wiedorn MO, Knoska J, Pena G, Awel S (2019) Evaluation of serial crystallographic structure determination within megahertz pulse trains. Struct Dyn. https://doi.org/10.1063/1.5124387
76. Patel J, Round A, Bielecki J, Doerner K, Kirkwood H, Letrun R, Schulz J, Sikorski M, Vakili M, De Wijn R, Peele A, Mancuso AP, Abbey B (2022) Towards real-time analysis of liquid jet alignment in serial femtosecond crystallography. J Appl Crystallogr 55:944–952. https://doi.org/10.1107/S1600576722005891
77. Henrich B, Becker J, Dinapoli R, Goettlicher P, Graafsma H, Hirsemann H, Klanner R, Krueger H, Mazzocco R, Mozzanica A, Perrey H, Potdevin G, Schmitt B, Shi X, Srivastava AK, Trunk U, Youngman C (2011) The adaptive gain integrating pixel detector AGIPD a detector for the European XFEL. Nucl Instrum Methods Phys Res Sect Accel Spectrometers Detect Assoc Equip 633:S11–S14. https://doi.org/10.1016/j.nima.2010.06.107
78. Kuster M, Boukhelef D, Donato M, Dambietz J-S, Hauf S, Maia L, Raab N (2014) Detectors and calibration concept for the European XFEL. Synchrotron Radiat News. https://doi.org/10.1080/08940886.2014.930809
79. Barty A, Kirian RA, Maia FRNC, Hantke M, Yoon CH, White TA, Chapman H (2014) Cheetah: software for high-throughput reduction and analysis of serial femtosecond X-ray diffraction data. J Appl Crystallogr 47:1118–1131. https://doi.org/10.1107/S1600576714007626
80. Hadian-Jazi M, Sadri A, Barty A, Yefanov O, Galchenkova M, Oberthuer D, Komadina D (2021) Data reduction for serial crystallography using a robust peak finder. J Appl Crystallogr 54:1360–1378. https://doi.org/10.1107/S1600576721007317

81. McRee DE (1999) Practical Protein Crystallography. 2nd edition. Elsevier. https://doi.org/10.1016/B978-0-12-486052-0.X5000-3
82. Kabsch W (2012) XDS. In: International Tables for Crystallography Vol. F, ch. 11.6, pp. 304–310. https://doi.org/10.1107/97809553602060000835
83. Leslie AGW, Powell HR (2007) Processing diffraction data with Mosflm. Evol Methods Macromol Crystallogr. https://doi.org/10.1007/978-1-4020-6316-9_4
84. White TA, Kirian RA, Martin AV, Aquila A, Nass K, Barty A, Chapman HN (2012) CrystFEL: a software suite for snapshot serial crystallography. J Appl Crystallogr. https://doi.org/10.1107/S0021889812002312
85. Brehm W, Diederichs K (2014) Breaking the indexing ambiguity in serial crystallography. Acta Crystallogr D Biol Crystallogr. https://doi.org/10.1107/s1399004713025431
86. Liu H, Spence JCH (2014) The indexing ambiguity in serial femtosecond crystallography (SFX) resolved using an expectation maximization algorithm. IUCrJ. https://doi.org/10.1107/s2052252514020314
87. Li C, Li X, Kirian R, Spence JCH, Liu H, Zatsepin NA (2019) SPIND: a reference-based auto-indexing algorithm for sparse serial crystallography data. IUCrJ 6: 72–84. https://doi.org/10.1107/S2052252518014951
88. Kirian RA, Wang X, Weierstall U, Schmidt KE, Spence JCH, Hunter M, Fromme P, White T, Chapman HN, Holton J (2010) Femtosecond protein nanocrystallography—data analysis methods. Opt Express 18:5713–5723. https://doi.org/10.1364/OE.18.005713
89. Evans P (2006) Scaling and assessment of data quality. Acta Crystallogr D Biol Crystallogr 62:72–82. https://doi.org/10.1107/S0907444905036693
90. Adams PD, Afonine PV, Bunkóczi G, Chen VB, Davis IW, Echols N, Headd JJ, Hung L-W, Kapral GJ, Grosse-Kunstleve RW, McCoy AJ, Moriarty NW, Oeffner R, Read RJ, Richardson DC, Richardson JS, Terwilliger TC, Zwart PH (2010) PHENIX: a comprehensive Python-based system for macromolecular structure solution. Acta Crystallogr D Biol Crystallogr 66: 213–221. https://doi.org/10.1107/S0907444909052925
91. Loh N-TD, Elser V (2009) Reconstruction algorithm for single-particle diffraction imaging experiments. Phys Rev. https://doi.org/10.1103/PhysRevE.80.026705
92. Yoon CH, Schwander P, Abergel C, Andersson I, Andreasson J, Aquila A, Bajt S (2011) Unsupervised classification of single-particle X-ray diffraction snapshots by spectral clustering. Opt Express 19: 16542–16549. https://doi.org/10.1364/OE.19.016542
93. Bobkov SA, Teslyuk AB, Kurta RP, Gorobtsov OY, Yefanov OM, Ilyin VA, Senin RA, Vartanyants IA (2015) Sorting algorithms for single-particle imaging experiments at X-ray free-electron lasers. J Synchrotron Radiat 22:1345–1352. https://doi.org/10.1107/S1600577515017348
94. Dempster AP, Laird NM, Rubin DB (1977) Maximum likelihood from incomplete data via the EM algorithm. J R Stat Soc. https://doi.org/10.1111/j.2517-6161.1977.tb01600.x
95. Fienup JR (1978) Reconstruction of an object from the modulus of its fourier transform. Opt Lett. https://doi.org/10.1364/ol.3.000027
96. Fienup JR (1982) Phase retrieval algorithms: a comparison. Appl Opt. https://doi.org/10.1364/ao.21.002758
97. Chen C-C, Jianwei Miao CWW, Lee TK (2007) Application of optimization technique to noncrystalline X-ray diffraction microscopy: guided hybrid input-output method. Phys Rev B. https://doi.org/10.1103/physrevb.76.064113
98. Marchesini S, He H, Chapman HN, Hau-Riege SP, Noy A, Howells MR, Weierstall U, Spence JCH (2003) X-ray image reconstruction from a diffraction pattern alone. Phys Rev B. https://doi.org/10.1103/physrevb.68.140101
99. Thibault P, Dierolf M, Bunk O, Menzel A, Pfeiffer F (2009) Probe retrieval in ptychographic coherent diffractive imaging. Ultramicroscopy 109:338–343. https://doi.org/10.1016/j.ultramic.2008.12.011

Small-Angle X-Ray Scattering for Macromolecular Complexes

11

Stephanie Hutin, Mark D. Tully, and Martha Brennich

Abstract

Small angle X-ray scattering (SAXS) is a versatile technique that can provide unique insights in the solution structure of macromolecules and their complexes, covering the size range from small peptides to complete viral assemblies. Technological and conceptual advances in the last two decades have tremendously improved the accessibility of the technique and transformed it into an indispensable tool for structural biology. In this chapter we introduce and discuss several approaches to collecting SAXS data on macromolecular complexes, including several approaches to online chromatography. We include practical advice on experimental design and point out common pitfalls of the technique.

Keywords

SAXS · BioSAXS · SEC-SAXS · automation · HPLC

S. Hutin · M. D. Tully
Structural Biology Group, European Synchrotron Radiation Facility, Grenoble, Grenoble, France

M. Brennich (✉)
European Molecular Biology Laboratory, Grenoble, Grenoble, France
e-mail: brennich@embl.fr

11.1 Introduction

Understanding the function and functional mechanism of macromolecular complexes is tightly linked to knowledge of their structure. Ideally, we would like to completely characterize the structure in the natural environment of the complex, the living cell [1, 2]. Indeed, recent approaches in super resolution light microscopy enable direct studies of super-complexes, such as the nuclear pore complex [3, 4], meiotic chromosome axes [5] or the pericentriolar material (PCM) [6]. However, the complex environment of the cell can also limit our degree of control, turning direct conclusions on causal relationships on the molecular level difficult to impossible. In addition, the resolution of fluorescence microscopy is intrinsically limited by its need of fluorescent labels [7]. Often, reducing the complexity of the system by isolating the object of interest can resolve these difficulties, thereby allowing functional studies on purified complexes in solution to be performed.

Small-angle X-ray scattering (SAXS) can provide structural information of macromolecules in solution on the nanometer scale. It is well suited to study the conformation and/or conformational ensembles of macromolecular complexes. Recent examples include studies of the Tat:AFF4:P-TEFb complex involved in proviral transcriptional activation of HIV-1 [8], the binding of vaccinia virus (VACV) DNA polymerase E9 to

M. C. Vega, F. J. Fernández (eds.), *Advanced Technologies for Protein Complex Production and Characterization*, Advances in Experimental Medicine and Biology 1453,
https://doi.org/10.1007/978-3-031-52193-5_11

DNA [9], the structure of host factor Q-beta phage (Hhq) of *Escherichia coli* in complex with mRNA [10], sub-complexes of the multisynthetase complex [11], ternary complexes of the C2 domain of coagulation factor VIII (fVIII) and inhibitory antibodies [12], or mannan-binding lectin (MBL) and MBL-associated serine protease-1 (MASP-1) complexes of the lectin pathway of the complement system [13].

Typical questions for a SAXS experiment would be "What is the solution structure of my complex?", "How do complex formation and dissociation depend on the concentration of the partners?" and "Which factors can alter the complex or its formation?" There are many SAXS analysis tools available to address this kind of questions, for example FoXSDock [14], SASREF [15], DAMMIF [16], DENFERT [17], or OLIGOMER. However, none of these can provide meaningful answers if the quality of the SAXS data is not good enough. It is therefore imperative to design SAXS experiments well and to collect high quality data to avoid misinterpretations based on experimental artefacts.

In this chapter, we will discuss different approaches for collecting SAXS data for complexes, when it is appropriate to use them, and which pitfalls need to be avoided. Figure 11.1 illustrates the typical SAXS workflow.

11.2 Dilution Series SAXS

The classical BioSAXS experiment consists of recording a series of dilutions of the sample of interest at known concentrations; see, *e.g.*, [18]. A typical setup first measures buffer, sample at concentration 1, buffer, sample at concentration 2, buffer, sample at concentration 3, and a final buffer measurement. This setup has numerous advantages:

- High quality data from low concentration samples. For large complexes as little as 0.5 mg/mL is sufficient.
- Experiments can be performed on any SAXS setup, including in-house instruments, as low photon flux can be compensated by increased exposure time.
- Low sample volumes of 30–50 μL are required. This is of special interest for additive screenings.
- In general, macromolecular concentrations are well-defined and known, allowing more accurate mass estimates from forward scattering. This is of particular interest for flexible systems for which volume-based approaches (Porod volume, correlated volume, etc.) struggle.
- When the macromolecular concentration is controlled, it is possible to investigate its effect on the scattering signal and draw conclusions on the mixing state.
- The static nature of the sample means that complementary experiments (*e.g.*, UV-Vis spectroscopy, dynamic light scattering) can be performed on the same sample or under identical conditions with relatively low technical effort.
- It is often easier to achieve high macromolecular concentrations when not relying on online purification. This improves the signal-to-noise, in particular at higher diffraction angles where the signal is often weak.
- On high-flux instruments, *e.g.*, synchrotron beamlines, individual measurements are relatively fast (1–5 min), enabling easy access to slow kinetics [19].

For stable complexes, a dilution series experiment is therefore often the best way to obtain high quality SAXS data. For example, M. Karlsen and co-workers recently determined the low-resolution structure of the dimeric and tetrameric complexes of the BAR domain protein PICK1 [20] by deconvoluting SAXS data of the equilibrium dimer-tetramer system. Cordeiro et al. went one step further, combing SAXS and molecular dynamics simulations to obtain the low-resolution structure of PCNA-p15 complexes and their dissociation constant at the same time [21].

Even if the composition of a sample is not concentration dependent, it is still necessary to

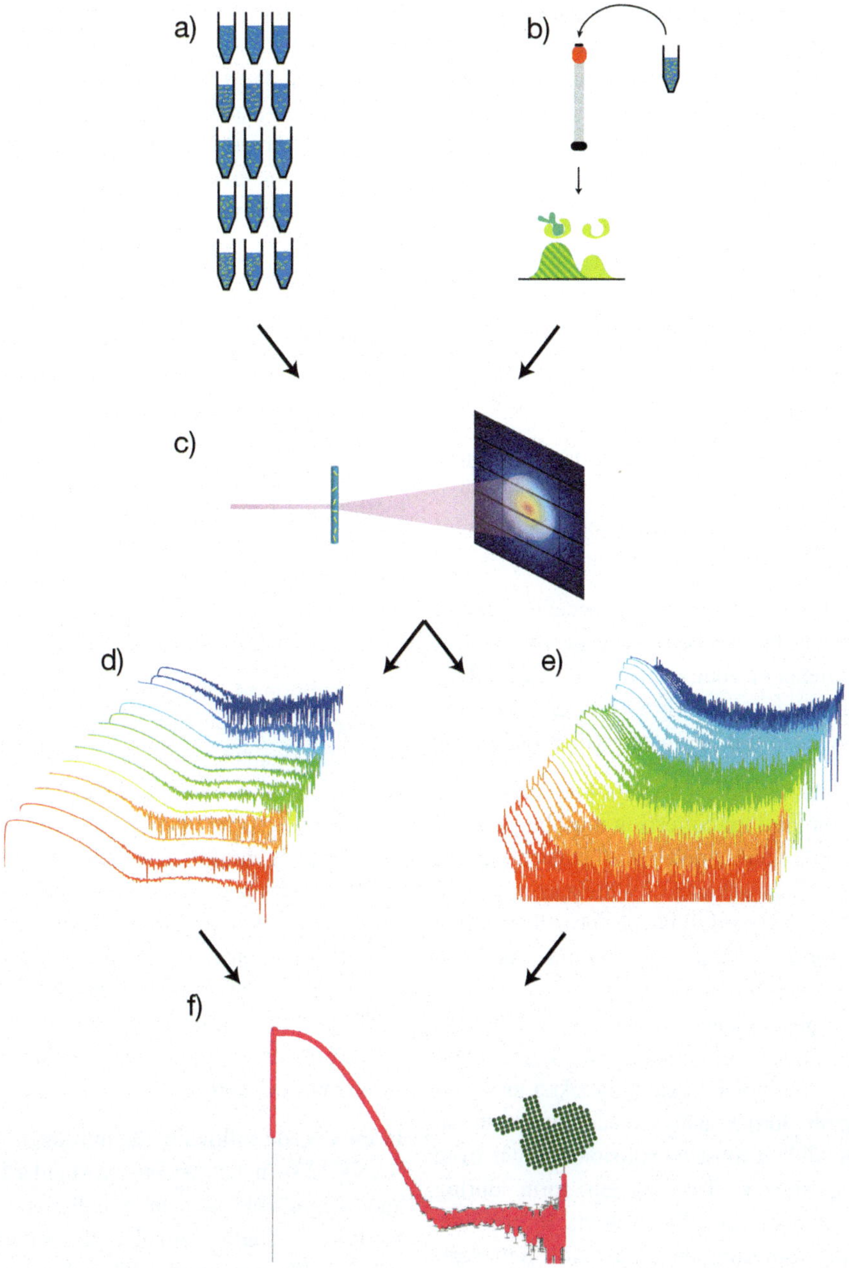

Fig. 11.1 Workflows of SAXS experiments. To obtain SAXS data of a complex from "static" samples, multiple samples at different absolute and relative concentrations are necessary (**a**). For each of these samples individual SAXS curves are collected (**c**, **d**). The information from these curves can be combined to an idealized SAXS curve of the complex suitable for SAXS-based modelling (**f**). For obtaining SAXS data of a complex via SEC-SAXS, a sample is applied on the column and data are directly collected on the eluent (**b**, **c**). The resulting set of continuous scattering curves (**e**) can be deconvoluted to obtain the SAXS curve of the complex (**f**)

measure several concentrations as additional phenomena, such as inter-particle scattering or aggregation can still affect the signal in a concentration dependent manner.

The standard strategy that is advised for investigating a stable complex or homo-oligomer requires the following measurements:

- SAXS concentration series of the individual components.
- SAXS concentration series of the complex.

For non-stable complexes, it is advisable to also vary the mixing ratios:

- SAXS concentration series of the individual components.
- Several SAXS concentration series at different mixing ratios of the components.

Note that for this kind of experiments, precise knowledge of the number and respective size of the constituents of your system is essential. This requirement is directly linked to the disadvantages to this approach, which include:

- Good background subtraction is essential, *i.e.,* the buffer used for correction must be identical to the one of the samples at all concentrations. This is ideally achieved by dialysis, although carefully designed and performed spiking, diafiltration or desalting column runs can serve as alternatives.
- Any "contaminants" such as aggregates, higher order oligomers or degradation products that are not explicitly taken into consideration during analysis and interpretation will invalidate most conclusions drawn from the experiment. These often form during freeze thawing of the protein. If possible, checking your sample with dynamic light scattering prior to measurement will alert you to any larger aggregates.

11.3 Inline Purification SAXS – Fighting Polydispersity

As mentioned above, obtaining the scattering curve of the monodisperse complex is essential for any further structural analysis. While such a curve can sometimes be obtained by deconvolution from SAXS data of mixtures even for non-stable complexes, we would ideally measure it directly to reduce the number of possible systematic errors. As association and dissociation of complexes and oligomers is generally not instantaneous, purification directly before data collection can provide scattering curves of monodisperse systems.

Chromatography techniques are particularly well suited to be coupled to SAXS data acquisition. As the sample passes through the porous matrix of the column, it separates by size, affinity or charge. The most common chromatography method used in combination with SAXS is size exclusion (SEC). Continuous data collection of the chromatography-coupled-SAXS experiment has several advantages over the more conventional concentration series SAXS experiments:

- Separation of large (*e.g.*, aggregates) or small (*e.g.*, excess buffer components) size contaminants.
- Elution of the sample in the running buffer facilitates correct background matching, if the column is equilibrated correctly.
- Peak shifts can directly reveal binding. In addition, calibrated SEC allows simultaneous determination of the Stokes radius based on the peak position is possible [22].
- Large structural variability (*e.g.*, domain movements) within a macromolecule/complex can be revealed.

In the decade following the introduction of SEC-SAXS [23], it has become a standard approach and is available in many small-angle scattering facilities, including small-angle neutron scattering (SANS) instruments [24–30].

There are however disadvantages to SEC-SAXS that also need to be considered:

- Higher protein concentrations are required due to column dilution effects, ideally upwards of 5 mg/mL.
- Experiments should be performed at synchrotrons or other high photon flux instruments.

In this experimental design, it is important to consider separation properties. Often, the apparent size of a complex is close to that of one of its constituents, even though their masses might differ significantly. They are therefore difficult to separate, and it is essential to eliminate excess of this constituent, *e.g.*, by adding its partner in excess.

Choosing the right size exclusion column for your experiment is essential for the success of a SEC-SAXS experiment. Ideally, the elution peak should be well centered between the void volume and the total column volume to avoid overlaps with aggregates and/or excess small molecules. This implies that more than one column type might be necessary to study a complex and all its constituents. Additionally, some macromolecules might be able to interact with the column medium, resulting in abnormal elution or even conformational changes. In general, concentration, flow rate, and volume dictate the quality of separation.

Sometimes, even online purification cannot completely separate different species, either because they are too close to each other in size or because of fast dynamics. Just as for dilution-series measurements, it is possible to deconvolute SAXS data to obtain idealized SAXS curve of individual components.

The program COSMICS uses the fact that the information content of SEC-SAXS data can be increased by emphasizing changes in different parts of the curve (*e.g.*, Kratky representation) and applies the multivariate curve resolution alternating least squares (MCR-ALS) chemometric method to obtain "pure" curves [31, 32].

The SEC-SAXS package of US-SOMO decomposes the chromatogram into individual peaks of pre-defined shape [33]. In addition, it can also reduce the influence of capillary spoiling due to material deposition on the final data.

The finite size of individual peaks allows evolving factor analysis (EFA) to deconvolute overlapping peaks [27, 34]. A SAXS specific implementation can be found in BioXTAS RAW [35].

One important requirement for all these approaches is that the SAXS signal of each individual contributor must be constant. Large scale flexibility (*e.g.*, movement of complex partners relative to each other) therefore renders these approaches unfeasible.

Of course, online purification is not limited to SEC. For very large multi-subunit macromolecular complexes or complexes that dissociate under shear stress, differential ultracentrifugation can be better suited [36].

For complexes that cannot be separated by size, or which are prone to aggregation during the concentration steps, separation by physical and chemical properties might be more appropriate. For example, ion exchange chromatography separates based on surface charge. As the bulk protein concentration is low prior to elution, high concentrations can be achieved without losses to aggregation [37].

Of particular interest for complexes is online affinity chromatography, especially if one of the constituents is insoluble on its own [38].

Since all these approaches rely on buffer variation for the separation, correct background subtraction is non-trivial, but manageable. All these approaches are less routinely available at SAXS instruments and require communication with local responsible about necessary equipment.

11.4 Buffer Considerations

Although SAXS measurements can be performed in almost any buffer, a few considerations should be kept in mind when selecting the buffer for a given experiment:

- Contrast: the strength of the SAXS signal depends on the difference in electron density between the buffer and the macromolecules. Try to keep the concentrations of all buffer components as low as possible.
- The salt concentration in the buffer should be ideally between 20 and 300 mM. High salt concentrations reduce the contrast and consequently increase the noise level. Low salt concentrations increase undesired inter-particle scattering.
- Preferentially use potassium salts in experiments with RNAs.

- Viscosity and surface tension: keep viscosity low because at many facilities the sample needs to be handled by a pipetting robot.
- Glycerol content should be kept as low as possible (max. 20% (v/v)). Note that glycerol-containing buffers slow down column equilibration. This is relevant for both SEC-SAXS experiments and for buffer matching prior to experiments. Up to 10 column volumes may be required to equilibrate a column sufficiently for SEC-SAXS.
- Only use reducing agents if absolutely required. Which agent is appropriate depends on the sample and needs to be tested.
- Do not use protease inhibitors in the final sample.

As the buffer composition might affect the structure and/or stability of a complex, additional control experiments might be necessary if complementary techniques require different buffer conditions (*e.g.*, no salt). These measurements might result in suboptimal SAXS data but might still help to confirm or exclude structural changes.

11.5 Analysis of Macromolecular Complexes

Assuming the data collection of a macromolecular complex was successful the data sets need to be analyzed. Primary processing creates idealized scattering curves and determines SAXS invariants [39]. A variety of software tools help with all these steps for both dilution series and SEC-SAXS such as ATSAS [40], ScÅtter or BioXTAS RAW [35].

Modeling and model validation are generally more specific for the problem at hand. Before attempting any SAXS-based modeling, one should know the answers to the following questions:

- Does the SAXS curve represent a (structurally) monodisperse system? Or are different oligomeric states present?
- Is there only one kind of component (*e.g.*, only protein)? Or are there different kinds of components (*e.g.*, protein/DNA complexes)?
- Are there flexible regions?
- Does complex formation affect the shape of the constituents?

Flexible systems generally require atomistic ensemble approaches to modeling, *e.g.*, EOM [41], MES [42] or SASSIE [43]. Protein/DNA complexes do not require special precautions when building atomistic models, but for building *ab initio* models the difference in scattering contrast between protein and nucleic acid needs to be considered, *e.g.*, MONSA [44].

For both atomistic and *ab initio* modeling it is possible to include oligomer formation.

11.6 Presenting "Complex" Data for Publication

When presenting SAXS data of complexes for publication it is recommended to follow the IUCr's publication guidelines for small-angle scattering data from biomolecules in solution [45]. Especially for non-stable complexes, it is important to precisely explain how sample monodispersity was ensured.

Additionally, one should not only present the complex data itself, but – if attainable – also SAXS data of the constituents. If different models for a complex (*e.g.*, different interaction interfaces) can be constructed, comparisons between rejected models and the recorded data should also be presented.

11.7 Case Study

Vaccinia is the prototype member of the Poxviridae. This family contains important pathogens, such as the smallpox-causing virus Variola, with large double-stranded DNA genomes that replicate exclusively in the cytoplasm of the host cell, where the replication is solely dependent on virus-encoded proteins [46]. E9 represents the catalytic subunit of the viral DNA polymerase. Over the years, a number of genetic and biochemical studies have characterized this protein [47], but only very recently a high-resolution structure of full-length E9 was solved at 2.7 Å resolution [9]. The protein displays a classical palm, thumb, finger, exonuclease, and N-terminal domains of a family B polymerase in an open conformation. The high-resolution structure permitted the identification of important poxvirus-specific structural insertions [9]. Furthermore, E9 is the target of several antivirals [48, 49], hence it is of high medical interest not only to understand the structure but also acquire knowledge of the structure of the DNA-bound state to understand resistance mutations against polymerase inhibitors.

Related published family B polymerase structures in complex with DNA oligonucleotides indicate that considerable domain movements occur upon DNA binding leading to closed structures compared to the apo forms. To study the domain movements of E9, SEC-SAXS experiments on an isomorphous exo^{minus} mutant (without exonuclease activity) bound to a 29mer DNA hairpin were performed and compared to models of E9 in elongation mode based on the yeast pol δ structure bound to DNA (PDB ID 3IAY).

The E9 exo^{minus}/29-mer hairpin DNA complex mimicking template and primer strand was formed by mixing the protein with a 20% molar excess of the DNA. 50 μL of this mix were injected onto a SEC-column (Superdex 200 Increase 5/150 GL column, GE Healthcare) in-line with the flow cell for SAXS measurements at the BioSAXS BM29 beamline at the European Synchrotron Research Facility (ESRF) in Grenoble, France. The column was equilibrated with 3 column volumes of 20 mM Tris-HCl, pH 7.5, 100 mM NaCl prior to the experiment. 1000 frames of 1 s exposure time per frame were collected at a flow rate of 0.3 mL min^{-1}, covering 1.7 column volumes. The individual frames were processed automatically and independently within the EDNA framework [50], yielding radially averaged curves of normalized intensity *vs.* scattering angle $q = 4\pi \sin\theta/\lambda$. The frames that corresponded to the elution of the protein/DNA complex, the protein alone and the DNA were individually merged and analyzed further using the tools of the ATSAS package [51]. Comparison to apo E9 exo^{minus} showed that formation of the complex between E9 exo^{minus} and the 29-mer DNA hairpin reduced the radius of gyration from 3.83 nm for the apo form to 3.45 nm, directly implying DNA binding in a cavity of E9.

Different models of E9 exo^{minus} with bound DNA were built based on its high-resolution crystal structure and the structure of yeast pol δ bound to DNA, independently of the SAXS experiments. Comparison of the predicted scattering curves computed with CRYSOL [52] from each of these models to the SAXS data allowed to evaluate the models. It was not sufficient to adjust only the position of the thumb domain of E9, which is expected to show the strongest movement, but only when the individual domains of E9 are adjusted, the theoretical curve of the E9 exo^{minus}/DNA matched very well the observed scattering curve (lowest χ^2, Fig. 11.2).

In this experiment SAXS provided essential insights into the conformational changes of E9 upon DNA binding. Understanding these changes greatly enhances our possibilities for developing drugs targeting E9.

11.8 Conclusions

Small-angle X-ray scattering is a powerful and versatile tool for structural biology. Applied correctly, it can provide unique insights into structures and structural dynamics of bio-macromolecular complexes.

The success of a SAXS experiment depends to a large extent on the sample preparation. Due to the increased popularity of the technique, there is a plethora of recent literature with advice on sample (and buffer) preparation for successful experiments.

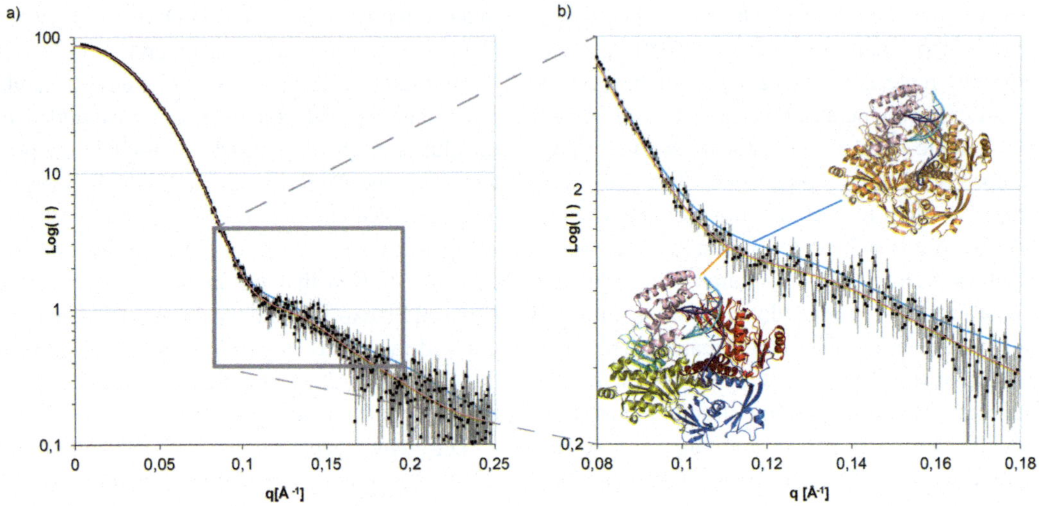

Fig. 11.2 Comparison of the scattering data of E9 exo^{minus}/29-mer hairpin DNA complex to different models. To obtain the best fit (orange curve) is necessary to adjust the position of all domains of E9. Adjusting only the thumb domain (pink domain) results in a systematic deviation of the predicted curve (blue curve). Adapted from [9].

Acknowledgments The authors thank N. Tarbouriech and W. Burmeister permission to show the data on E9 exo^{minus} and for advice on the manuscript.

References

1. Sydor AM, Czymmek KJ, Puchner EM, Mennella V (2015) Super-resolution microscopy: from single molecules to supramolecular assemblies. Trends Cell Biol 25:730–748. https://doi.org/10.1016/j.tcb.2015.10.004
2. Baddeley D, Bewersdorf J (2017) Biological insight from super-resolution microscopy: what we can learn from localization-based images. Annu Rev Biochem 87:965
3. Löschberger A, van de Linde S, Dabauvalle M-C (2012) Super-resolution imaging visualizes the eightfold symmetry of gp210 proteins around the nuclear pore complex and resolves the central channel with nanometer resolution. J Cell Sci 125:570–575. https://doi.org/10.1242/jcs.098822
4. Szymborska A, Marco A, Daigle N (2013) Nuclear pore scaffold structure analyzed by super-resolution microscopy and particle averaging. Science 341:655–658. https://doi.org/10.1126/science.1240672
5. Köhler S, Wojcik M, Xu K, Dernburg AF (2017) Superresolution microscopy reveals the three-dimensional organization of meiotic chromosome axes in intact Caenorhabditis elegans tissue. Proc Natl Acad Sci USA 114:E4734. https://doi.org/10.1073/pnas.1702312114
6. Mennella V, Agard DA, Huang B, Pelletier L (2014) Amorphous no more: subdiffraction view of the pericentriolar material architecture. Trends Cell Biol 24:188–197. https://doi.org/10.1016/j.tcb.2013.10.001
7. Balzarotti F, Eilers Y, Gwosch KC (2016) Nanometer resolution imaging and tracking of fluorescent molecules with minimal photon fluxes. Science: aak9913. https://doi.org/10.1126/science.aak9913
8. Schulze-Gahmen U, Echeverria I, Stjepanovic G (2016) Insights into HIV-1 proviral transcription from integrative structure and dynamics of the Tat: AFF4: P-TEFb: TAR complex. elife 5:e15910
9. Tarbouriech N, Ducournau C, Hutin S (2017) The vaccinia virus DNA polymerase structure provides insights into the mode of processivity factor binding. Nat Commun 8. https://doi.org/10.1038/s41467-017-01542-z
10. Peng Y, Curtis JE, Fang X, Woodson SA (2014) Structural model of an mRNA in complex with the bacterial chaperone Hfq. PNAS 111:17134–17139. https://doi.org/10.1073/pnas.1410114111
11. Fu Y, Kim Y, Jin KS (2014) Structure of the ArgRS–GlnRS–AIMP1 complex and its implications for mammalian translation. PNAS 111:15084–15089. https://doi.org/10.1073/pnas.1408836111
12. Walter JD, Werther RA, Polozova MS (2013) Characterization and solution structure of the factor VIII C2 domain in a ternary complex with classical and non-classical inhibitor antibodies. J Biol Chem 288: 9905–9914. https://doi.org/10.1074/jbc.M112.424564
13. Kjaer TR, Le LTM, Pedersen JS (2015) Structural insights into the initiating complex of the lectin

pathway of complement activation. Structure 23:342–351. https://doi.org/10.1016/j.str.2014.10.024
14. Schneidman-Duhovny D, Hammel M, Tainer JA, Sali A (2016) FoXS, FoXSDock and MultiFoXS: single-state and multi-state structural modeling of proteins and their complexes based on SAXS profiles. Nucleic Acids Res 44:W424–W429. https://doi.org/10.1093/nar/gkw389
15. Petoukhov MV, Svergun DI (2005) Global rigid body modeling of macromolecular complexes against small-angle scattering data. Biophys J 89:1237–1250. https://doi.org/10.1529/biophysj.105.064154
16. Franke D, Svergun DI (2009) DAMMIF, a program for rapid ab-initio shape determination in small-angle scattering. J Appl Crystallogr 42:342–346. https://doi.org/10.1107/S0021889809000338
17. Koutsioubas A, Perez J (2013) Incorporation of a hydration layer in thedummy atom'ab initio structural modelling of biological macromolecules. J Appl Crystallogr 46:1884–1888
18. Koch MHJ, Vachette P, Svergun DI (2003) Small-angle scattering: a view on the properties, structures and structural changes of biological macromolecules in solution. Q Rev Biophys 36:147–227. https://doi.org/10.1017/S0033583503003871
19. Ibrahim Z, Martel A, Moulin M (2017) Time-resolved neutron scattering provides new insight into protein substrate processing by a AAA+ unfoldase. Sci Rep 7. https://doi.org/10.1038/srep40948
20. Karlsen ML, Thorsen TS, Johner N (2015) Structure of dimeric and tetrameric complexes of the BAR domain protein PICK1 determined by small-angle X-ray scattering. Structure 23:1258–1270. https://doi.org/10.1016/j.str.2015.04.020
21. Cordeiro TN, Chen P, Biasio A (2017) Disentangling polydispersity in the PCNA–p15PAF complex, a disordered, transient and multivalent macromolecular assembly. Nucleic Acids Res 45:1501–1515. https://doi.org/10.1093/nar/gkw1183
22. Irvine GB (2001) Determination of molecular size by size-exclusion chromatography (gel filtration). In: Current protocols in cell biology. Wiley, New York
23. Mathew E, Mirza A, Menhart N (2004) Liquid-chromatography-coupled SAXS for accurate sizing of aggregating proteins. J Synchrotron Radiat 11:314–318. https://doi.org/10.1107/S0909049504014086
24. David G, Pérez J (2009) Combined sampler robot and high-performance liquid chromatography: a fully automated system for biological small-angle X-ray scattering experiments at the Synchrotron SOLEIL SWING beamline. J Appl Crystallogr 42:892–900. https://doi.org/10.1107/S0021889809029288
25. Watanabe Y, Inoko Y (2009) Size-exclusion chromatography combined with small-angle X-ray scattering optics. J Chromatogr A 1216:7461–7465
26. Graewert MA, Franke D, Jeffries CM (2015) Automated pipeline for purification, biophysical and X-ray analysis of biomacromolecular solutions. Sci Rep 5. https://doi.org/10.1038/srep10734
27. Meisburger SP, Taylor AB, Khan CA, Zhang S, Fitzpatrick PF, Ando N (2016) Domain movements upon activation of phenylalanine hydroxylase characterized by crystallography and chromatography-coupled small-angle X-ray scattering. J Am Chem Soc 138:6506–6516. https://doi.org/10.1021/jacs.6b01563
28. Jordan A, Jacques M, Merrick C (2016) SEC-SANS: size exclusion chromatography combined in situ with small-angle neutron scattering. J Appl Crystallogr 49: 2015–2020. https://doi.org/10.1107/S1600576716016514
29. Brennich ME, Round AR, Hutin S (2017) Online size-exclusion and ion-exchange chromatography on a SAXS beamline. J Vis Exp. https://doi.org/10.3791/54861
30. Ryan TM, Trewhella J, Murphy JM (2018) An optimized SEC-SAXS system enabling high X-ray dose for rapid SAXS assessment with correlated UV measurements for biomolecular structure analysis. J Appl Crystallogr 51. https://doi.org/10.1107/S1600576717017101
31. Tauler R, Smilde A, Kowalski B (1995) Selectivity, local rank, three-way data analysis and ambiguity in multivariate curve resolution. J Chemom 9:31–58. https://doi.org/10.1002/cem.1180090105
32. Herranz-Trillo F, Groenning M, Maarschalkerweerd A (2017) Structural analysis of multi-component amyloid systems by chemometric SAXS data decomposition. Structure 25:5–15. https://doi.org/10.1016/j.str.2016.10.013
33. Brookes E, Vachette P, Rocco M, Pérez J (2016) US-SOMO HPLC-SAXS module: dealing with capillary fouling and extraction of pure component patterns from poorly resolved SEC-SAXS data. J Appl Crystallogr 49:1827–1841. https://doi.org/10.1107/S1600576716011201
34. Gampp H, Maeder M, Meyer CJ, Zuberbühler AD (1985) Calculation of equilibrium constants from multiwavelength spectroscopic data—III: model-free analysis of spectrophotometric and ESR titrations. Talanta 32:1133–1139. https://doi.org/10.1016/0039-9140(85)80238-1
35. Hopkins JB, Gillilan RE, Skou S (2017) BioXTAS RAW: improvements to a free open-source program for small-angle X-ray scattering data reduction and analysis. J Appl Crystallogr 50:1545–1553. https://doi.org/10.1107/S1600576717011438
36. Hynson RMG, Duff AP, Kirby N (2015) Differential ultracentrifugation coupled to small-angle X-ray scattering on macromolecular complexes. J Appl Crystallogr 48:769–775. https://doi.org/10.1107/S1600576715005051
37. Hutin S, Brennich M, Maillot B, Round A (2016) Online ion-exchange chromatography for small-angle X-ray scattering. Acta Crystallogr D Struct Biol 72
38. Weinhäupl K, Brennich M, Kazmaier U, Lelievre J, Ballell L, Goldberg A, Schanda P, Fraga H (2018) The antibiotic cyclomarin blocks arginine-phosphate–

induced millisecond dynamics in the N-terminal domain of ClpC1 from Mycobacterium tuberculosis. J Biol Chem 293:8379–8393. https://doi.org/10.1074/jbc.RA118.002251
39. Brennich M, Pernot P, Round A (2017) How to analyze and present SAS data for publication. In: Biological small angle scattering: techniques, strategies and tips. Springer, Singapore, pp 47–64
40. Franke D, Petoukhov MV, Konarev PV, Panjkovich A, Tuukkanen A, Mertens HDT, Kikhney AG, Hajizadeh NR, Franklin JM, Jeffries CM, Svergun DI (2017) ATSAS 2.8: a comprehensive data analysis suite for small-angle scattering from macromolecular solutions. J Appl Crystallogr 50:1212–1225. https://doi.org/10.1107/S1600576717007786
41. Tria G, Mertens HDT, Kachala M, Svergun DI (2015) Advanced ensemble modelling of flexible macromolecules using X-ray solution scattering. IUCrJ 2:207–217. https://doi.org/10.1107/S205225251500202X
42. Pelikan M, Hura GL, Hammel M (2009) Structure and flexibility within proteins as identified through small angle X-ray scattering. Gen Physiol Biophys 28:174–189
43. Curtis JE, Raghunandan S, Nanda H, Krueger S (2012) SASSIE: a program to study intrinsically disordered biological molecules and macromolecular ensembles using experimental scattering restraints. Comput Phys Commun 183:382–389. https://doi.org/10.1016/j.cpc.2011.09.010
44. Svergun DI (1999) Restoring low resolution structure of biological macromolecules from solution scattering using simulated annealing. Biophys J 76:2879–2886
45. Trewhella J, Duff AP, Durand D, Gabel F, Guss JM, Hendrickson WA, Hura GL, Jacques DA, Kirby NM, Kwan AH, Pérez J, Pollack L, Ryan TM, Sali A, Schneidman-Duhovny D, Schwede T, Svergun DI, Sugiyama M, Tainer JA, Vachette P, Westbrook J, Whitten AE (2017) 2017 publication guidelines for structural modelling of small-angle scattering data from biomolecules in solution: an update. Acta Crystallogr D Struct Biol 73:710–728. https://doi.org/10.1107/S2059798317011597
46. Moss B (2013) Poxviridae. In: Fields BN, Knipe DM, Howley PM (eds) Fields virology, 6th edn. Lippincott Williams & Wilkins, Philadelphia, pp 2129–2159
47. Czarnecki MW, Traktman P (2017) The vaccinia virus DNA polymerase and its processivity factor. Virus Res 234:193–206
48. Gammon DB, Snoeck R, Fiten P (2008) Mechanism of antiviral drug resistance of vaccinia virus: identification of residues in the viral DNA polymerase conferring differential resistance to antipoxvirus drugs. J Virol 82:12520–12534
49. Duraffour S, Andrei G, Topalis D (2012) Mutations conferring resistance to viral DNA polymerase inhibitors in camelpox virus give different drug-susceptibility profiles in vaccinia virus. J Virol 86: 7310–7325
50. Brennich ME, Kieffer J, Bonamis G (2016) Online data analysis at the ESRF bioSAXS beamline, BM29. J Appl Crystallogr 49:203–212. https://doi.org/10.1107/S1600576715024462
51. Petoukhov MV, Franke D, Shkumatov AV, Tria G, Kikhney AG, Gajda M, Gorba C, Mertens HDT, Konarev PV, Svergun DI (2012) New developments in the ATSAS program package for small-angle scattering data analysis. J Appl Crystallogr 45:342–350. https://doi.org/10.1107/S0021889812007662
52. Svergun D, Barberato C, Koch MHJ (1995) CRYSOL – a program to evaluate X-ray solution scattering of biological macromolecules from atomic coordinates. J Appl Crystallogr 28:768–773. https://doi.org/10.1107/S0021889895007047

Sample Preparation for Electron Cryo-Microscopy of Macromolecular Machines

12

Aurélien Deniaud, Burak V. Kabasakal, Joshua C. Bufton, and Christiane Schaffitzel

Abstract

High-resolution structure determination by electron cryo-microscopy underwent a step change in recent years. This now allows study of challenging samples which previously were inaccessible for structure determination, including membrane proteins. These developments shift the focus in the field to the next bottlenecks which are high-quality sample preparations. While the amounts of sample required for cryo-EM are relatively small, sample quality is the key challenge. Sample quality is influenced by the stability of complexes which depends on buffer composition, inherent flexibility of the sample, and the method of solubilization from the membrane for membrane proteins. It further depends on the choice of sample support, grid pre-treatment and cryo-grid freezing protocol. Here, we discuss various widely applicable approaches to improve sample quality for structural analysis by cryo-EM.

Keywords

Electron cryo-microscopy · Sample preparation · Sample stability · Sample optimization · Grid preparation · Membrane proteins

12.1 Introduction

Electron cryo-microscopy (cryo-EM) emerged in 2013 as a technology capable of providing structural insights into biological macromolecules at near-atomic resolution. Technical advances in cryo-EM and in image processing made this revolution in resolution possible [1–3] (see Chap. 13 by J.L. Carrascosa). The breakthrough is mainly based on the implementation of direct electron detectors that record movies, rather than images, with unprecedented speed and quality. Increased computation power and new image processing software allow correction for sample movements during imaging and help to deal more efficiently with sample heterogeneity. Using this new technology even smaller specimens can be structurally characterized by cryo-EM. The highest resolution reported to date (2018) is the 1.8 Å cryo-EM structure of 334 kDa glutamate dehydrogenase [4].

A. Deniaud
Univ. Grenoble Alpes, CNRS, CEA, IRIG - Laboratoire de Chimie et Biologie des Métaux, Grenoble, France
e-mail: aurelien.deniaud@cea.fr

B. V. Kabasakal
School of Biochemistry, University of Bristol, Bristol, UK

Turkish Accelerator and Radiation Laboratory, Gölbaşı, Ankara, Türkiye

J. C. Bufton · C. Schaffitzel (✉)
School of Biochemistry, University of Bristol, Bristol, UK
e-mail: christiane.berger-schaffitzel@bristol.ac.uk

M. C. Vega, F. J. Fernández (eds.), *Advanced Technologies for Protein Complex Production and Characterization*, Advances in Experimental Medicine and Biology 1453,
https://doi.org/10.1007/978-3-031-52193-5_12

In the last 5 years, cryo-EM structures deposited in the EM data base (https://www.ebi.ac.uk/pdbe/emdb/) reached an average resolution of 11.2 Å (Fig. 12.1a). At this resolution it is not possible to recognize amino acid side chains and this precludes building of atomic models without prior structural knowledge. However, it is very interesting to notice that the average resolution improved more than threefold between 2014 (22.3 Å) and 2018 (6.9 Å), highlighting the potential of cryo-EM for many different samples (Fig. 12.1b) and the increasing access to state-of-the-art cryo-microscopes and detectors.

The reason for limited resolution of many cryo-EM structures is inherent to the technique: In single particle cryo-EM, the purified, imaged macromolecular particles are *assumed* to be identical. Accordingly, it is possible to computationally average over many identical, but noisy images of particles – their identity being determined in a sorting process called classification [5–7]. Averaging improves the signal-to-noise ratio and thus the resolution in the resulting 2D class averages. High-resolution 2D class averages showing secondary structure features can most readily be obtained from homogeneous subsets of particles, which are present in high-quality samples. 2D classification also provides a means to identify sample heterogeneity and to computationally purify the sample.

Sample heterogeneity can be caused by the presence of contaminants. These are usually simple to identify by sample quality control, for instance by size-exclusion chromatography, Coomassie-stained SDS PAGE gels and mass spectrometry. Similarly, the sub-stoichiometric presence or absence of subunits/interaction partners in macromolecular complexes is a source of sample heterogeneity. This type of heterogeneity usually can be addressed by a clever choice of tags for affinity purification fused to minority components of a complex, improved protein expression systems and/or additional purification steps.

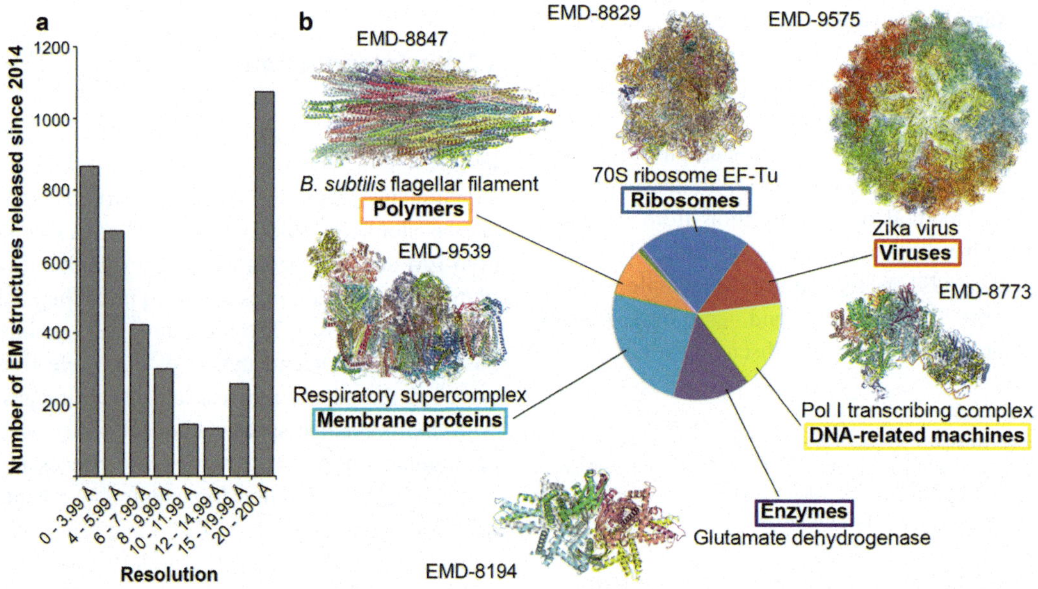

Fig. 12.1 Analysis of EM structures. (**a**) Numbers of EM structures in the EM data bank (EMDB) by resolution range. Only EM structures released *since 2014* were taken into account. *De novo* atomic model building is usually possible only for structures with a resolution of better than 4 Å. (**b**) Classification of 584 high-resolution cryo-EM structures according to biological function or specific properties in terms of architecture or cellular location. Only structures with a resolution better than 4 Å resolution were considered. One example is presented for each category with EMDB entry number (EMD-8194 [4], EMD-9539 [91], EMD-9575 [92], EMD-8773 [93], EMD-8847 [94], EMD-8829 [95])

A second source of heterogeneity, which is more difficult to address, is the dynamics of macromolecular complexes. In fact, macromolecular machines often undergo a number of conformational changes to perform their function. Such a sample then consists of a mixture of different conformational states. This leads to limited resolution of the resulting cryo-EM map if the different conformations cannot be sorted out biochemically or computationally.

Sample heterogeneity caused by flexibility/dynamics of the sample cannot be detected readily by standard protein biochemistry. Sample flexibility has to be revealed by additional quality control of the sample, such as dynamic light scattering or negative stain EM analyses (Fig. 12.2). Negative stain EM provides high contrast images and therefore allows assessment of contaminants, sample concentration and homogeneity [8]. 2D class averages from negative stain EM can reveal different conformations of a sample. Negative stain EM analyses can also indicate if the sample adopts one preferred conformation or if a specific conformation of the sample can be stabilized, *e.g.*, by addition of conformation-specific antibodies, inhibitors, substrate analogues and/or mutagenesis. These approaches are adopted from X-ray crystallography where the crystal lattice further purifies the proteins and forces the sample to adopt one specific conformation [9].

State-of-the-art image processing can deal with conformational flexibility of samples in cases where the sample adopts a series of discrete states. It is then possible to solve the cryo-EM structures of the different states. To date, maximum likelihood-based classification is commonly used in cryo-EM to sort for differences in composition and/or conformation in the sample [10, 11]. It identifies structurally homogeneous subset(s) of particles and leads to 3D reconstructions that can be refined to higher resolution. However, 2D and 3D classification cannot resolve conformational flexibility which is due to continuous motion of parts of the sample. Currently, very flexible parts of the structure are masked out, and the structure determination is then limited to the rigid parts of the sample which adopt a defined state [12, 13]. However, masking out more flexible, dynamic parts of macromolecules leaves them poorly defined, requiring new biochemical approaches and additional manipulations to improve sample homogeneity to be able to visualize such flexible regions at high-resolution.

Here, we review innovative methodological approaches to improve cryo-EM sample quality, in particular sample stability and homogeneity (Fig. 12.2). Biochemical methods to analyze and optimize sample purity and stability are discussed in detail. In addition, we focus on existing tools and new developments for cryo-grid preparation, a critical and often underestimated step in cryo-EM. We place special emphasis on membrane proteins and their complexes which are particularly difficult to prepare and characterize structurally. Notably, high-resolution structures of several biologically important, much sought-after membrane protein complexes, which previously could not be solved by crystallography, were recently determined by cryo-EM (Fig. 12.3). We highlight new trends in membrane protein sample preparation for cryo-EM, which contributed to a wealth of novel high-resolution cryo-EM structures.

12.2 Approaches for High-Quality Sample Preparation

At first sight, sample preparation for single particle cryo-EM is straightforward and much simpler compared to crystallography or nuclear magnetic resonance spectroscopy (NMR) as neither crystals nor labeling is required. However, samples must be imaged in a high vacuum in EM, which is incompatible with biological samples in aqueous buffer solutions. Vitrification, which was first established by Jacques Dubochet (Nobel prize in Chemistry 2017), overcomes this limitation of EM by flash-freezing the sample and imaging at liquid nitrogen temperatures [14]. In practice, a small volume of sample is applied onto a carbon-coated grid, made from a metal mesh such as copper, gold, nickel, molybdenum, rhodium or tungsten with different mesh sizes (200–400 mesh grids are commonly used) and

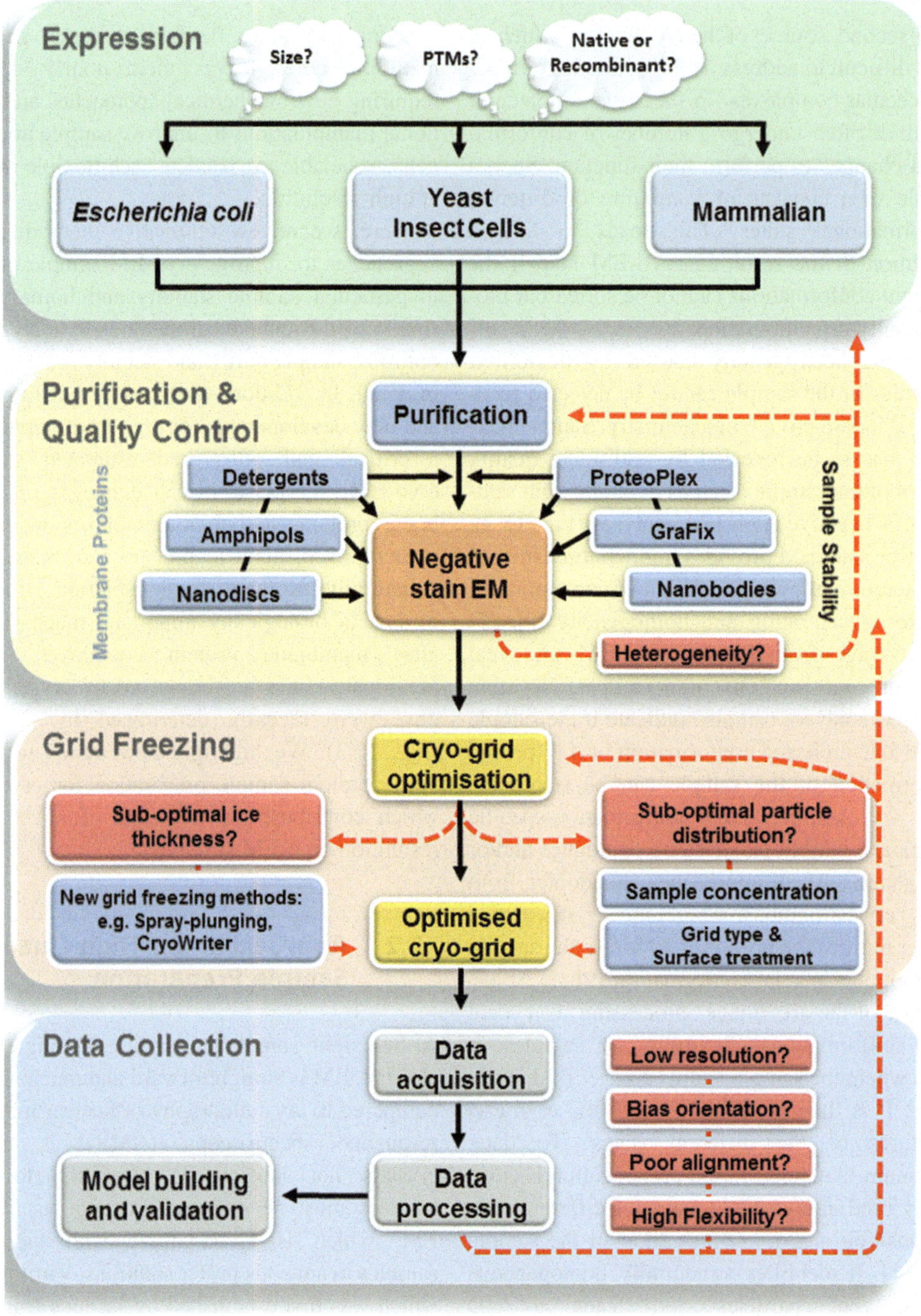

Fig. 12.2 Schematic overview of a typical cryo-EM sample preparation workflow. The properties of the specimen have to be considered first to identify an appropriate expression system to produce structural biology-grade material. Then, purification and quality control (QC) by negative stain EM is carried out to verify sample homogeneity and stability. Often, multiple rounds of purification optimization and QC using the techniques discussed in the main text are required. For homogeneous samples, freezing conditions that yield an ideal ice thickness must be determined. This can be achieved by varying sample concentration, grid type, grid pre-treatments, application/plunging method, and buffer conditions (see main text). Screening of the cryo-grids prepared under different conditions will indicate if further optimization is required before continuing with data acquisition. Subsequent data analysis (including 2D class averaging, initial model generation, 3D classification) could lead to a satisfactory

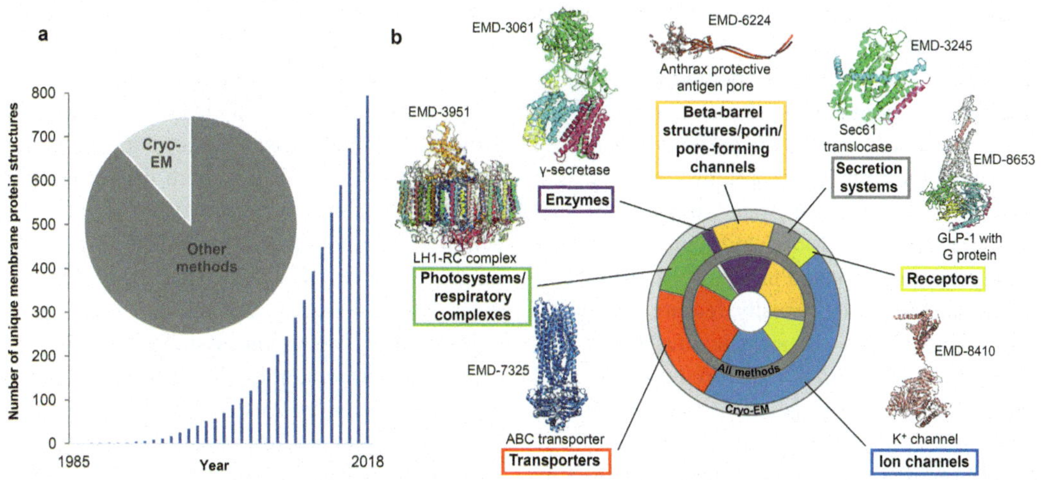

Fig. 12.3 Structural analysis of membrane proteins. (**a**) Numbers of unique membrane protein structures published by year (total of 796 PDB entries). Inset: contribution of cryo-EM to the unique membrane protein structures (97 entries). (**b**) Outer ring: Classification of 97 unique cryo-EM structures of membrane proteins (all published since 2013, all with <10 Å resolution) according to their biological function. Examples of membrane protein cryo-EM structures are shown for each category with EMDB entry number (EMD-3061 [23], EMD-6224 [43], EMD-3245 [44], EMD-8410 [45], EMD-8653 [46], EMD-7325 [47], EMD-3951 [48]). The inner ring represents unique structures determined by all methods, including crystallography and NMR (796 in total), using the same colors for the classes. Unclassified structures are shown in white. Data taken from Stephen White's webpage 'membrane proteins of known 3D structure' (http://blanco.biomol.uci.edu/mpstruc)

with a thin carbon support film on one side. Subsequently, most of the sample is blotted away with filter paper, and then the grid with a thin aqueous film of the sample is flash frozen in liquid ethane. The freezing must occur at high speed (faster than 10^4 K/sec) to avoid formation of ice crystals and to favor formation of amorphous, vitreous ice [15]. Vitrification prevents biological specimens from drying out in the high vacuum of the EM, preserves the native structure of the macromolecules and keeps them immobilized during imaging.

It can be rather time-consuming and require significant amounts of samples to identify optimal grid-freezing conditions: Cryo-grids suitable for data collection require good particle concentration and distribution in *thin* ice. The ideal ice thickness is just a bit thicker than the macromolecules embedded, ensuring the best possible signal-to-noise ratio (Fig. 12.2). The contrast/signal of macromolecules in cryo-EM is generally low because it is mostly provided by C, O, N and H atoms which are also present in the solvent. Due to the similar density of macromolecules and solvent, it is easier to visualize very large macromolecular complexes in vitreous ice.

12.2.1 Selecting the Optimal Expression System

In the past, purification of macromolecular complexes was mostly performed from native

Fig. 12.2 (continued) reconstruction. However, it may also indicate that the ice conditions are not ideal or that the sample or cryo-grid preparation requires further optimization. Once a suitable 3D reconstruction of the sample is obtained from image processing the final steps of atomic model building and validation can begin

sources [16]. For Mega Dalton-sized complexes such as ribosomes and their complexes, it is still the only viable preparation method (Fig. 12.1b). Bacterial expression systems, in particular *Escherichia coli,* remain the preferred sources of individual recombinant proteins and smaller complexes because they are simple, fast and cost-effective [17]. More "difficult to express" eukaryotic multi-domain proteins or large macromolecular complexes usually require eukaryotic expression systems, the most popular being *Saccharomyces cerevisiae*, baculovirus-insect cell expression and mammalian cell-based systems [18, 19].

The optimal expression system is defined by specific features of the sample. These may include sequence and size of the proteins as well as activity, folding (the requirement of specific chaperones) and post-translational modifications (PTMs). PTMs, such as acetylation, phosphorylation, addition of lipids or carbohydrates, can be essential to protein activity and can alter the proteins' physicochemical nature. Moreover, overexpression of certain proteins can be tolerated in one system but be toxic in another. While *S. cerevisiae* is the least expensive and the most convenient to use eukaryotic protein expression system, baculovirus-insect cell systems have a more sophisticated chaperone machinery for folding mammalian proteins and can achieve more complex PTMs. For mammalian proteins, the proper conformation and modifications are most likely to be achieved by expression in mammalian cells. More distant expression systems may provide higher yields, but not add all PTMs which are found in the native protein in the cell. Therefore, overexpression in the endogenous or a closely related host organism is usually the preferred approach for production of a correctly assembled, fully active protein complex.

Examples for "difficult to express" proteins include eukaryotic kinases and membrane protein complexes: Human SMG1 kinase (~420 kDa), a member of the phosphatidylinositol 3-kinase-related kinase family involved in genome stability and nonsense-mediated mRNA decay, can be expressed in its active, auto-phosphorylated form in human HEK293T cells. Its yields are further increased when the interacting protein SMG9 is co-expressed [20, 21]. In contrast, SMG1 overexpression using baculovirus-insect cell expression yielded only insoluble, aggregated proteins (unpublished). Eukaryotic membrane protein complexes often require the presence of specific lipids and glycosylation for stability and activity which is difficult to achieve in recombinant expression systems: Human γ-secretase, comprising four subunits each with a transmembrane domain, was produced in human HEK293F cells by co-expression of all four subunits from a single expression vector [22, 23]. In the resulting high-resolution cryo-EM structure of γ-secretase, lipids were identified bound to specific sites on the transmembrane domains of the proteins and eleven glycosylated residues were present in the nicastrin subunit as judged by the EM density (Fig. 12.4) [23].

12.2.2 Sample Quality Control and Buffer Optimization

Biochemical purification procedures traditionally use buffers that mimic the cellular environment of the sample, *i.e.*, a close-to physiological pH and additives such as glycerol or sucrose to enhance stability. However, glycerol, sucrose and detergents are known to interfere with cryo-grid preparation as they reduce the contrast of the sample in vitreous ice. To obtain optimal cryo-EM data for a purified specimen it is therefore advised to optimize its buffer and assess the need for additives.

For crystallography, protein homogeneity is usually assessed under multiple conditions by dynamic light scattering [24] or by Thermofluor/differential scanning fluorimetry (DSF) [25]. For Thermofluor/DSF analysis, a dye is added to the sample which is quenched in aqueous solution and becomes hyper-fluorescent in an apolar environment, *e.g.*, when exposed to hydrophobic residues of a protein. However, DSF is not suitable for analysis of large complexes because it measures protein stability as the midpoint of the unfolding transition, assuming cooperative unfolding. This assumption is not valid for most

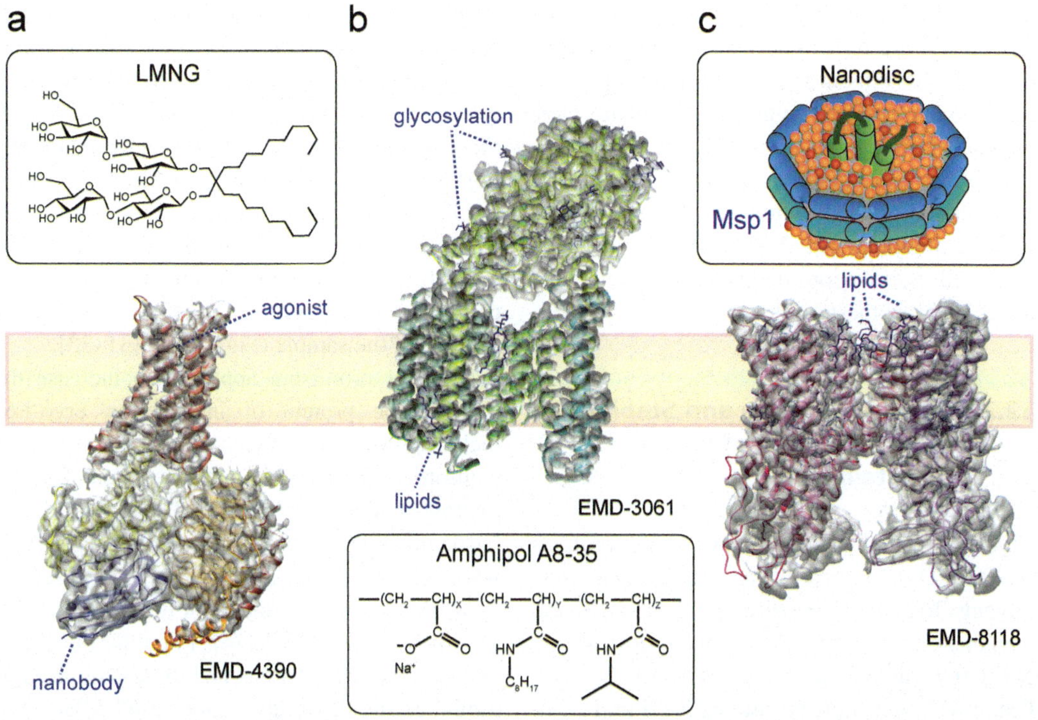

Fig. 12.4 Low CMC and detergent-free systems for cryo-EM analysis of membrane proteins. (**a**) Chemical structure of LMNG and cryo-EM structure of agonist-bound adenosine A_{2A} receptor in LMNG (red, agonist: black), heterotrimeric G protein (yellow: mini-G_S, orange-red: βγ subunits) and nanobody (blue) at 4.1 Å resolution [33]. (**b**) Schematic illustration of amphipol A8–35 [51] and cryo-EM structure of γ-secretase with amphipol [23]. The four subunits are depicted in different green colors. Representative glycan and lipid residues (dark blue) are labeled. (**c**) Schematic illustration of a nanodisc with an embedded integral membrane protein [54] and cryo-EM structure of TRPV1 channel in a nanodisc [96]. The four subunits are depicted in pink-purple colors; lipids (dark blue) are labeled. All EM densities are depicted in transparent grey

complexes as they usually do not unfold cooperatively, but in a multi-state process of complex dissociation and unfolding of the subunits.

A DSF-based method, named ProteoPlex, screens for conditions of monodispersity and increased stability of large macromolecular complexes [26]. ProteoPlex comprises a software for DSF data analysis in order to identify buffer conditions leading to a single melting temperature transition of a complex [26]. Switching from multiple unfolding transitions to a single transition is a sign of sample monodispersity, and an increased melting temperature indicates better sample stability. To achieve this, a large panel of buffer conditions can be screened: pHs, salt concentration, diverse additives, as well as available specific ligands, inhibitors, nanobodies or antibody fragments.

Notably, the pH optima identified using DSF often differ significantly from physiological pH (7.0–8.0); they range from pH 4 to 9 and peak around pH 6.5 [27]. Most cryo-EM structures found in the EMDB have been solved at a pH between 7 and 8.5, indicating that the stability of these specimens was not assessed systematically. Likely, these structures were determined at non-optimal buffer conditions.

Based on the analyses of ~80 macromolecular complexes, Chari and Stark predicted that the best buffer conditions for crystallization and cryo-EM of macromolecules are likely to be very similar [26, 27]. For example, using pH 6 instead of pH 8 led to a five-fold increase in purification yields of AAA+ ATPase p97. In negative stain EM intact, homogenous oligomers of p97 were observed rather than partial aggregates which were obtained from previous purifications [26]. This exemplifies that ProteoPlex is very useful for not only sample quality control and optimization of buffer conditions for EM analyses, but also improved purification under conditions where the complex is most stable.

12.2.3 Improving Size and Stability: Nanobodies and Fusion Proteins

Cryo-EM is challenging for small proteins (<100 kDa), in particular small membrane proteins. Because of the difficulty in crystallizing membrane proteins or solving their structure by NMR (699 unique structures since 1985), cryo-EM provides an attractive alternative (Fig. 12.3a). 97 unique cryo-EM structures of membrane proteins from different classes were solved since 2013 (Fig. 12.3b), demonstrating the power of the technology. One means to overcome the problems associated with a small sample is to increase the protein size, for instance with the help of specific Fab antibody fragments (Fabs) or nanobodies (Fig. 12.4a). Nanobodies consist of a single monomeric variable antibody domain and have a molecular weight of ~15 kDa.

For example, the cryo-EM structure of a ~130 kDa heterodimeric ATP binding cassette (ABC) transporter TmrAB from *Thermus thermophilus* has been reported at ~8 Å resolution using a specific 50 kDa Fab fragment [28]. The Fab fragment not only increased size but also aided the determination of particle orientations [28]. Using the same strategy, the cryo-EM structure of the HIV integrase dimer (65 kDa) in complex with two specific Fabs (50 kDa each) was reported at 10 Å resolution [29].

Nanobodies are commonly used for purification and detection of proteins. Moreover, when a complex is flexible and fails to achieve high-resolution EM reconstructions, a very attractive strategy is to use antibodies/nanobodies that stabilize a specific conformation. For instance, using a specific nanobody, a 8.7 Å cryo-EM reconstruction has been achieved for the A/V-type ATPase rotary motor from *Thermus thermophilus* [30]. The presence of the nanobody improved the local resolution of the map, *i.e.*, the rigidity of the region where it bound to the ATPase. Similarly, nanobodies have been used to analyze or reduce the flexibility of G protein-coupled receptors (GPCRs) as well as to increase the size of the sample (Fig. 12.4a) [31–33].

Another promising approach to increase the size of the protein of interest for cryo-EM includes fusion of the protein of interest to a well-ordered, symmetric oligomer. For instance, maltose-binding protein (MBP, 40 kDa) was covalently linked to glutamine synthetase (GS) which forms dodecamers [34]. The resulting cryo-EM structure yielded 4 Å resolution for GS and 6–10 Å for MBP [34]. For this, the linker between the two proteins had to be optimized to minimize the flexibility of the fused MBP. The resolution was lowest at the junction between the proteins. Consequently, this approach may be limited by the requirement to optimize the linker for each new protein fusion, thus minimizing flexibility and enforcing 12-fold symmetry [34].

12.2.4 Improving Stability: Polyproteins and Crosslinking

Polyproteins can help overcome the problem of unstable complexes that dissociate during purification or grid preparation. Polyproteins are covalently linked individual proteins and are used in nature by many viruses for most efficient protein production. For structural biology they have been used to achieve the correct stoichiometry of proteins forming a complex and to stabilize the resulting complexes [35, 36]. The cryo-EM structure of the *E. coli* ribosomal targeting complex in the 'closed' conformation critically relied on stabilizing the complex between the signal recognition particle (SRP) and its receptor FtsY. This was achieved by covalently linking the SRP protein Ffh with FtsY into a single polypeptide chain without compromising activity [37].

Alternatively, purified complexes can be stabilized by mild chemical crosslinking. The simplest approach is to use batch reactions. The optimal concentration of crosslinker is determined in SDS-PAGE as the minimal concentration required for a complete shift of all subunits of a complex into a higher-molecular weight band. For instance, the lysine-specific crosslinker BS3 (bis(sulfosuccinimidyl)suberate) was successfully used to stabilize Target of Rapamycin Complex 2 (TORC2) which was purified from source. The corresponding cryo-EM structure of TORC2 reached 7.9 Å resolution [38].

GraFix is a more elaborate method to combine separation by size in a density gradient centrifugation with mild chemical crosslinking using glutaraldehyde [39]. About 100 μg sample are loaded onto a glycerol and glutaraldehyde gradient. During centrifugation the sample is exposed to increasing concentration of the glutaraldehyde crosslinker and simultaneously size-fractionated in the density gradient. This favors intramolecular crosslinks and removes any complexes with intermolecular crosslinks due to their increased size. For cryo-EM, glycerol is removed from the buffer using a desalting column [39]. 'On-column' crosslinking follows a similar principle: β_2 adrenergic receptor-G protein complexed with β-arrestin was crosslinked during size-exclusion chromatography, leading to enrichment of a 'tight' conformation [32].

It should be noted that chemical crosslinking leads to reduced conformational heterogeneity, favoring the most compact state of the sample. Sometimes, the resulting compact structure does not represent the conformation the sample usually adopts in solution. It is therefore advisable to compare the crosslinked and native sample, *e.g.*, using negative stain EM.

12.2.5 Stabilizing Membrane Proteins

Since 2013, structures of many important membrane protein complexes were determined by cryo-EM [40] (Fig. 12.3b). Ground-breaking examples are the transient receptor potential channel TRPV1 [41], human γ-secretase [22, 23] (Fig. 12.4b, c), and the 1 MDa mammalian NADH:ubiquinone oxidoreductase (respiratory complex I) [42]. Examples of high-resolution cryo-EM structures exist for virtually all classes of membrane proteins (Fig. 12.3b): Enzymes, photosystems, pore-forming channels, secretion systems, receptors, ion channels and transporters [23, 43–48]. Many of the high-resolution membrane protein structures additionally show tightly bound lipids, metal ions, glycan residues and bound ligands (Fig. 12.4).

12.2.5.1 New Amphiphilic Detergents

Solving high-resolution structures of membrane protein complexes requires the latest technology and optimized sample preparation techniques, including the choice of suitable detergents. To date, it is unclear which detergents work best for cryo-EM studies of membrane proteins. Analyzing recent cryo-EM structures of membrane proteins and detergents utilized for solubilization, indicates that a large number of detergents are compatible with cryo-EM; including n-Dodecyl-β-Maltoside, Cymal7, Digitonin and Tween-20 [49]. For cryo-EM, detergents with low critical micellar concentration (CMC) such as Lauryl Maltose Neopentyl Glycol (LMNG) and detergent-free systems (amphipols, nanodiscs, SMA or saposins) became very popular in recent years (Fig. 12.4).

High detergent concentrations (above the CMC) are required for solubilization and purification of membrane proteins. These high concentrations often prevent thin ice formation during grid preparation. Moreover, the detergent micelles around membrane proteins diminish the contrast in vitreous ice. Finally, free detergent micelles can be mistaken as protein particles, complicating image processing.

To address the need for high detergent concentration, **GraDeR,** a variation of GraFix, has been established [50]. Two inverse gradients are used for ultracentrifugation, with increasing concentrations of glycerol and decreasing LMNG concentrations. The LMNG structure resembles a lipid with two aliphatic chains (Fig. 12.4a). LMNG has a very low CMC (in H_2O ~0.01 nM/0.001% w/v) and an extremely slow off-rate, allowing to temporarily lower the LMNG concentration below the CMC without observing protein aggregation. Centrifuging membrane proteins into LMNG concentrations below the CMC avoids the presence of detergent micelles and facilitates grid preparation [50]. As a proof of principle, GraDeR was successfully applied to bacterial V-type ATPase, a *Caenorhabditis elegans* gap junction channel and mammalian F_oF_1 ATPase, significantly improving the quality of the cryo-grid preparations [50].

Similarly, LMNG alone has facilitated cryo-EM sample preparation. For instance, the cryo-EM structure of the adenosine A_{2A} receptor bound to an agonist (NECA), a heterotrimeric G protein (comprising an engineered mini-G_S and βγ subunits) and nanobody Nb35 was solved in LMNG at 4.1 Å resolution [33] (Fig. 12.4a).

12.2.5.2 Amphipathic Polymers: A Substitute for Detergent

Amphipols (amphipathic polymers) have been designed to stabilize integral membrane proteins [51]. They are constituted of polyacrylate polymers which are very soluble due to hydrophilic carboxylate groups in their side chains (Fig. 12.4b). Hydrophobic side chains bind to the hydrophobic surface of membrane proteins and thus stabilize the membrane protein's native structure [51]. Because of multiple associations between the amphipols and the membrane protein, amphipols can replace detergents completely, avoiding the problems associated with detergent micelles.

In the case of γ-secretase, replacing the detergent digitonin by the amphipol A8–35 was critical for achieving a high-resolution cryo-EM structure [22, 23] (Fig. 12.4b). Other examples include the 3.9 Å resolution cryo-EM structure of the membrane-embedded part of Vacuolar-type ATPase from yeast [52] and a 3.4 Å cryo-EM structure of polycystic kidney disease-like channel PKD2L1 [53] which were both achieved using amphipols.

12.2.5.3 Nanodiscs Mimic the Lipid Bilayer

Removing proteins from their natural lipid environment with detergents results in loss of native interactions, *e.g.*, with important lipid molecules and other membrane proteins. This can lead to dissociation of membrane protein complexes. Moreover, detergents do not properly mimic the membrane because the detergent micelle forms a spherical vesicle and not a bilayer. In contrast, nanodiscs offer a native-like lipid bilayer environment [54]. Nanodiscs are composed of two identical membrane scaffolding proteins, MSPs, which each contain an amphipathic helix forming a belt around the hydrophobic regions of the lipid bilayer (Fig. 12.4c). The diameter of the nanodisc is determined by the length of the MSP helix, which can be varied and range from 6 to 17 nm. By optimizing the lipid, membrane protein and MSP protein ratio, it is possible to reconstitute membrane proteins in nanodiscs and to remove all detergents [54].

The first membrane protein cryo-EM structures in nanodiscs were the bacterial SecYE translocon in complex with the ribosome [55], the TRPV1 channel [41] (Fig. 12.4c) and the RyR1 ryanodine receptor [56]. Notably, other high resolution RyR1 cryo-EM structures were solved using detergent (Tween-20) [57] and mixed lipid-detergent micelles (Chaps/lipids) [58], and they are virtually identical to the RyR1 structure in nanodiscs [56]. This suggests that this membrane protein complex is particularly stable and compatible with various solubilization methods.

12.2.5.4 Saposin-Based Nanoparticles and Styrene-Maleic Acid (SMA) Copolymers

Detergent-free saposin-lipoprotein nanoparticles ('Salipro') are composed of lipids and saposin [59]. Saposin-like proteins are highly stable due to amphipathic helices and disulfide bridges [60]. Saposin A surrounds the lipids and the transmembrane domains of the membrane protein and thus keeps them in solution. The Salipro system is adaptable to various sizes of proteins. For example, Saposin has been used to solve the cryo-EM structure of the bacterial PeptTSo2 peptide transporter at 6.5 Å resolution [59].

SMA copolymers comprise styrene and maleic acid groups. These polymers are remarkable because they have been shown to solubilize lipid-protein complexes directly from raw membranes or even cells [61]. Thus, SMAs avoid the detergent-solubilization step which is inherent to all strategies described above and which potentially disturbs the native membrane protein structure. This makes this copolymer particularly attractive for membrane protein preparations and structural biology. Recently, SMA lipid particles (SMALPs) have been used to solve the 3.4 Å cryo-EM structure of the bacterial alternative complex III [62]. The complex participates in electron transport by catalyzing the oxidation of membrane-bound quinol and the reduction of cytochrome *c* [62].

12.3 Cryo-grid Preparation and Ice Thickness

Successful cryo-grid preparation depends on many parameters, including the choice of grid, pre-treatment of grids, blotting conditions, the nature of the sample and buffer composition. In practice, grid preparations need to be optimized for new samples, and often many grids are screened to identify suitable areas on the grid for data collection. As discussed above, the presence of detergents, glycerol or sucrose in the buffer is known to cause problems with ice quality [15]. For instance, detergents lower the surface tension and make it more difficult to produce thin ice. Holey carbon grids with smaller holes sometimes help to overcome this problem [63].

Other parameters that affect ice thickness include the blotting time and the post-blot incubation time before the grid is flash-frozen in liquid ethane. These parameters can be tightly controlled with semi-automated blotting and plunging devices. Moreover, the filter paper, that is used to remove excess fluid from the cryo-grid, can touch the grid from both sides, from the sample application side or from the 'back' side. A longer incubation time before freezing the grid allows water to evaporate; the extent of evaporation depends on the temperature and humidity of the environment. Notably, evaporation also changes the buffer composition (pH and ionic strength) which can affect the protein structure.

The sample support film (*e.g.*, homemade holey carbon film or commercial C-flat/Quantifoil holey carbon film), the size of the holes and its hydrophilicity are important parameters for grid preparation. Sometimes, remaining plastic on holey carbon film has to be removed using ethyl acetate [63] or using other protocols involving chloroform, acetone and/or ethanol. Hydrophilicity of the support film is required to allow complete wetting of the grid and formation of the ice layer. Hydrophilicity can be modified by exposure to UV radiation, glow discharge, plasma cleaning, poly-L-lysine or detergent treatment [64]. The extent of partition of the sample into the grid holes depends on the hydrophilicity of the support. Macromolecules often preferentially adsorb on the carbon film. This dilutes the concentration of particles in the holes. One solution is to apply the sample twice on the grid, hoping that the first application saturates the carbon film.

Affinity grids use a specimen-specific support film to immobilize particles, leading to a higher concentration of particles on the film and avoiding the interaction of particles with the water-air interface. Such functionalized surfaces include for example Ni-NTA-derivatized monolayers [65], antibody-coated carbon films [66] or crystalline streptavidin monolayers [67]. Of these, Ni-NTA and streptavidin-coated lipid monolayers were reported to lead to preferred orientations of particles on the grid.

When particles interact with the water-air interface they often adopt a preferred orientation or even denature, compromising structure determination. Approaches to overcome preferred orientation of particles include the addition of low amounts of detergents which concentrate at the water-air interface and thus protect proteins from the surface effects. Changing the glow-discharge parameters or any other treatment that affects the hydrophobicity of the support also can change particle distribution and orientation and thus can help to improve the particle view distribution in vitreous ice [63]. Collecting images of tilted specimen is the least preferable solution to the preferred view problem as it often prevents reaching high resolution.

12.3.1 New Grid Materials to Reduce Particle Movement

Particle movement during irradiation of the specimen is caused by stage drift and/or beam-induced charging. In order to restore high-resolution information, motion correction is applied during image processing [68]. Motion correction showed that commonly used amorphous holey carbon films are not very stable during irradiation, as they were deformed and were poor conductors [69].

Consequently, new support materials were tested to identify films that are more stable during imaging. Graphene and gold films were shown to strongly reduce or nearly eliminate sample movement [69, 70]. Gold (Au) grids with a gold support film gave best results which is likely due to gold being highly conductive and very radiation-hard, *i.e.*, more resistant to radiation-caused alterations [69]. Accordingly, Au grids are now commercially available and commonly used for high-resolution cryo-EM. Recent examples of their application include the 3.85 Å cryo-EM structure of ASCT2, a human neutral amino acid transporter [71] and the human HCN1 hyperpolarization-activated channel at 3.5 Å resolution [72].

Notably, the most pronounced particle movement occurs within the exposure to the first 3–5 electrons per $Å^2$ [69]. Unfortunately, it is currently not possible to computationally correct for this movement. Therefore, the corresponding movie frames are discarded, despite the fact that these frames contain the most valuable high-resolution information as they are least damaged by irradiation. In the future, better grid supports which even further limit particle movement, combined with new software may allow processing of these images and thus could lead to a further step change in resolution achievable by cryo-EM.

12.3.2 Cryo-grid Preparation: From Pipetting-Blotting-Plunging Devices to New Automated Cryo-Grid Preparation Procedures

Optimizing cryo-grid preparation can involve screening sample concentrations, buffer optimization, different grid types and blotting conditions. **Blotting-plunging** robots (such as FEI Vitrobot, Leica GP, Gatan CP3) have been developed to make cryo-grid freezing more reproducible (see above). However, blotting with filter paper removes >99.9% of the sample, and it can lead to sample aggregation or denaturation. Moreover, the process of grid screening is low-throughput, leaving this process rather unpredictable. To address this bottleneck, new cryo-grid preparation devices are developed, such as 'Spotiton' [73], a Microsprayer Chip [74] and 'CryoWriter' [75].

Spotiton is a piezoelectric inkjet dispensing device, incorporated into a purpose-built vitrification robot [73]. It spots a very small volume (~2–16 nl) of sample onto a novel *self-blotting* grid on which the sample is spread to a thin film, avoiding the use of filter paper [76]. Self-blotting grids are composed of copper and rhodium (Cu/Rh) or palladium (Cu/Pd) and are covered with nanowires on the copper surface. The opposite side (Rh or Pd) is smooth (no nanowires) and holds the support film, holey carbon or gold. The sample is spotted with the dispensing device onto the nanowire side of the support and spreads across the support [76]. The nanowires act as blotting paper, rapidly removing the sample when it comes in contact, and leaving a thin liquid film which subsequently is flash-frozen.

Spraying-plunging methods use microfluidic devices with silicon−/polydimethylsiloxane-based chips [74, 77]. A chip comprises a micromixer with two inlets for the reactants (originally built for time-resolved EM) and a microreaction channel from which the reaction mix is sprayed onto the grid. This device can also be used for normal samples, rather than reaction mixtures. By adjusting gas pressure and the distance between the sprayer and the grid, the ice thickness can be controlled. Currently, rather large sample volumes (9 µl/grid) are required. A proof-of-principle study with apoferritin showed that cryo-grids prepared with the microsprayer can lead to a 3 Å cryo-EM structure [74].

The **CryoWriter** avoids the blotting and the spraying step. CryoWriter uses a microcapillary to directly apply and spread a very small volume of sample (~3–20 nl) onto the grid, such that the drop spans from the tip to the grid surface [75]. During application, the grid is moved to spread the sample over the surface and to fill the holes in the support film. Real-time monitoring of the thickness of the sample film allows to adjust the volumes applied and to change the sample film thickness prior to flash-freezing [75]. Sample thinning is achieved by controlled water evaporation. This is possible because the device is not humidity-controlled. When the monitoring system (a laser diode and a photo detector) indicates that the desired liquid film thickness is reached, the grid is rapidly vitrified using a robotic arm.

12.4 Imaging In-Focus: The Volta Phase Plate

The use of phase plates for imaging by electron microscopy in-focus was already suggested in 1947 by Hans Boersch [78]. For light microscopy, Fritz Zernike realized in-focus imaging. He developed a phase plate that shifts the phase of scattered light by 90° relative to the phase of un-scattered light [78]. For cryo-EM, in-focus imaging has only recently been achieved: Radostin Danev and colleagues discovered that a charged carbon film can be used to introduce the required phase shift for scattered electrons [79]. This 'Volta Phase Plate' yields an impressive enhancement of contrast for in-focus images [80]. The introduction of the phase plate is now rapidly transforming single particle cryo-EM and cryo-tomography [80, 81]. Contrast-enhanced cryo-EM using the Volta Phase Plate is particularly interesting for small proteins, including small membrane proteins. For example, 64 kDa hemoglobin was solved at 3.2 Å resolution [82] and several GPCR complexes in the range of 150 kDa were reported. Examples include agonist-bound calcitonin receptor with G_s protein at 4.1 Å resolution [83] and agonist-bound glucagon-like peptide-1 receptor complex at 3.3 Å [84].

12.5 Challenges and Outlook

Due to technological advances in recent years cryo-EM became a more automated, higher throughput and more routine technique, allowing high-resolution structural biology. Future advances will include next-generation direct electron detectors with higher sensitivity and faster readout as well as developments in image processing software addressing the problems associated with the initial movie frames. In parallel, new tools and routines for improved sample production and cryo-grid preparations are required. In fact, sample quality is the bottleneck in many projects.

In particular, structures of small proteins and membrane protein complexes can now be solved by cryo-EM [82–84]. The challenges for membrane protein samples are multi-fold, including the choice of expression system to produce sufficient amounts of protein/complex, mild but efficient solubilization using detergents or detergent-free systems, purification of homogeneous protein sample and last but not least cryo-grid preparation (Fig. 12.2). It is key to find the best detergent and buffer conditions where the protein/complex is stable. If sufficient sample is available, buffer optimization is ideally achieved using high-throughput screening techniques, such as ProteoPlex [26]. Currently, there is no general answer to the biochemical problem of how to stabilize unstable and flexible specimens. Possible solutions, including antibody fragments [28, 29] or nanobodies [31], polyproteins [36, 37] or mild chemical cross-linking [38, 39] are discussed above. Establishing a protocol for cryo-grid preparation is a challenge which currently is addressed by new grid materials [69, 70, 76] and novel sample application approaches [73–75] substituting for the pipetting-blotting-plunging technique.

For membrane proteins, the optimal solution to many of the problems associated with protein extraction from the membrane would be *in situ* analysis at high-resolution in their native membrane. In fact, Electron cryo-Tomography (cryoET) underwent a resolution revolution as well, thanks to direct electron detectors, Volta phase plate and improved sample preparation techniques, such as cryo-focused ion beam (FIB) milling [85]. Similarly, improved software for 3D classification and subtomogram averaging tools contribute to increasing resolution of tomograms [86–88]. CryoET analyses of the ~56 MDa core of the nuclear pore complex at ~21 Å resolution [89, 90] impressively highlight the great potential of this technique.

Acknowledgments The authors would like to thank the members of the Berger and Schaffitzel team and Dr. Ufuk Borucu for critically reading the manuscript. CS acknowledges funding by the BBSRC (BB/P000940/1), the MRC (MR/P019471/1) and the Wellcome Trust (210701/Z/18/Z). AD acknowledges funding by the CEA DRF-Impulsion program (FIB-Bio grant).

References

1. Kühlbrandt W (2014) Cryo-EM enters a new era. elife 3:e03678
2. Fernandez-Leiro R, Scheres SH (2016) Unravelling biological macromolecules with cryo-electron microscopy. Nature 537(7620):339–346
3. Frank J (2016) Generalized single-particle cryo-EM--a historical perspective. Microscopy (Oxf) 65(1):3–8. https://doi.org/10.1093/jmicro/dfv358
4. Merk A, Bartesaghi A, Banerjee S, Falconieri V, Rao P, Davis MI, Pragani R, Boxer MB, Earl LA, Milne JLS, Subramaniam S (2016) Breaking Cryo-EM resolution barriers to facilitate drug discovery. Cell 165(7):1698–1707. https://doi.org/10.1016/j.cell.2016.05.040
5. van Heel M, Frank J (1981) Use of multivariate statistics in analysing the images of biological macromolecules. Ultramicroscopy 6(2):187–194. https://doi.org/10.1016/S0304-3991(81)80197-0
6. Lyumkis D, Brilot AF, Theobald DL, Grigorieff N (2013) Likelihood-based classification of cryo-EM images using FREALIGN. J Struct Biol 183(3):377–388. https://doi.org/10.1016/j.jsb.2013.07.005
7. Nogales E, Scheres SH (2015) Cryo-EM: a unique tool for the visualization of macromolecular complexity. Mol Cell 58(4):677–689
8. Scarff CA, Fuller MJG, Thompson RF, Iadaza MG (2018) Variations on negative stain electron microscopy methods: tools for tackling challenging systems. J Vis Exp 132. https://doi.org/10.3791/57199
9. Vedadi M, Niesen FH, Allali-Hassani A, Fedorov OY, Finerty PJ Jr, Wasney GA, Yeung R, Arrowsmith C, Ball LJ, Berglund H, Hui R, Marsden BD, Nordlund P, Sundstrom M, Weigelt J, Edwards AM (2006) Chemical screening methods to identify ligands that promote protein stability, protein crystallization, and structure determination. Proc Natl Acad Sci U S A 103(43):15835–15840. https://doi.org/10.1073/pnas.0605224103
10. Scheres SH (2016) Processing of structurally heterogeneous Cryo-EM data in RELION. Methods Enzymol 579:125–157
11. Punjani A, Rubinstein JL, Fleet DJ, Brubaker MA (2017) cryoSPARC: algorithms for rapid unsupervised cryo-EM structure determination. Nat Methods 14(3):290–296. https://doi.org/10.1038/nmeth.4169
12. Nguyen THD, Galej WP, Bai XC, Oubridge C, Newman AJ, Scheres SHW, Nagai K (2016) Cryo-EM structure of the yeast U4/U6.U5 tri-snRNP at 3.7 A resolution. Nature 530(7590):298–302. https://doi.org/10.1038/nature16940
13. von Loeffelholz O, Natchiar SK, Djabeur N, Myasnikov AG, Kratzat H, Menetret JF, Hazemann I, Klaholz BP (2017) Focused classification and refinement in high-resolution cryo-EM structural analysis of ribosome complexes. Curr Opin Struct Biol 46:140–148
14. Lepault J, Booy FP, Dubochet J (1983) Electron microscopy of frozen biological suspensions. J Microsc 129(Pt 1):89–102. https://doi.org/10.1111/j.1365-2818.1983.tb04163.x
15. Thompson RF, Walker M, Siebert CA, Muench SP, Ranson NA (2016) An introduction to sample preparation and imaging by cryo-electron microscopy for structural biology. Methods 100:3–15. https://doi.org/10.1016/j.ymeth.2016.02.017
16. Mesa P, Deniaud A, Montoya G, Schaffitzel C (2013) Directly from the source: endogenous preparations of molecular machines. Curr Opin Struct Biol 23(3):319–325
17. Chen R (2012) Bacterial expression systems for recombinant protein production: *E. coli* and beyond. Biotechnol Adv 30(5):1102–1107. https://doi.org/10.1016/j.biotechadv.2011.09.013
18. Zhao Y, Bishop B, Clay JE, Lu W, Jones M, Daenke S, Siebold C, Stuart DI, Jones EY, Aricescu AR (2011) Automation of large scale transient protein expression

in mammalian cells. J Struct Biol 175(2):209–215. https://doi.org/10.1016/j.jsb.2011.04.017
19. Pelosse M, Crocker H, Gorda B, Lemaire P, Rauch J, Berger I (2017) MultiBac: from protein complex structures to synthetic viral nanosystems. BMC Biol 15(1):99
20. Yamashita A, Izumi N, Kashima I, Ohnishi T, Saari B, Katsuhata Y, Muramatsu R, Morita T, Iwamatsu A, Hachiya T, Kurata R, Hirano H, Anderson P, Ohno S (2009) SMG-8 and SMG-9, two novel subunits of the SMG-1 complex, regulate remodeling of the mRNA surveillance complex during nonsense-mediated mRNA decay. Genes Dev 23(9):1091–1105. https://doi.org/10.1101/gad.1767209
21. Deniaud A, Karuppasamy M, Bock T, Masiulis S, Huard K, Garzoni F, Kerschgens K, Hentze MW, Kulozik AE, Beck M, Neu-Yilik G, Schaffitzel C (2015) A network of SMG-8, SMG-9 and SMG-1 C-terminal insertion domain regulates UPF1 substrate recruitment and phosphorylation. Nucleic Acids Res 43(15): 7600–7611. https://doi.org/10.1093/nar/gkv668
22. Lu P, Bai XC, Ma D, Xie T, Yan C, Sun L, Yang G, Zhao Y, Zhou R, Scheres SHW, Shi Y (2014) Three-dimensional structure of human gamma-secretase. Nature 512(7513):166–170. https://doi.org/10.1038/nature13567
23. Bai XC, Yan C, Yang G, Lu P, Ma D, Sun L, Zhou R, Scheres SHW, Shi Y (2015) An atomic structure of human gamma-secretase. Nature 525(7568):212–217. https://doi.org/10.1038/nature14892
24. Proteau A, Shi R, Cygler M (2010) Application of dynamic light scattering in protein crystallization. Curr Protoc Protein Sci Chapter 17: Unit 17 10
25. Ericsson UB, Hallberg BM, Detitta GT, Dekker N, Nordlund P (2006) Thermofluor-based high-throughput stability optimization of proteins for structural studies. Anal Biochem 357(2):289–298. https://doi.org/10.1016/j.ab.2006.07.027
26. Chari A, Haselbach D, Kirves JM, Ohmer J, Paknia E, Fischer N, Ganichkin O, Moller V, Frye JJ, Petzold G, Jarvis M, Tietzel M, Grimm C, Peters JM, Schulman BA, Tittmann K, Markl J, Fischer U, Stark H (2015) ProteoPlex: stability optimization of macromolecular complexes by sparse-matrix screening of chemical space. Nat Methods 12(9):859–865. https://doi.org/10.1038/nmeth.3493
27. Stark H, Chari A (2016) Sample preparation of biological macromolecular assemblies for the determination of high-resolution structures by cryo-electron microscopy. Microscopy (Oxf) 65(1):23–34
28. Kim J, Wu S, Tomasiak TM, Mergel C, Winter MB, Stiller SB, Robles-Colmanares Y, Stroud RM, Tampe R, Craik CS, Cheng Y (2015) Subnanometre-resolution electron cryomicroscopy structure of a heterodimeric ABC exporter. Nature 517(7534): 396–400. https://doi.org/10.1038/nature13872
29. Wu S, Avila-Sakar A, Kim J, Booth DS, Greenberg CH, Rossi A, Liao M, Li X, Alian A, Griner SL, Juge N, Yu Y, Mergel CM, Chaparro-Riggers J, Strop P, Tampe R, Edwards RH, Stroud RM, Craik CS, Cheng Y (2012) Fabs enable single particle cryoEM studies of small proteins. Structure 20(4): 582–592. https://doi.org/10.1016/j.str.2012.02.017
30. Davies RB, Smits C, Wong ASW, Stock D, Christie M, Sandin S, Stewart AG (2017) Cryo-EM analysis of a domain antibody bound rotary ATPase complex. J Struct Biol 197(3):350–353. https://doi.org/10.1016/j.jsb.2017.01.002
31. Westfield GH, Rasmussen SG, Su M, Dutta S, DeVree BT, Chung KY, Calinski D, Velez-Ruiz G, Oleskie AN, Pardon E, Chae PS, Liu T, Li S, Woods VL Jr, Steyaert J, Kobilka BK, Sunahara RK, Skiniotis G (2011) Structural flexibility of the G alpha s alpha-helical domain in the beta2-adrenoceptor Gs complex. Proc Natl Acad Sci U S A 108(38):16086–16091. https://doi.org/10.1073/pnas.1113645108
32. Shukla AK, Westfield GH, Xiao K, Reis RI, Huang LY, Tripathi-Shukla P, Qian J, Li S, Blanc A, Oleskie AN, Dosey AM, Su M, Liang CR, Gu LL, Shan JM, Chen X, Hanna R, Choi M, Yao XJ, Klink BU, Kahsai AW, Sidhu SS, Koide S, Penczek PA, Kossiakoff AA, Woods VL Jr, Kobilka BK, Skiniotis G, Lefkowitz RJ (2014) Visualization of arrestin recruitment by a G-protein-coupled receptor. Nature 512(7513):218–222. https://doi.org/10.1038/nature13430
33. Garcia-Nafria J, Lee Y, Bai X, Carpenter B, Tate CG (2018) Cryo-EM structure of the adenosine A_{2A} receptor coupled to an engineered heterotrimeric G protein. elife 7. https://doi.org/10.7554/eLife.35946.001
34. Coscia F, Estrozi LF, Hans F, Malet H, Noirclerc-Savoye M, Schoehn G, Petosa C (2016) Fusion to a homo-oligomeric scaffold allows cryo-EM analysis of a small protein." Sci Rep. Sci Rep 6:30909. https://doi.org/10.1038/srep30909
35. Vijayachandran LS, Viola C, Garzoni F, Trowitzsch S, Bieniossek C, Chaillet M, Schaffitzel C, Busso D, Romier C, Poterszman A, Richmond TJ, Berger I (2011) Robots, pipelines, polyproteins: enabling multiprotein expression in prokaryotic and eukaryotic cells. J Struct Biol 175(2):198–208. https://doi.org/10.1016/j.jsb.2011.03.007
36. Crepin T, Swale C, Monod A, Garzoni F, Chaillet M, Berger I (2015) Polyproteins in structural biology. Curr Opin Struct Biol 32:139–146
37. von Loeffelholz O, Jiang Q, Ariosa A, Karuppasamy M, Huard K, Berger I, Shan SO, Schaffitzel C (2015) Ribosome-SRP-FtsY cotranslational targeting complex in the closed state. Proc Natl Acad Sci U S A 112(13):3943–3948. https://doi.org/10.1073/pnas.1424453112
38. Karuppasamy M, Kusmider B, Oliveira TM, Gaubitz C, Prouteau M, Loewith R, Schaffitzel C (2017) Cryo-EM structure of *Saccharomyces cerevisiae* target of rapamycin complex 2. Nat Commun 8(1):1729. https://doi.org/10.1038/s41467-017-01862-0
39. Kastner B, Fischer N, Golas MM, Sander B, Dube P, Boehringer D, Hartmuth K, Deckert J, Hauer F,

Wolf E, Uchtenhagen H, Urlaub H, Herzog F, Peters JM, Poerschke D, Luhrmann R, Stark H (2008) GraFix: sample preparation for single-particle electron cryomicroscopy. Nat Methods 5(1):53–55. https://doi.org/10.1038/nmeth1139
40. Mio K, Sato C (2018) Lipid environment of membrane proteins in cryo-EM based structural analysis. Biophys Rev 10(2):307–316
41. Liao M, Cao E, Julius D, Cheng Y (2013) Structure of the TRPV1 ion channel determined by electron cryo-microscopy. Nature 504(7478):107–112. https://doi.org/10.1038/nature12822
42. Vinothkumar KR, Zhu J, Hirst J (2014) Architecture of mammalian respiratory complex I. Nature 515(7525): 80–84. https://doi.org/10.1038/nature13686
43. Jiang JS, Pentelute BL, Collier RJ, Zhou ZH (2015) Atomic structure of anthrax protective antigen pore elucidates toxin translocation. Nature 521(7553): 545–U323. https://doi.org/10.1038/nature14247
44. Voorhees RM, Hegde RS (2016) Structure of the Sec61 channel opened by a signal sequence. Science 351(6268):88–91. https://doi.org/10.1126/science.aad4992
45. Tao X, Hite RK, MacKinnon R (2017) Cryo-EM structure of the open high-conductance ca(2+)-activated K(+) channel. Nature 541(7635):46–51. https://doi.org/10.1038/nature20608
46. Zhang Y, Sun B, Feng D, Hu H, Chu M, Qu Q, Tarrasch JT, Li S, Sun Kobilka T, Kobilka BK, Skiniotis G (2017) Cryo-EM structure of the activated GLP-1 receptor in complex with a G protein. Nature 546(7657):248–253. https://doi.org/10.1038/nature22394
47. Kim Y, Chen J (2018) Molecular structure of human P-glycoprotein in the ATP-bound, outward-facing conformation. Science 359(6378):915–919. https://doi.org/10.1126/science.aar7389
48. Qian P, Siebert CA, Wang P, Canniffe DP, Hunter CN (2018) Cryo-EM structure of the *Blastochloris viridis* LH1-RC complex at 2.9 A. Nature 556(7700): 203–208. https://doi.org/10.1038/s41586-018-0014-5
49. Vinothkumar KR (2015) Membrane protein structures without crystals, by single particle electron cryomicroscopy. Curr Opin Struct Biol 33:103–114
50. Hauer F, Gerle C, Fischer N, Oshima A, Shinzawa-Itoh K, Shimada S, Yokoyama K, Fujiyoshi Y, Stark H (2015) GraDeR: membrane protein complex preparation for single-particle cryo-EM. Structure 23(9): 1769–1775. https://doi.org/10.1016/j.str.2015.06.029
51. Popot JL, Althoff T, Bagnard D, Baneres JL, Bazzacco P, Billon-Denis E, Catoire LJ, Champeil P, Charvolin D, Cocco MJ, Cremel G, Dahmane T, de la Maza LM, Ebel C, Gabel F, Giusti F, Gohon Y, Goormaghtigh E, Guittet E, Kleinschmidt JH, Kühlbrandt W, Le Bon C, Martinez KL, Picard M, Pucci B, Sachs JN, Tribet C, van Heijenoort C, Wien F, Zito F, Zoonens M (2011) Amphipols from A to Z. Annu Rev Biophys 40:379–408. https://doi.org/10.1146/annurev-biophys-042910-155219
52. Mazhab-Jafari MT, Rohou A, Schmidt C, Bueler SA, Benlekbir S, Robinson CV, Rubinstein JL (2016) Atomic model for the membrane-embedded V_O motor of a eukaryotic V-ATPase. Nature 539(7627): 118–122. https://doi.org/10.1038/nature19828
53. Su Q, Hu F, Liu Y, Ge X, Mei C, Yu S, Shen A, Zhou Q, Yan C, Lei J, Zhang Y, Liu X, Wang T (2018) Cryo-EM structure of the polycystic kidney disease-like channel PKD2L1. Nat Commun 9(1):1192. https://doi.org/10.1038/s41467-018-03606-0
54. Denisov IG, Sligar SG (2016) Nanodiscs for structural and functional studies of membrane proteins. Nat Struct Mol Biol 23(6):481–486
55. Frauenfeld J, Gumbart J, Sluis EO, Funes S, Gartmann M, Beatrix B, Mielke T, Berninghausen O, Becker T, Schulten K, Beckmann R (2011) Cryo-EM structure of the ribosome-SecYE complex in the membrane environment. Nat Struct Mol Biol 18(5): 614–621. https://doi.org/10.1038/nsmb.2026
56. Efremov RG, Leitner A, Aebersold R, Raunser S (2015) Architecture and conformational switch mechanism of the ryanodine receptor. Nature 517(7532): 39–43. https://doi.org/10.1038/nature13916
57. Yan Z, Bai X, Yan C, Wu J, Li Z, Xie T, Peng W, Yin C, Li X, Scheres SHW, Shi Y, Yan N (2015) Structure of the rabbit ryanodine receptor RyR1 at near-atomic resolution. Nature 517(7532):50–55. https://doi.org/10.1038/nature14063
58. Zalk R, Clarke OB, des Georges A, Grassucci RA, Reiken S, Mancia F, Hendrickson WA, Frank J, Marks AR (2015) Structure of a mammalian ryanodine receptor. Nature 517(7532):44–49. https://doi.org/10.1038/nature13950
59. Frauenfeld J, Loving R, Armache JP, Sonnen AF, Guettou F, Moberg P, Zhu L, Jegerschold C, Flayhan A, Briggs JA, Garoff H, Low C, Cheng Y, Nordlund P (2016) A saposin-lipoprotein nanoparticle system for membrane proteins. Nat Methods 13(4): 345–351. https://doi.org/10.1038/nmeth.3801
60. Bruhn H (2005) A short guided tour through functional and structural features of saposin-like proteins. Biochem J 389(Pt 2):249–257. https://doi.org/10.1042/BJ20050051
61. Knowles TJ, Finka R, Smith C, Lin YP, Dafforn T, Overduin M (2009) Membrane proteins solubilized intact in lipid containing nanoparticles bounded by styrene maleic acid copolymer. J Am Chem Soc 131(22):7484–7485. https://doi.org/10.1021/ja810046q
62. Sun C, Benlekbir S, Venkatakrishnan P, Wang Y, Hong S, Hosler J, Tajkhorshid E, Rubinstein JL, Gennis RB (2018) Structure of the alternative complex III in a supercomplex with cytochrome oxidase. Nature 557(7703):123–126. https://doi.org/10.1038/s41586-018-0061-y
63. Cheng Y, Grigorieff N, Penczek PA, Walz T (2015) A primer to single-particle cryo-electron microscopy. Cell 161(3):438–449. https://doi.org/10.1016/j.cell.2015.03.049

64. Grassucci RA, Taylor D, Frank J (2008) Visualization of macromolecular complexes using cryo-electron microscopy with FEI Tecnai transmission electron microscopes. Nat Protoc 3(2):330–339. https://doi.org/10.1038/nprot.2007.474
65. Kelly D, Dukovski D, Walz T (2010) A practical guide to the use of monolayer purification and affinity grids. Methods Enzymol 481:83–107. https://doi.org/10.1016/S0076-6879(10)81004-3
66. Yu G, Vago F, Zhang D, Snyder JE, Yan R, Zhang C, Benjamin C, Jiang X, Kuhn RJ, Serwer P, Thompson DH, Jiang W (2014) Single-step antibody-based affinity cryo-electron microscopy for imaging and structural analysis of macromolecular assemblies. J Struct Biol 187(1):1–9. https://doi.org/10.1016/j.jsb.2014.04.006
67. Crucifix C, Uhring M, Schultz P (2004) Immobilization of biotinylated DNA on 2-D streptavidin crystals. J Struct Biol 146(3):441–451. https://doi.org/10.1016/j.jsb.2004.02.001
68. Scheres SH (2014) Beam-induced motion correction for sub-megadalton cryo-EM particles. elife 3:e03665. https://doi.org/10.7554/eLife.03665
69. Russo CJ, Passmore LA (2014) Electron microscopy: ultrastable gold substrates for electron cryomicroscopy. Science 346(6215):1377–1380. https://doi.org/10.1126/science.1259530
70. Russo CJ, Passmore LA (2014) Controlling protein adsorption on graphene for cryo-EM using low-energy hydrogen plasmas. Nat Methods 11(6):649–652. https://doi.org/10.1038/nmeth.2931
71. Garaeva AA, Oostergetel GT, Gati C, Guskov A, Paulino C, Slotboom DJ (2018) Cryo-EM structure of the human neutral amino acid transporter ASCT2. Nat Struct Mol Biol 25(6):515–521. https://doi.org/10.1038/s41594-018-0076-y
72. Lee CH, MacKinnon R (2017) Structures of the human HCN1 hyperpolarization-activated channel. Cell 168(1–2):111–120 e111. https://doi.org/10.1016/j.cell.2016.12.023
73. Razinkov I, Dandey V, Wei H, Zhang Z, Melnekoff D, Rice WJ, Wigge C, Potter CS, Carragher B (2016) A new method for vitrifying samples for cryoEM. J Struct Biol 195(2):190–198. https://doi.org/10.1016/j.jsb.2016.06.001
74. Feng X, Fu Z, Kaledhonkar S, Jia Y, Shah B, Jin A, Liu Z, Sun M, Chen B, Grassucci RA, Ren Y, Jiang H, Frank J, Lin Q (2017) A fast and effective microfluidic spraying-plunging method for high-resolution single-particle cryo-EM. Structure 25(4):663–670 e663. https://doi.org/10.1016/j.str.2017.02.005
75. Arnold SA, Albiez S, Bieri A, Syntychaki A, Adaixo R, McLeod RA, Goldie KN, Stahlberg H, Braun T (2017) Blotting-free and lossless cryo-electron microscopy grid preparation from nanoliter-sized protein samples and single-cell extracts. J Struct Biol 197(3):220–226. https://doi.org/10.1016/j.jsb.2016.11.002
76. Wei H, Dandey VP, Zhang Z, Raczkowski A, Rice WJ, Carragher B, Potter CS (2018) Optimizing "self-wicking" nanowire grids. J Struct Biol 202(2):170–174. https://doi.org/10.1016/j.jsb.2018.01.001
77. Lu Z, Shaikh TR, Barnard D, Meng X, Mohamed H, Yassin A, Mannella CA, Agrawal RK, Lu TM, Wagenknecht T (2009) Monolithic microfluidic mixing-spraying devices for time-resolved cryo-electron microscopy. J Struct Biol 168(3):388–395. https://doi.org/10.1016/j.jsb.2009.08.004
78. Glaeser RM (2016) How good can cryo-EM become? Nat Methods 13(1):28–32
79. Danev R, Buijsse B, Khoshouei M, Plitzko JM, Baumeister W (2014) Volta potential phase plate for in-focus phase contrast transmission electron microscopy. Proc Natl Acad Sci U S A 111(44):15635–15640. https://doi.org/10.1073/pnas.1418377111
80. Danev R, Baumeister W (2016) Cryo-EM single particle analysis with the Volta phase plate. elife 5
81. Mahamid J, Pfeffer S, Schaffer M, Villa E, Danev R, Cuellar LK, Forster F, Hyman AA, Plitzko JM, Baumeister W (2016) Visualizing the molecular sociology at the HeLa cell nuclear periphery. Science 351(6276):969–972. https://doi.org/10.1126/science.aad8857
82. Khoshouei M, Radjainia M, Baumeister W, Danev R (2017) Cryo-EM structure of haemoglobin at 3.2 A determined with the Volta phase plate. Nat Commun 8:16099. https://doi.org/10.1038/ncomms16099
83. Liang YL, Khoshouei M, Radjainia M, Zhang Y, Glukhova A, Tarrasch J, Thal DM, Furness SGB, Christopoulos G, Coudrat T, Danev R, Baumeister W, Miller LJ, Christopoulos A, Kobilka BK, Wootten D, Skiniotis G, Sexton PM (2017) Phase-plate cryo-EM structure of a class B GPCR-G-protein complex. Nature 546(7656):118–123. https://doi.org/10.1038/nature22327
84. Liang YL, Khoshouei M, Glukhova A, Furness SGB, Zhao P, Clydesdale L, Koole C, Truong TT, Thal DM, Lei S, Radjainia M, Danev R, Baumeister W, Wang MW, Miller LJ, Christopoulos A, Sexton PM, Wootten D (2018) Phase-plate cryo-EM structure of a biased agonist-bound human GLP-1 receptor-Gs complex. Nature 555(7694):121–125. https://doi.org/10.1038/nature25773
85. Schaffer M, Mahamid J, Engel BD, Laugks T, Baumeister W, Plitzko JM (2017) Optimized cryo-focused ion beam sample preparation aimed at *in situ* structural studies of membrane proteins. J Struct Biol 197(2):73–82. https://doi.org/10.1016/j.jsb.2016.07.010
86. Wan W, Briggs JA (2016) Cryo-electron tomography and subtomogram averaging. Methods Enzymol 579:329–367
87. Baker LA, Grange M, Grünewald K (2017) Electron cryo-tomography captures macromolecular complexes in native environments. Curr Opin Struct Biol 46:149–156

88. Oikonomou CM, Jensen GJ (2017) Cellular electron cryotomography: toward structural biology *in situ*. Annu Rev Biochem 86:873–896. https://doi.org/10.1146/annurev-biochem-061516-044741
89. Kosinski J, Mosalaganti S, von Appen A, Teimer R, DiGuilio AL, Wan W, Bui KH, Hagen WJ, Briggs JA, Glavy JS, Hurt E, Beck M (2016) Molecular architecture of the inner ring scaffold of the human nuclear pore complex. Science 352(6283):363–365. https://doi.org/10.1126/science.aaf0643
90. Lin DH, Stuwe T, Schilbach S, Rundlet EJ, Perriches T, Mobbs G, Fan Y, Thierbach K, Huber FM, Collins LN, Davenport AM, Jeon YE, Hoelz A (2016) Architecture of the symmetric core of the nuclear pore. Science 352(6283):aaf1015. https://doi.org/10.1126/science.aaf1015
91. Wu M, Gu J, Guo R, Huang Y, Yang M (2016) Structure of mammalian respiratory supercomplex $I_1III_2IV_1$. Cell 167(6):1598–1609 e1510
92. Zhang S, Kostyuchenko VA, Ng TS, Lim XN, Ooi JS, Lambert S, Tan TY, Widman DG, Shi J, Baric RS, Lok SM (2016) Neutralization mechanism of a highly potent antibody against Zika virus. Nat Commun 7:13679
93. Han Y, Yan C, Nguyen THD, Jackobel AJ, Ivanov I, Knutson BA, He Y (2017) Structural mechanism of ATP-independent transcription initiation by RNA polymerase I. Elife 6:e27414. https://doi.org/10.7554/eLife.27414
94. Wang F, Burrage AM, Postel S, Clark RE, Orlova A, Sundberg EJ, Kearns DB, Egelman EH (2017) A structural model of flagellar filament switching across multiple bacterial species. Nat Commun 8(1):960
95. Fislage M, Zhang J, Brown ZP, Mandava CS, Sanyal S, Ehrenberg M, Frank J (2018) Cryo-EM shows stages of initial codon selection on the ribosome by aa-tRNA in ternary complex with GTP and the GTPase-deficient EF-Tu H84A. Nucleic Acids Res 46(11):5861–5874
96. Gao Y, Cao E, Julius D, Cheng Y (2016) TRPV1 structures in nanodiscs reveal mechanisms of ligand and lipid action. Nature 534(7607):347–351

Characterization of Complexes and Supramolecular Structures by Electron Microscopy

13

José L. Carrascosa

Abstract

Recent advancements in cryo-electron microscopy (cryo-TEM) have enabled the determination of structures of macromolecular complexes at near-atomic resolution, establishing it as a pivotal tool in Structural Biology. This high resolution allows for the detection of ligands and substrates under physiological conditions. Enhancements in detectors and imaging devices, like phase plates, improve signal quality, facilitating the reconstruction of even smaller macromolecular complexes. The 100-kDa barrier has been surpassed, presenting new opportunities for pharmacological research and expanding the scope of crystallographic analyses in the pharmaceutical industry. Cryo-TEM produces vast data sets from minimal samples, and refined classification methods can identify different conformational states of macromolecular complexes, offering deeper insights into the functional characteristics of macromolecular systems. Additionally, cryo-TEM is paving the way for time-resolved microscopy, with rapid freezing techniques capturing snapshots of vital structural changes in biological complexes. Finally, in Structural Cell Biology, advanced cryo-TEM, through tomographic procedures, is revealing conformational changes related to the specific subcellular localization of macromolecular systems and their interactions within cells.

Keywords

Cryo-electron microscopy (cryo-TEM) · Cryo-electron tomography (CET) · Electron microscopy (EM) · Direct electron detectors (DED)

13.1 Introduction

Electron microscopy (EM) has been a widely used structure determination method in Biomedical applications, but only recently it has emerged as a key technique for Structural Biology to retrieve structures at molecular and atomic resolution levels. This has been possible due to the recent incorporation of technical advances that provided the elements to overcome the classical limitations of this technique. The requirement to prepare samples for analysis under vacuum, which required extensive fixation procedures, as well as the use of heavy metal stains imposed by the poor specific contrast of biomaterials, was overcome by the incorporation of fast freezing for sample preparation. A second source of improvements came from the data acquisition systems, which incorporated computer-controlled

J. L. Carrascosa (✉)
Department of Structure of Macromolecules, Centro Nacional de Biotecnología (CNB, CSIC), Madrid, Spain
e-mail: jlcarras@cnb.csic.es

M. C. Vega, F. J. Fernández (eds.), *Advanced Technologies for Protein Complex Production and Characterization*, Advances in Experimental Medicine and Biology 1453,
https://doi.org/10.1007/978-3-031-52193-5_13

automation and the implementation of highly efficient direct electron detectors, which made possible data acquisition of electron images at a high speed and with a significant signal to noise ratio. Last, but not least, the progressive improvement in data processing, including sophisticated data standardization, statistical analyses, and three-dimensional reconstruction protocols have facilitated the collection of large amounts of images and their optimized combination to render three-dimensional experimental structures.

The incorporation of these complementary optimized tools dramatically changed the way EM was advancing during the last decades to recently reach an explosive "resolution revolution" allowing retrieving structural determination at atomic resolution from two-dimensional projection electron microscopic images. The success of cryo-electron microscopy (cryo-TEM) has been acknowledged by the 2017 Nobel Prize in Chemistry awarded to three main figures in the development of these methods for the high-resolution structure determination of biomolecules in solution: J. Frank, R. Henderson and J. Dubochet. Deposition of structures solved by EM in the Electron Microscopy Data Bank (EMDB at the Protein Data Bank Europe, https://www.ebi.ac.uk/pdbe/emdb/) has increased steadily, reaching more than 5200 in the late 2017. Up to 2014, average resolution for these structures was relatively stable at 1.5 nm. Nevertheless, this value exhibited a spectacular jump since then, reaching 0.8 nm in 2017. Even more significative, the maximum resolution trend has followed an impressive profile, reaching below 0.2 nm in 2017.

A very interesting aspect derived from the EMDB deposited structures is the size distribution of the maps: In 2016, they ranged from slightly less than 0.1 MDa (17) up to structures larger than 10 MDa (17). The larger number of structures solved that year corresponded to complexes between 1 and 10 MDa (196), with 137 corresponding to 0.5–1 MDa, 109 to 0.25 to 0.5 MDa and 41 to 0.1–0.25 MDa. This wide variety of sizes amenable to be solved by EM highlights one of the most interesting values of EM in Structural Biology: a large number of complexes of very different sizes can be tackled by EM irrespectively of their complexity and size (between 0.1 and several tens of MDa).

From the distribution of released maps at the EMDB as a function of the technique used, it is clear that a vast majority (77%) corresponds to structures of macromolecular complexes solved using single particle three-dimensional reconstruction methods [1]. Another 10% corresponds to complexes exhibiting helical symmetry or arranged as two-dimensional crystals, while the last 15% is derived from electron tomography. The success of single particle reconstruction in cryo-TEM has been exemplified by the production between 2014 and 2016 of more than 200 structures with resolution better than 0.4 nm. A critical call of attention on the possibilities of cryo-TEM in Structural Biology of macromolecules and macromolecular complexes was the publication in 2016 of a paper [2], where the structure of glutamate dehydrogenase (334 KDa) was solved at 0.18 nm, thus breaking the 0.2 nm resolution barrier by EM. In the same paper, the group of S. Subramaniam presented the structure of isocitrate dehydrogenase (with a mass of 93 kDa), demonstrating that cryo-TEM was able to retrieve structures for complexes below 100 KDa. Although the resolution obtained in this case was 0.38 nm, it was sufficient to identify conformational changes induced by binding of allosteric small-molecule inhibitors.

Although structural determination of macromolecular complexes is key for understanding their function, it is becoming more and more evident that many of their properties, including those of biomedical interest, might be closely related to their localization in the cell and to their relationship with other molecular machineries. Thus, a more comprehensive analysis will be progressively required, and the proteome and interactome will have to merge with cell structure at the highest possible resolution to define not only structure but also interactions at the physiological level. This is where tomography methods based on electron microscopy data find one most interesting application area. Cryoelectron tomography (CET) combines the potential of three-dimensional reconstruction from projections with a near native preservation of biological samples. From the maps at the EMDB, around 15% corresponds to data obtained by CET. CET also benefits from the recent technical advances mentioned above (cryo-preparation, direct electron detectors, improved reconstruction programs) and it presently offers resolution levels that allow to visualize large macromolecular complexes in their functional cellular environments. Resolutions of 3–4 nm are standard in most of the CET released maps, and the implementation of subtomogram averaging to get higher resolution is widely used (since 2012, resolutions below 1 nm are currently obtained), thus supporting CET promise for visualizing machines at work inside cells [3, 4].

In this chapter, we will briefly review the technical advances which support the TEM resolution revolution. Then, we will highlight the main advantages of cryo-TEM for structural determination of macromolecular complexes, underlying the possibility to study related conformations and structural transformations, as well as the possibility to correlate cryo-TEM data with other methods (X-ray crystallography, mass spectrometry, modelling). Finally, we will briefly describe the bases and main applications of electron tomography in the context of placing macromolecular machines under physiological conditions in the cellular environment.

13.2 Recent Technical Advances

Success of cryo-TEM today is the result of a steady accumulation of methodological improvements in sample preparation, data production and acquisition, and image processing. Nevertheless, after 2012, this trend experienced a sudden boost (the resolution revolution). This change of derivative has been due to a small number of recent technical advances in data acquisition and processing which allowed to alleviate one major problem related to cryo-TEM images: frozen, unstained biological samples are very sensitive to radiation damage and possess a very low contrast. To avoid destruction of the sample, very low electron dosages are used, then yielding very noisy images. Progressive automation in data acquisition [5] proved to be instrumental to generate large amounts of images required to average and reconstruct the cryo-TEM data. Recently, the use of phase plates [6], which are devices inserted in the back focal plane of the objective lens, have shown great potential to improve the contrast in images of vitrified samples. Even in the best imaging conditions, an additional bottleneck is recording the data. The standard use of CCDs resulted in a low efficiency, limiting attainable resolution from such noisy images. A main milestone was the incorporation of direct electron detectors (DEDs). These devices, due to their efficient recording of single electrons without the need of intermediate tools, have improved the sensitivity, quality and contrast of the images, offering the possibility of overcoming problems derived from sample movement and distortion during acquisition by using movie-like fast frame acquisition [7].

As a result of automation and the use of DEDs, large data sets are generated in cryo-TEM. To take full advantage of these data, improved processing packages are used to generate optimized data from the movies generated by DEDs, standardize, classify, and combine data from individual images to generate the three-dimensional volumes. A number of different packages are currently used, each one with specific advantages: FREALIGN [8], EMAN [9], XMIPP [10], and RELION [11]. Also, there are suites accessing different packages, such as SCIPION [12].

Taken together, the new automated microscopes equipped with phase plates and DEDs, and the availability of improved data processing packages, make cryo-TEM a very powerful tool for structural analysis of macromolecular complexes, not only yielding structures, but also offering interesting added value for tackling dynamic processes or addressing conformational heterogeneities.

13.3 Analysis of Macromolecular Complexes

13.3.1 Purification Requirements

A main advantage for using cryo-TEM analysis is that it is not required to have large amounts of sample to prepare microscope grids. Each grid requires around 5 μl of a sample at moderate concentration (0.1 to 0.5 mg/ml), and although it is required to do several rounds of microscopy tests to define optimal conditions for visualization (selection of appropriate buffer, concentration of sample, removal of undesirable chemical contaminants), the final amount of protein required (several tens of micrograms) is always quite moderate, well below the amounts needed for other methods as NMR or X-ray crystallography (several tens of milligrams).

Another aspect to take into account is the fact that cryo-TEM does not require a highly purified sample. This does not mean that very heterogeneous samples (lysates, first clarification steps in purification protocols) might be suitable for direct study but, rather, that samples for cryo-TEM do not need to be 100% pure to be analyzed. This is so because electron microscope images show individual projection views of each component in the sample. This allows to detect and identify the main components in the sample (by correlating the images with the biochemical results), thus allowing to select those particles that represent the major biochemical component. This allows for addressing the study of those complexes that are difficult to produce and to purify, as it is usually the case for those which cannot be overexpressed *ex vivo* and demand purification from naturally existing cells and tissues. A comprehensive approach to the importance of sample preparation for successful cryo-TEM of macromolecular complexes, and a review of innovative approaches to improve sample quality is given in Chap. 12 by A. Deniaud, B.V. Kabasaki, J. C. Buffon and C. Schaffitzel.

13.3.2 Structure Determination: From Molecular Up to Atomic Resolution

Structure determination by single-particle cryo-TEM presents a number of advantages, besides the pure structure description. A clarifying example for the added value of cryo-TEM is that related to the study of CorA, the major Mg^{2+} uptake system in prokaryotes [13]. This pentameric complex is ~200-kDa in size and it is a Mg^{2+}-dependent channel which is the key pathway for electrophoretic Mg^{2+} uptake. X-ray crystallographic studies of the CorA crystals obtained in the presence and absence of Mg^{2+} did not reveal significant changes in the channel conformation, thus leaving open the interpretation on how the gating mechanism could take place.

The group of S. Subramaniam made a single-particle cryo-TEM study to determine the structures of detergent-solubilized CorA from *Thermotoga maritima* under conditions that stabilize its closed (Mg^{2+}-bound) and open (Mg^{2+}-free) states [14]. Figure 13.1 shows the results obtained for the five-fold symmetric, Mg^{2+}-bound state. Typical cryo-TEM fields show characteristic projection views of the complexes (Fig. 13.1a). Further analysis of these views of the particles reveals that they can be grouped in classes (Fig. 13.1b). This classification step is a great advantage of the method, as it enables detection of eventual contaminants (*in silico* purification), and also provides a collection of two-dimensional projections of the particles which allows for their combination to produce a three-dimensional reconstruction. In this case, the resolution (3.8 Å) was sufficient to allow tracing of the five polypeptide subunits, revealing α-helical regions, β sheets, and even strong densities for larger side chains (Fig. 13.1c).

13.3.3 Localization of Ligands, Substrates and Small Molecules

The possibilities of cryo-TEM for structure determination are related to the achievable resolution limit and to whether that resolution would be sufficient to detect the location of small molecules such as ligands, substrates, etc. It is clear that the extension of the method towards drug investigation in pharmaceutical studies is much dependent on these capabilities. The studies of da Fonseca on the interaction of substrates with the proteasome in the framework of its use to fight malaria are a good example on how cryo-TEM studies might be instrumental for the design of

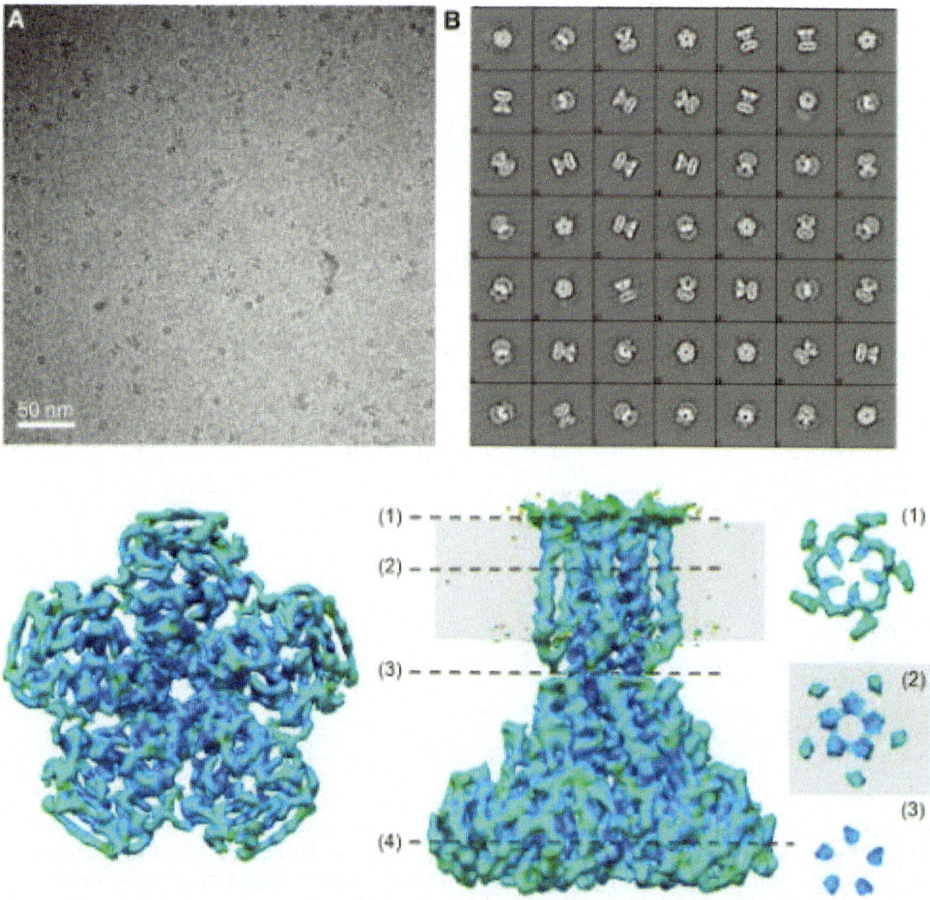

Fig. 13.1 Cryo-EM of TmCorA in the Presence of Mg2 + figure title. (**a**) Representative cryo-EM image with CorA particles visible in different orientations is shown. (**b**) Selected 2D class averages used to produce an initial model are shown. (Bottom) Top and side views of final map, colored according to local resolution. Slices through the map at indicated positions are shown on the right. (Reproduced with permission from [14])

improved therapeutic strategies against an important human disease [15].

The eukaryotic proteasome is a large protein complex of about 750 kDa, which plays a fundamental role in cell homeostasis by degrading specifically tagged proteins. It has been considered as a good target for different diseases, as it is involved in degradation of key cell cycle regulators. Structures of different subcomplexes related to the proteasome have been solved by X-ray crystallography. Nevertheless, the analysis of the interaction of inhibitors with the proteasome has been limited by the difficulty in obtaining crystals under conditions required for drug interaction. This is an ideal area where cryo-TEM has proven to be quite appropriate, as samples for cryo-TEM can be prepared under conditions currently used for studies on drug interaction. The fact that fast freezing is the only manipulation required makes this procedure ideal to work under a wide variety of pH, salt and ionic conditions at concentrations that are typical for biochemical assays.

The structure of the 20S core of the human proteasome was solved by cryo-TEM at about 3.5 Å resolution [16]. Figure 13.2a,b shows planes of the three-dimensional reconstruction of the core, built by seven α and seven β subunits

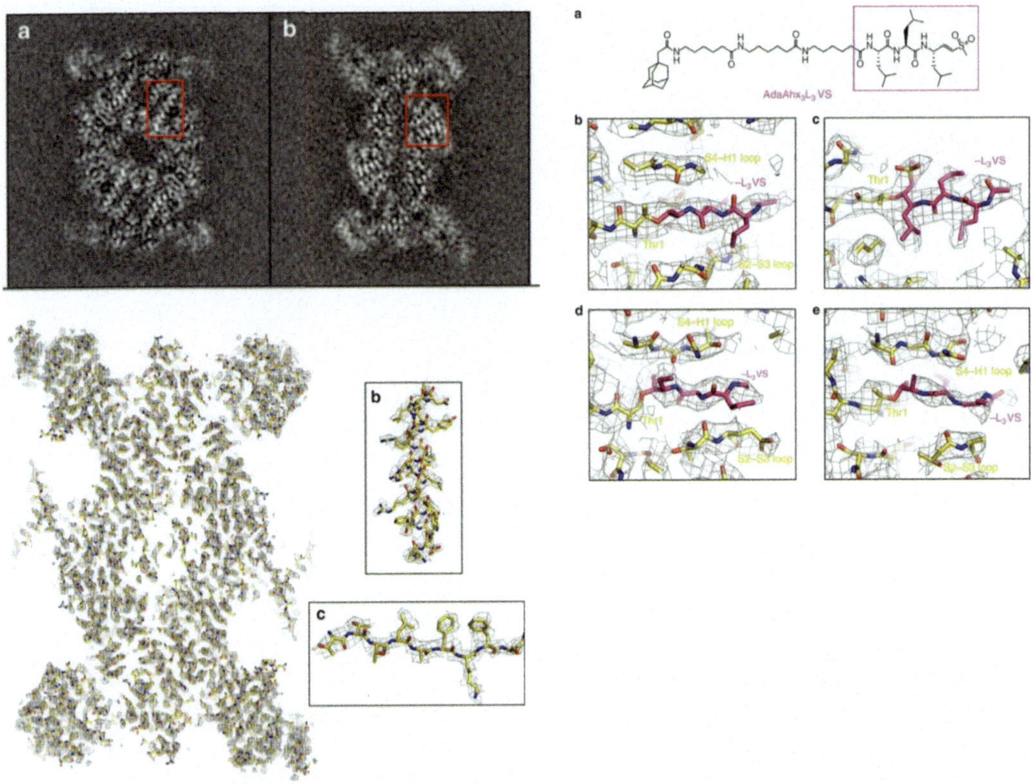

Fig. 13.2 Cryo-EM map of the human 20S–AdaAhx$_3$L$_3$VS complex. (Left, top) Individual sections 1-Å thick of the 3D map are represented as grey-scale. Regions showing the pattern of α-helical secondary structure and the separation of sheet forming β strands are boxed. (Left, bottom) Section of the map showing the global agreement between the map densities and the coordinates of the 20S proteasome complex. Close-up representations of an α-helix and a β strand are shown, illustrating substantial recovery of side-chain information. (Right) Structural formula of the inhibitor AdaAhx$_3$L$_3$VS. The map of the human 20S–AdaAhx$_3$L$_3$VS complex, Fourier low-pass filtered to 3.4 Å, is shown as a mesh. Clear densities are seen extending from the N-terminal Thr residues of the β_5, β_2 and β_1 subunits that are consistent with the -L$_3$VS moiety of the AdaAhx$_3$L$_3$VS molecule, with the vinyl sulfone group and the side chains of the three leucine residues shown clearly resolved at the β_5 site. (Modified and reproduced with permission from [16])

arranged in pseudo-seven-fold rings, two copies of which stack into a barrel shaped assembly, revealing secondary structure elements (α-helices and β-strands) of the structural peptides of the proteasome core. When this structure was compared to the crystal structure derived from a mouse apo20S core, the agreement found allowed to completely identify the polypeptide chain (Fig. 13.2c) in the cryo-EM volume. Interestingly enough, this proteasome core was complexed with AdaAhx3L3VS, which is a highly potent proteasome inhibitor that irreversibly binds the 20S core proteolytic active sites. The resolution attained by cryo-EM in the complex allowed locating densities corresponding to the inhibitor, which extended out of the active sites (Fig. 13.2d). The possibility to detect the location and conformation of the inhibitor at the different active sites of the proteasome demonstrates that cryo-EM can be a most useful tool for structural studies of protein–ligand interactions.

13.3.4 Study of Related Conformations and Snapshots of Dynamic Processes

The existence of different conformational states of a functional complex is probably more frequent than initially expected. Most of the knowledge we have accumulated on structure of macromolecules and macromolecular complexes has been derived from protein crystallography, which represents a single, "frozen" view of the possible conformational space required to get crystallization. One of the clear advantages of cryo-EM is that it does not impose any condition to "fix" one of the possible conformations, as all of them might be collected simultaneously during image acquisition. If a sufficient number of views are collected, it is possible to search the conformational space for any given complex, thus allowing to study not only those more frequent conformations, but also other less abundant ones, which can be also more populated under certain functional conditions. Obviously, this type of study requires to apply powerful classification methods to a large number of views to sort the specimen projections into distinct classes which, in turn, have to be translated into three-dimensional classification [11]. This added value for structural analysis makes cryo-EM a unique tool to study structural heterogeneity and its relationship to functional states.

A particular example of the potential of cryo-EM for detecting different conformations of a protein complex is the study of membrane receptors, where the existence of structural changes relevant to their function, together with the added difficulty that they are membrane proteins, pose a great technical challenge. The study by Meyerson et al. [17] of a type of ionotropic glutamate receptors (iGluRs), which are major mediators of excitatory synaptic transmission in the central nervous system, is quite representative. AMPA (α-amino-3-hydroxy-5-methyl-4-isoxazole propionic acid) receptor is one of the three subfamilies of these receptors, which have been the subject of many studies over the years due to their role in brain function and development. AMPA is a tetramer of four homologous subunits, with an extracellular NTD domain, a transmembrane domain (TMD) and an intracellular CTD domain. The combination of biochemical studies and crystallographic analysis of the isolated domains of these complexes have resulted in working models on how the transition between the different states of these channels takes place. Nevertheless, a complete picture on the structural relationship among the different states was missing. The incorporation of cryo-EM to these studies has given a detailed view of the structural bases of the glutamate receptor activation and desensitization. Figure 13.3 shows that a modified form of AMPA (GluA2) in a desensitized state can adopt many different conformations in solution (Fig. 13.3a). Three-dimensional classification allowed to define three main classes which, although at a relatively poor resolution (2–3 nm), revealed different orientations of the ATD domains (Fig. 13.3b), and the comparison with the close and active states of the receptor revealed striking differences in the way both the ATD and LBD domains are arranged in the different states, including symmetry transitions. These data allowed to finally integrate previous biochemical and structural data from different origins [17].

Another interesting possibility derived from the way single-particle reconstruction is obtained in cryo-EM is to tackle dynamic processes by studying snapshots along a functional cycle. This sort of time-resolved microscopy does not only mean that samples have to be taken along a period of time (made possible by the way frozen samples are prepared) but, also, that analysis of the projection views within a sample where a reaction is taking place allows to get different three-dimensional states which can, in turn, be correlated with defined functional steps.

A classical study reviewed by the group of H. Stark [18] showed how the tRNA moves in the ribosome during the translocation in protein synthesis. In this study, samples were taken by fast freezing at different times along the translocation reaction, taking advantage of the relative slow rate of certain steps. A large number of images were processed (more than 2×10^6), yielding around 50 different three-dimensional structures, which were then refined to show 8 key states (snapshots). By docking atomic structures of the different ribosomal components

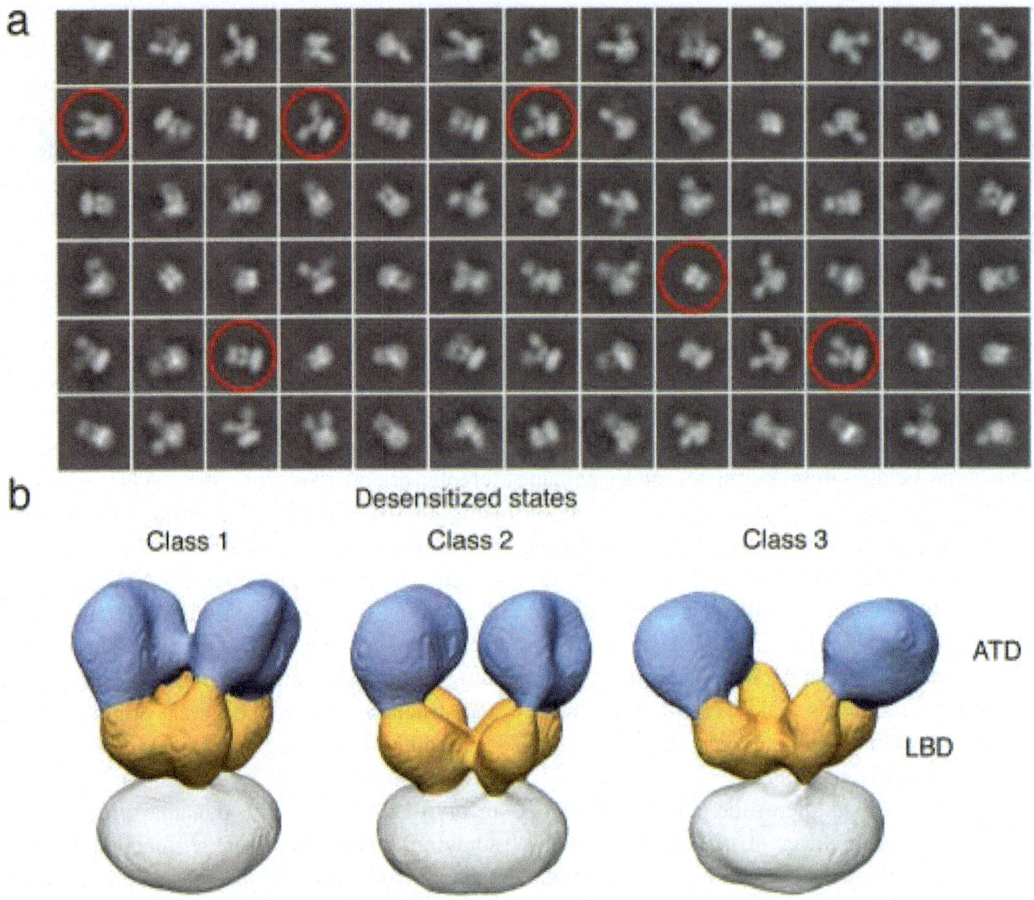

Fig. 13.3 Conformational ensemble of desensitized GluA2 by Cryo-EM. (**a**) Representative desensitized state GluA2em 2D class averages from initial classification of 35,083 projection images. Selected class-averages that illustrate the range of observed conformations are highlighted. (**b**) Segmented isosurface representations of three distinct desensitized state GluA2em structures, with the ATD and LBD layers identified in blue and orange, respectively. (Modified and reproduced with permission from [17])

into the experimental volumes obtained by cryo-EM, it was possible to define the movement of the two tRNAs moving into different positions in the ribosome (Fig. 13.4), also revealing the tight coupling of this tRNA movement with different conformational changes in the ribosome. The use of cryo-EM also allows studying the behavior of functional complexes under different parameters, as temperature, pH, concentration of specific components, inhibitors, etc., thus offering a wide application spectrum with unique possibilities with respect to other alternative and complementary methods.

13.4 *In Situ* Structural Determination by Electron Tomography

During recent years it has become increasingly evident that understanding biological functions demands not only the study of individual cell components, but also to understand their interactions and the way they are organized in space and time. Cellular processes are organized in a precise spatial framework, and the functional states of the different cellular machineries are

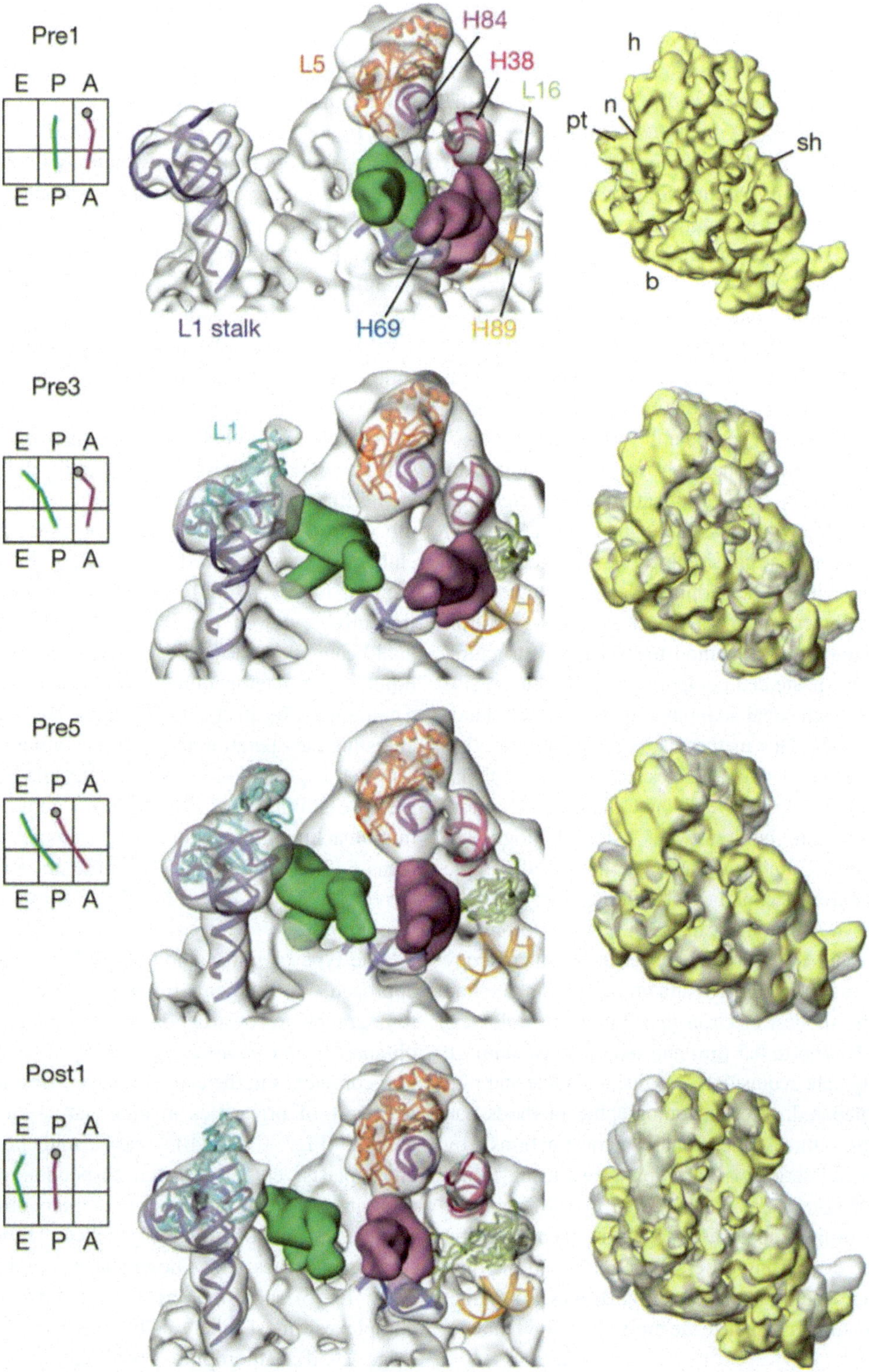

Fig. 13.4 Dynamics of ribosome–tRNA interactions and of the 30S subunit. tRNA positions in selected states of (retro-)translocation (left-hand boxes) and contacts with 50S subunit regions for representative sub-states (middle panels) are depicted along with the actual state of the 30S subunit (right-hand panels) in yellow, overlaid with the preceding state in grey. Note the anticlockwise 30S body rotation from pre1 to pre5 and the switch back to the non-rotated conformation upon transition to post1. 30S landmarks are: *b* body, *h* head, *n* neck, *pt.* platform, *sh* shoulder. Protein L1 in the L1 stalk is depicted in light blue (no clear density for L1 in pre1). (Modified and reproduced with permission from [18])

dependent on complex interactions among them, which, in turn, are related to their location in the cellular framework. Different approaches have tackled this fundamental problem from diverse perspectives providing comprehensive insights into the cellular proteome and interactome.

One key aspect of the knowledge of the organization of the cellular machinery is the definition of the structural context, *i.e.*, the cellular map of compartments and components at the highest possible resolution. To this end, correlation of light microscopy and electron microscopy presents a great potential to combine the whole cell perspective in vivo with the high intrinsic resolution of cryo-EM [19]. The combination of fast freezing, which offers the possibility to preserve the cell structure at near physiological level, together with recent technical developments to get tomographic data using cryo-EM, have positioned cryo-electron tomography as a method to bridge the structural analysis of the cell components with their precise three-dimensional mapping in the cellular landscape [20]. This method is based on the fact that electron microscopy images are two-dimensional projections, from which three-dimensional information can be retrieved using tomographic procedures. Implementation of improved sample preparation techniques, including fast freezing and cryo-substitution, makes it possible to prepare cells and even tissues for electron microscopy observation. The two limitations of the method are the modest electron penetration in biological samples (up to 0.5 μm) and the radiation damage during data acquisition. Provided suitable samples are prepared, electron tomographic methods can retrieve three-dimensional reconstructions by collecting series of projection views using tilting at different angles (reviewed in [4]). Merging of these projections in a tomographic reconstruction can be done using different methods, and the final volume can be analyzed and processed to extract and to further visualize the main structural features.

13.4.1 Sample Preparation for Cryo-Electron Tomography

Acceptable cell preservation for relatively thin samples can be obtained either by direct plunge freezing [21] or, in the case of thicker samples, by more complex setups as high-pressure freezing [22]. In any case, the limiting factor to get cryo-electron tomographic reconstructions is electron penetration in the frozen, hydrated sample, which cannot be higher than 0.5 μm. This limits the application of the method to those areas close to the border of the cells or, more frequently, to the need to obtain thin sections of the frozen sample. To this end, cryo-sectioning or cryo-substitution followed by conventional thin sectioning has been used with some success [22, 23]. A more attractive way to obtain sections of frozen cells with accurate thickness and without the compression artifacts derived from mechanical sectioning procedures, or the requirement of using plastic embedding, is to produce lamellas by cryo-focused ion beam milling [24]. In this method (recently reviewed by Schaffer et al. [25]), a dual-beam focused ion beam machine is used at liquid nitrogen temperature to mill a sample using a gallium ion beam at defined angles. By carefully adjusting the operation using the scanning microscope, it is possible to produce a lamella by progressive thinning of the frozen sample. This micromachining procedure yields samples where areas with a thickness of 100 nm can be transferred to the cryo-EM for cryo-electron tomography.

The type of information attainable using this approach is shown in Fig. 13.5. Cryo-tomographic reconstructions of frozen lamellas obtained from *Chlamydomonas* [25] show molecular complexes in the cell cytoplasm. Besides the overview of organelles in their native environment (Fig. 13.5a,b), as mitochondria, chloroplasts, etc., a detailed inspection of the tomograms reveals the presence of many different complexes bound to cell organelles. Mitochondrial membranes show bowl-shaped aggregates (Fig. 13.5c), and the cristae are covered by complexes, most likely ATP synthase dimers (Fig. 13.5d). Chloroplast thylakoids also show ATP synthase monomers, together with other larger complexes (Fig. 13.5e). Clathrin- and COPI-coated vesicles are also visible near the Golgi (Fig. 13.5f–h). Although the nominal resolution of these tomograms is not better than several nm, the possibility to detect and identify

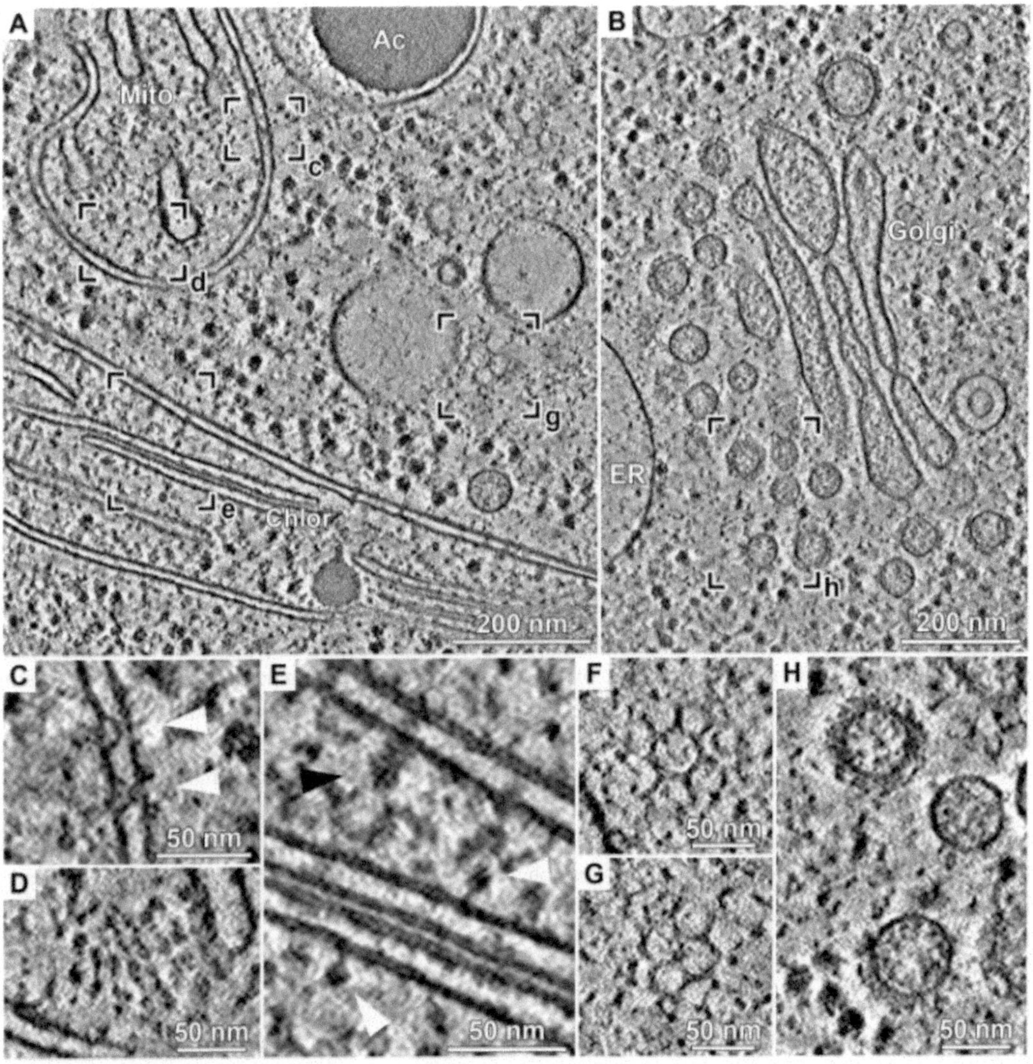

Fig. 13.5 Optimized cryo-focused ion beam sample preparation for in situ structural studies of membrane proteins. Visualization of molecular complexes within *Chlamydomonas* cell FIB lamellas. (**a**–**b**) Overview slices from tomographic volumes. *Mito* mitochondrion, *Chlor* chloroplast, *Ac* acidocalcisome, *ER* endoplasmic reticulum. The framed regions in (**a**) and (**b**) correspond to the indicated close-up views in C–H. These magnified views show different tomographic slices than those depicted in (**a**) and (**b**, **c**) Circular bowl-shaped complexes (arrowheads), seen as semicircles in this side view, positioned between the mitochondrion's inner and outer membranes. (**d**) Top view of rows of ATP synthase dimers bound to a crista membrane. (**e**) Single ATP synthases bound to thylakoid membranes (white arrowheads) and a large complex that spans the outer and inner membranes of the chloroplast envelope (black arrowhead). (**f**–**g**) Clathrin coats in the cytoplasm (**f** is from a region of the tomogram in **a** that is not shown). H: Vesicles near the cis-Golgi with complete (top), partial (bottom) or no (middle) COPI coats. The lamellas were 140 nm (**a**) and 95 nm (**b**) thick, and each covered with 60 nm of condensed water vapor followed by 5 nm of sputtered Pt. Tomograms were acquired with the VPP, a target defocus of −0.5 μm, and an object pixel size of 0.342 nm. (Modified and reproduced with permission from [25])

complexes within the cytoplasm of native cells is a major step forward in the definition of the cell cartography.

13.4.2 Improvement of Tomogram Interpretation and Resolution

The possibility to get structural information of complexes at molecular resolution from cryo-electron tomograms depends on different factors. The quality of lamella preparation, as well as the reproducibility of the thickness is a first factor which influences the resolution potentially attainable in the tomograms. Taking into account the poor penetration power of electrons, the thicker the lamella the poorer signal to noise ratio it would yield in the tomogram. This problem might be alleviated by using energy filtering to remove inelastic scattering, although the outcome will depend on the low intrinsic contrast of frozen, unstained material and the relative concentration of material in the different cell areas.

Also related to the interaction of electrons with the frozen material, the heavy energy transfer from the electron beam into the sample implies the use of very limited electron dosages which leads to very low contrast images which demand to use defocusing for generating phase contrast. The recent incorporation of Phase Plates (PP) has greatly improved generation of contrast and the production of images with increased signal to noise ratios. There are two main types of phase plates, the Zernike PP [26, 27], and the more recent Volta PP [28]. Although phase plates can also be used in data acquisition for single particle cryo-EM reconstruction, they show their highest potential in tomography, where, in combination with energy filters, seem to be the best option for automated tomographic series acquisition.

The full exploitation of the information content in cryo-electron tomography requires a set of steps integrated in complex workflows (reviewed in [29]). These procedures include the use of specific image processing methods, which allow for contrast transfer function correction, denoising and reconstruction. Then, interpretation of the tomographic volume requires additional steps. Selection of the regions of interest is normally done by segmentation of particular areas or sub-volumes, which then can be used for searching specific macromolecular complexes.

One particular aspect of interest is the possibility to improve the resolution attainable from tomographic data by averaging those structural components that are present in multiple, identical forms within the cell cytoplasm [29, 30]. To this end, template matching is very useful for automatic search of specific macromolecular complexes, using cross correlation of known structures with the experimental tomographic volume. These procedures lead to a map of the positions of the specific complexes of interest within the tomogram and provide the coordinates of the spatial distribution of these complexes. From these coordinates, subtomograms containing each specific complex can be computationally extracted and then averaged using the same rationale as that used for single particle reconstruction (see above).

Subtomogram averaging is a major tool to expand the use of tomography to perform structural biology *in situ*. The extensive use of improved classification and refinement methods, already developed for single particle reconstruction, have led to obtaining different relevant complexes structures in their physiological environment at nanometric resolution. Examples of such complexes are given in Fig. 13.6, ranging from very large nuclear pore complexes to

Fig. 13.6 (continued) reference stabilizes. (Bottom) Examples of recent structures solved by subtomogram averaging, shown approximately to scale. (**a**) Ribosomes on the ER membrane. (**b**) COPI coated vesicles. (**c**) The glycoprotein spike of HIV [11]. (**d**) The human nuclear pore. (**e**) A microtubule doublet from a *Chlamydomonas flagellum*. Panels were adapted from the original references. (Modified and reproduced with permission from [30])

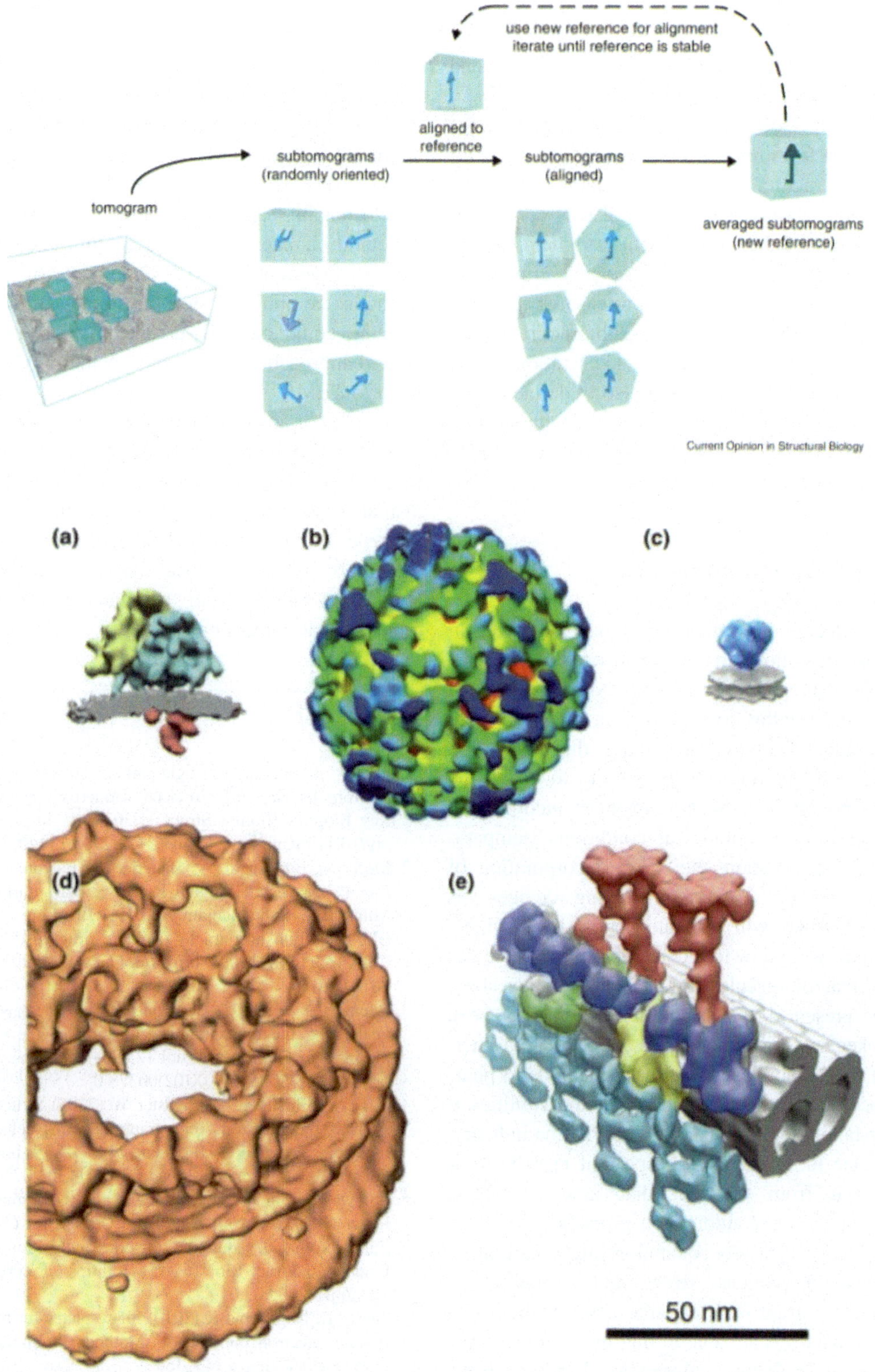

Fig. 13.6 The potential of subtomogram averaging to elucidate structural biology *in situ*. (Top) An overview of subtomogram averaging. Subtomograms are extracted from the tomogram. They are rotationally and translationally aligned against a reference. The aligned subtomograms are then averaged to generate a new reference. The new reference is then used for alignment of the subtomograms again. This procedure is repeated until the

relatively small viral spikes or ribosomes attached to the endoplasmic reticulum [30].

The use of nanometric resolution volumes obtained from tomograms can be also useful for fitting structural data obtained either by single particle reconstruction of isolated complexes (or some of their components), or to compare them with those structures obtained after purification. A recent study of the proteasome [31] is a clear example of the power of subtomogram averaging to define not only the localization of the 26S Proteasome in situ, but also to define different forms of the complex (with or without substrate) which correlate their specific function to unique positions at the cellular level.

13.5 Conclusions

The successful application of new advances in cryo-TEM allows to retrieve structures of macromolecular complexes at near atomic resolution, thus positioning cryo-TEM as a major tool for Structural Biology. Achieving this resolution level also provides the possibility to detect the presence of ligands and substrates incorporated under truly physiological conditions (temperature, concentration, etc.). The incorporation of improved detectors and imaging devices, as phase plates, will generate better signal which, in turn, will allow to reconstruct macromolecules and macromolecular complexes of smaller sizes. The present 100 KDa barrier has already been broken and this might attract a great deal of interest for the application of these methods to pharmacological studies, extending the possibilities of crystallographic analyses in the Pharma industry.

One most interesting aspect of cryo-TEM is derived from the fact that new cryo-TEM instruments and automation procedures produce very large data sets containing huge amounts of individual particles from minute amount of samples. Improved classification methods on these data sets are able to find the eventual presence of different conformational states of the complexes under study. The analysis of these structures at atomic detail might be key to give novel insights into functional features and relationships of macromolecular machineries.

Cryo-TEM also offers the possibility of the approximation towards time-resolved microscopy. Fast freezing of samples allows, in principle, to obtain snapshots of biologically relevant structural changes of complexes correlated to a certain function. The development of new methods combining stop-flow and ultra-fast freezing could open in the near future a new way to approach the study of dynamic processes by cryo-TEM.

The use of advanced cryo-TEM is also instrumental in Structural Cell Biology. Tomographic procedures using the new technical possibilities available today are yielding new insights into the existence of conformational changes correlated to the precise subcellular localization of macromolecular machineries, and their interactions with other cellular components.

References

1. Frank J (2002) Single-particle imaging of macromolecules by Cryo-electron microscopy. Annu Rev Biophys Biomol Struct 31:303–319. https://doi.org/10.1146/annurev.biophys.31.082901.134202
2. Merk A, Bartesaghi A, Banerjee S, Falconieri V, Rao P, Davis MI, Pragani R, Boxer MB, Earl LA, Milne JLS, Subramaniam S (2016) Breaking Cryo-EM resolution barriers to facilitate drug discovery. Cell 165:1698–1707. https://doi.org/10.1016/j.cell.2016.05.040
3. Baumeister W (2002) Electron tomography: towards visualizing the molecular organization of the cytoplasm. Curr Opin Struct Biol 12:679–684. https://doi.org/10.1016/S0959-440X(02)00378-0
4. Lučić V, Förster F, Baumeister W (2005) Structural studies by electron tomography: from cells to molecules. Annu Rev Biochem 74:833–865. https://doi.org/10.1146/annurev.biochem.73.011303.074112
5. Suloway C, Shi J, Cheng A, Pulokas J, Carragher B, Potter CS, Zheng SQ, Agard DA, Jensen GJ (2009) Fully automated, sequential tilt-series acquisition with Leginon. J Struct Biol 167:11–18. https://doi.org/10.1016/j.jsb.2009.03.019
6. Danev R, Nagayama K (2010) Phase plates for transmission electron microscopy. Methods in Enzymology. Elsevier, In, pp 343–369
7. McMullan G, Faruqi AR, Henderson R (2016) Direct electron detectors. Methods in Enzymology. Elsevier, In, pp 1–17

8. Grigorieff N (2007) FREALIGN: high-resolution refinement of single particle structures. J Struct Biol 157:117–125. https://doi.org/10.1016/j.jsb.2006.05.004
9. Ludtke SJ, Baldwin PR, Chiu W (1999) EMAN: Semiautomated software for high-resolution single-particle reconstructions. J Struct Biol 128:82–97. https://doi.org/10.1006/jsbi.1999.4174
10. Sorzano COS, Marabini R, Velázquez-Muriel J, Bilbao-Castro JR, Scheres SHW, Carazo JM, Pascual-Montano A (2004) XMIPP: a new generation of an open-source image processing package for electron microscopy. J Struct Biol 148:194–204. https://doi.org/10.1016/j.jsb.2004.06.006
11. Scheres SHW (2012) RELION: implementation of a Bayesian approach to cryo-EM structure determination. J Struct Biol 180:519–530. https://doi.org/10.1016/j.jsb.2012.09.006
12. de la Rosa-Trevín JM, Quintana A, Del Cano L, Zaldívar A, Foche I, Gutiérrez J, Gómez-Blanco J, Burguet-Castell J, Cuenca-Alba J, Abrishami V, Vargas J, Otón J, Sharov G, Vilas JL, Navas J, Conesa P, Kazemi M, Marabini R, Sorzano COS, Carazo JM (2016) Scipion: a software framework toward integration, reproducibility and validation in 3D electron microscopy. J Struct Biol 195:93–99. https://doi.org/10.1016/j.jsb.2016.04.010
13. Maguire ME (2006) Magnesium transporters: properties, regulation and structure. Front Biosci 11: 3149. https://doi.org/10.2741/2039
14. Matthies D, Dalmas O, Borgnia MJ, Dominik PK, Merk A, Rao P, Reddy BG, Islam S, Bartesaghi A, Perozo E, Subramaniam S (2016) Cryo-EM structures of the Magnesium Channel CorA reveal symmetry break upon gating. Cell 164:747–756. https://doi.org/10.1016/j.cell.2015.12.055
15. Li H, Bogyo M, Da Fonseca PCA (2016) The cryo-EM structure of the *plasmodium falciparum* 20S proteasome and its use in the fight against malaria. FEBS J 283:4238–4243. https://doi.org/10.1111/febs.13780
16. Da Fonseca PCA, Morris EP (2015) Cryo-EM reveals the conformation of a substrate analogue in the human 20S proteasome core. Nat Commun 6:7573. https://doi.org/10.1038/ncomms8573
17. Meyerson JR, Kumar J, Chittori S, Rao P, Pierson J, Bartesaghi A, Mayer ML, Subramaniam S (2014) Structural mechanism of glutamate receptor activation and desensitization. Nature 514:328–334. https://doi.org/10.1038/nature13603
18. Fischer N, Konevega AL, Wintermeyer W, Rodnina MV, Stark H (2010) Ribosome dynamics and tRNA movement by time-resolved electron cryomicroscopy. Nature 466:329–333. https://doi.org/10.1038/nature09206
19. Sartori A, Gatz R, Beck F, Rigort A, Baumeister W, Plitzko JM (2007) Correlative microscopy: bridging the gap between fluorescence light microscopy and cryo-electron tomography. J Struct Biol 160:135–145. https://doi.org/10.1016/j.jsb.2007.07.011
20. Lučić V, Rigort A, Baumeister W (2013) Cryo-electron tomography: the challenge of doing structural biology in situ. J Cell Biol 202:407–419. https://doi.org/10.1083/jcb.201304193
21. Dubochet J, Adrian M, Chang J-J, Homo J-C, Lepault J, McDowall AW, Schultz P (1988) Cryo-electron microscopy of vitrified specimens. Q Rev Biophys 21:129–228. https://doi.org/10.1017/S0033583500004297
22. Marsh BJ, Mastronarde DN, Buttle KF, Howell KE, McIntosh JR (2001) Organellar relationships in the Golgi region of the pancreatic beta cell line, HIT-T15, visualized by high resolution electron tomography. Proc Natl Acad Sci U S A 98:2399–2406. https://doi.org/10.1073/pnas.051631998
23. Al-Amoudi A, Chang J-J, Leforestier A, McDowall A, Salamin LM, Norlén LPO, Richter K, Blanc NS, Studer D, Dubochet J (2004) Cryo-electron microscopy of vitreous sections. EMBO J 23:3583–3588. https://doi.org/10.1038/sj.emboj.7600366
24. Marko M, Hsieh C, Moberlychan W, Mannella CA, Frank J (2006) Focused ion beam milling of vitreous water: prospects for an alternative to cryo-ultramicrotomy of frozen-hydrated biological samples. J Microsc 222:42–47. https://doi.org/10.1111/j.1365-2818.2006.01567.x
25. Schaffer M, Mahamid J, Engel BD, Laugks T, Baumeister W, Plitzko JM (2017) Optimized cryo-focused ion beam sample preparation aimed at in situ structural studies of membrane proteins. J Struct Biol 197:73–82. https://doi.org/10.1016/j.jsb.2016.07.010
26. Danev R, Kanamaru S, Marko M, Nagayama K (2010) Zernike phase contrast cryo-electron tomography. J Struct Biol 171:174–181. https://doi.org/10.1016/j.jsb.2010.03.013
27. Marko M, Leith A, Hsieh C, Danev R (2011) Retrofit implementation of Zernike phase plate imaging for cryo-TEM. J Struct Biol 174:400–412. https://doi.org/10.1016/j.jsb.2011.01.005
28. Danev R, Baumeister W (2017) Expanding the boundaries of cryo-EM with phase plates. Curr Opin Struct Biol 46:87–94. https://doi.org/10.1016/j.sbi.2017.06.006
29. Asano S, Engel BD, Baumeister W (2016) In situ Cryo-electron tomography: a post-reductionist approach to structural biology. J Mol Biol 428:332–343. https://doi.org/10.1016/j.jmb.2015.09.030
30. Briggs JAG (2013) Structural biology in situ–the potential of subtomogram averaging. Curr Opin Struct Biol 23:261–267. https://doi.org/10.1016/j.sbi.2013.02.003
31. Asano S, Fukuda Y, Beck F, Aufderheide A, Förster F, Danev R, Baumeister W (2015) A molecular census of 26 *S* proteasomes in intact neurons. Science 347:439–442. https://doi.org/10.1126/science.1261197

Correction to: Advanced Technologies for Protein Complex Production and Characterization

M. Cristina Vega and Francisco J. Fernández

Correction to: M. C. Vega, F. J. Fernández (eds.), *Advanced Technologies for Protein Complex Production and Characterization*, Advances in Experimental Medicine and Biology 1453, https://doi.org/10.1007/978-3-031-52193-5

The book was inadvertently published with incorrect series volume number as 3243 instead of 1453.

The correction has been incorporated by including the updated series volume number as 1453 at all instances in the book.

The updated version of this book can be found at
https://doi.org/10.1007/978-3-031-52193-5

M. C. Vega, F. J. Fernández (eds.), *Advanced Technologies for Protein Complex Production and Characterization*, Advances in Experimental Medicine and Biology 1453,
https://doi.org/10.1007/978-3-031-52193-5_14

Index

M. C. Vega, F. J. Fernández (eds.), *Advanced Technologies for Protein Complex Production and Characterization*, Advances in Experimental Medicine and Biology 1453,
https://doi.org/10.1007/978-3-031-52193-5

GPSR Compliance

The European Union's (EU) General Product Safety Regulation (GPSR) is a set of rules that requires consumer products to be safe and our obligations to ensure this.

If you have any concerns about our products, you can contact us on ProductSafety@springernature.com

In case Publisher is established outside the EU, the EU authorized representative is:

Springer Nature Customer Service Center GmbH
Europaplatz 3
69115 Heidelberg, Germany

The manufacturer's authorised representative in the EU is Springer Nature Customer Service Centre GmbH, Europaplatz 3, 69115 Heidelberg, Germany. If you have any concerns regarding our products, please contact ProductSafety@springernature.com

Printed and bound by CPI Group (UK) Ltd, Croydon, CR0 4YY
01/10/2026
02225867-0020